The Milky Way

Other Books by the same Author

The Relevance of Physics

Brain, Mind and Computers
Lecomte du Nouy Prize and Medal, 1970

The Paradox of Olbers' Paradox:
A Case History of Scientific Thought

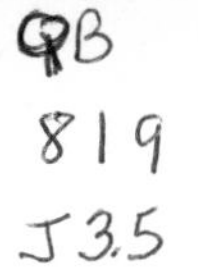

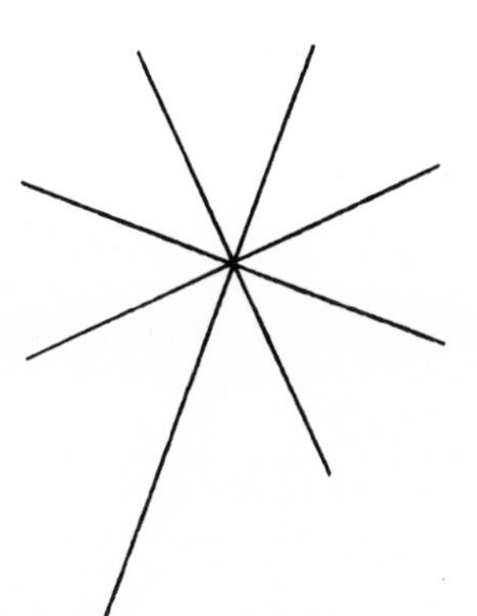

The Milky Way

An Elusive Road for Science

Stanley L. Jaki

Science History Publications
A DIVISION OF
Neale Watson Academic Publications, Inc.
NEW YORK · 1972

First published in the United States by

Science History Publications

A DIVISION OF

Neale Watson Academic Publications, Inc.
156 Fifth Avenue, New York 10010

First Paperback Edition 1975

Library of Congress Catalog Card Number 72-87334

International Standard Book Number 0-88202-022-6

Designed and Manufactured in the U. S. A.

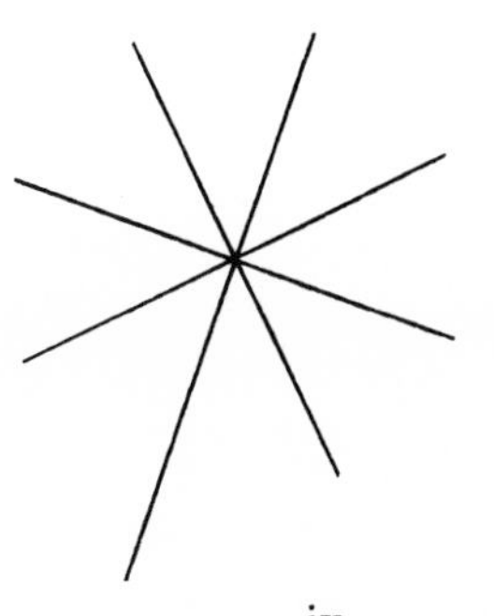

Contents

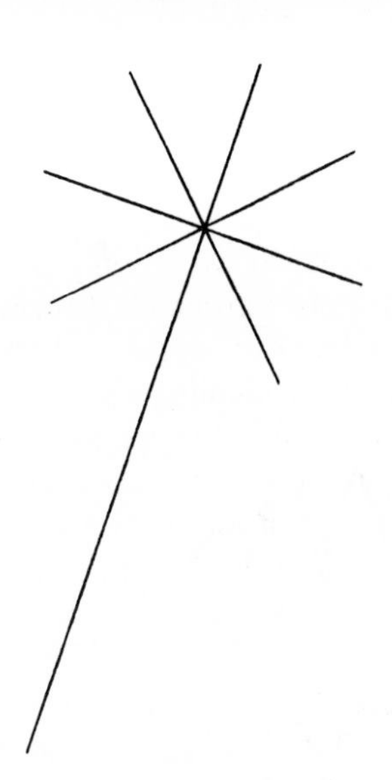

Introduction

In writing the Introductions to their works, authors of monographs often face the delicate task of justifying their undertakings. Such is especially true in the case of conspicuous topics. These form a prime target for research, with the result that it soon becomes well-nigh impossible to add a really new twist to an already old story. Conspicuous and grand as the Milky Way is, a monograph on the history of its investigation will avoid at least one risk —that of being an echo. It may seem astonishing, but no one has yet tried to write that story. The astonishment should be proportional to one's familiarity with the usually extensive literature on astronomical and cosmological topics of comparable importance.

The very circumstance which assures undisputed novelty to a history of the Milky Way, should also make evident its difficulty. Only one short, though pivotal section of that long story may be said to have received careful attention thus far. William Herschel's work and ideas on the Milky Way found in Michael A. Hoskin a most perceptive interpreter. He further pioneered by editing hitherto unappreciated manuscripts of Thomas Wright on the subject. But, with that, smooth going for the historian of the Milky Way reaches an abrupt end. The great bibliography of astronomy by Houzeau and Lancaster would only lead him to an insignificant article by Houzeau, published in 1879, which contained a list of some half dozen obscure authors of the 16th and 17th centuries that he copied from Lalande's *Bibliographie astronomique* (1803). The historian's discouragement would only increase after he had turned to the great classics of astronomical literature written prior to and during Newton's age. For, so far as the Milky Way was concerned, Copernicus initiated a stunning silence, Galileo gave rise to a mesmerizing myopia, and Newton created a dazzling distraction.

In the history of science, neglect of something obvious can, of course, be most instructive about the often illogical workings of the so-called scientific atmosphere—which can befog even the mind of a genius. The early history of the Milky Way is particularly noteworthy in this respect. The Milky Way, a great singularity in the skies, did not fit either the Aristotelian closed world with its spherical and perfect heavens, nor its successor, the infinite, homogeneous universe of stars. As a result, the champions of both world pictures either manhandled the phenomenon of the Milky Way, or shied away from it. This was all the more regrettable in that the whitish band of the Milky Way might have served as a powerful reminder of Olbers' Paradox, or the paradoxical absence of a sky bathed day and night in a whitish blaze, the fusion of the light of innumerable stars. Failure to face the problem of the Milky Way was indeed much akin to the centuries-long failure to recognize the full cosmological weight of Olbers' Paradox.

As for the earlier part of the story, the historian can spare himself a frustrating search if he turns early enough to such dust-gathering documents of early astronomical literature as the folios of Riccioli, Gassendi, Mersenne, and Stöffler. It is there that he finds mention of a number of authors who, between 1500 and 1640, made some statement about the Milky Way. Locating their works and others from the same period is another task.

While it is relatively easy to gather the dicta of classical antiquity on the Milky Way, medieval gropings concerning it form an area hitherto unexamined in any systematic manner. Here too, as in many other areas of medieval science, the myth of the dark ages can not withstand a careful look at the record. Actually, it was in modern times that scientific myths were created about the Milky Way. "Wright's wrong" was merrily perpetuated until quite recently, and, until now, no real attention has been paid to the nineteenth-century myth of the "one-island universe."

The breakdown of that myth occurred in 1924 when the Milky Way proved to be only one of an immense though most likely finite number of island galaxies. Modern relativistic cosmology uses galaxies as the building blocks of a world picture which defies Newton as well as Aristotle. Man's cultural history might indeed have followed a notably different course, had the Greeks, to say nothing of the Copernicans, Galileans, and Newtonians, stumbled on the true, roughly grindstone shape of the Milky Way. Conceptually, it was not at all impossible that the Greeks should have done so. They achieved several still more startling insights into the structure of nature. Here, however, discovery eluded them in striking anticipation of a long and baffling series of near misses. The major cause of this failure seems to be a persistent unwillingness to confront the implications of a major cosmic singularity. As a result, the Milky Way became a most elusive road for science.

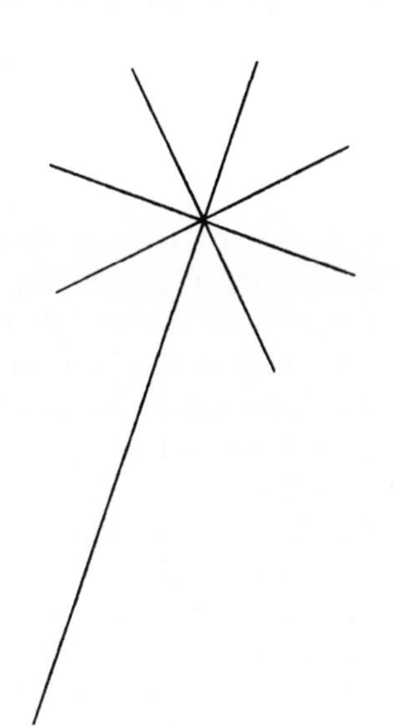

CHAPTER ONE

Greek Hunches

ORIGINS of nearly all major scientific topics are lost in tales of mythology and folklore. Such is certainly the case with man's speculations about the Milky Way.[1] Its whitish gleam suggested to ancient Egyptians, those most methodical storekeepers of grain, that it was the work of Isis, who spread large quantities of wheat across the sky. The gold-hoarding Incas believed it to be the golden dust of stars. Eskimos of the Arctic saw in it a snowy band. For Bushmen seeking relief from the chilly night of the veldt, the Milky Way evoked the ashes of campfires. The Arabs of the parched deserts fancied, typically enough, that the Milky Way was a great river. In the eyes of the fishermen of the Far Orient it appeared as a school of fish frightened by the new moon, a symbol of the hook. The Indians of the Great Lakes also projected something of their homeland into their vision of the Milky Way, when they pictured it as a muddy creek stirred up by a turtle swimming along the bottom of the sky. The Polynesians did much the same by calling the Milky Way the "long, blue, cloud-eating shark."

Over and above these local peculiarities, there was a universal vision of it as a Way or Road. To the early Hindus it represented the path of Aryaman, leading to his celestial throne. To the people of the Yellow River, it was the great yellow road. The Celts of old saw in it the route along which Gwydyon pursued his erring wife. Old Teutons spoke of it as the lane to Valhalla, the home of fallen heroes. The Iroquois called it the "Road of Souls," the pathway to the eternal kingdom of Ponemah. To the pilgrims of the Middle

Ages it was the trail to Rome. French peasants still refer to it as "le chemin du pèlerin," a simpler form of the older name, "le chemin de Saint Jacques de Galice." Legend has it[2] that this expression originated with Charlemagne. Speaking to his councilor, Eginhard, he noted that it chanced to point toward Santiago de Compostela in Galicia, once the foremost place of pilgrimage in medieval Europe.

The first scientific discussion of the Milky Way begins with a reference to mythology. That discussion appears in Aristotle's *Meteorologica.*[3] There Aristotle recalled that according to the Pythagoreans

> *(a)* the Milky Way is the path taken by one of the stars at the time of the legendary fall of Phaethon,

and that some Pythagoreans held that

> *(b)* the Milky Way is the circle in which the sun once moved. And the region is supposed to have been scorched or affected in some other such way as a result of the passage of these bodies [the sun and the stars].

In addition, Aristotle reported two earlier opinions about the Milky Way:

> *(c)* The schools of Anaxagoras and Democritus maintain that the Milky Way is the light of certain stars. The sun, they say, in its course beneath the earth, does not shine upon some of the stars; the light of those upon which the sun does shine is not visible to us, being obscured by its rays, while the Milky Way is the light peculiar to those stars which are screened from the sun's light by the earth [Illustration Ia].

Excluding the legend of Phaethon as a theory, Aristotle wrote:

> *(d)* There is still a third theory about the Milky Way. For some say that it is a reflection of our vision to the sun, just as a comet was supposed to be.

The refutation of these theories required no special insight by a logician of Aristotle's stature. He seemed to enjoy belaboring the point that all these theories failed to satisfy either the principles of logic, or some well-established information about the universe. Logic demanded that if the sun scorched the heavens before it shifted its course to the zodiac, then "the circle of the zodiac should also be so affected, indeed more so than the Milky Way: for all the planets, as well as the sun, move in it. But though the whole zodiac circle is visible to us (for we can see half of it at any time during the night) it shows no sign of being so affected, except when a part of it overlaps the Milky Way."

Again, logic demanded that if the opinion which Aristotle ascribed to Anaxagoras and Democritus were true, the Milky Way should be a circular spot moving across the sky at the rate of motion of the sun. In addition, Aristotle recalled certain estimates about the respective sizes of the sun and earth, and about the ratio of the sun-earth distance to the distance of the sphere of fixed stars. It remains, however, conjectural whether those estimates were substantially the same as those generally ascribed to Aristarchus (fl. 160 B.C.).[4] According to the latter, the earth-moon distance was about 80 earth radii and the earth-sun distance about 1500 earth radii. This provided the basis in Ptolemaic astronomy for setting the distance of fixed stars at little more than 20,000 earth radii. Aristotle himself referred generally "to astronomical researches" which "have now shown that the size of the sun is greater than that of the earth and that the stars are far farther away than the sun from the earth, just as the sun is farther than the moon from the earth." The conclusion thus followed readily: "the vertex of the cone formed by the rays of the sun will not fall very far from the earth, nor will the earth's shadow (which we call night) reach the stars. The sun must therefore shine on all the stars, nor can the earth screen any of them from it" (Illustration Ib).

The refutation of theory *(d)* was also based on its inconsistency with an obvious phenomenon of nature, specifically, the motion of the sun. A moving source of light, like the sun, could not produce a stationary image on the differently moving surface of the sphere of the stars. At this point Aristotle should have stopped. But it was his wont to claim all truth—an urge that trapped him badly more than once. In this case he also expressed himself rather clumsily both in his reporting of the theory and in his refutation of it. Probably no one will ever know whether the blame for this should lie with a copyist or with the obscurantist theory which Aristotle continued to refute at length with no real necessity. Contrary to his suggestion, the theory in question was not weakened by the fact that the Milky Way could be seen at night as a reflection in water or in similar reflecting surfaces. If Aristotle could admit that the sun's rays were reflected from the crystalline or ethereal walls of the heavens, a repetition of that reflection in water had to be a most logical possibility. Aristotle's own reference to several mirrors was the best pointer to the fact of multiple reflections. Aristotle merely flogged a dead horse when he noted that the reflection theory ran counter to the ever-changing relative distance between the sun and the stars. He hardly improved his own position by battling the reflection theory with a reference to changes in the time of the rising of Delphinus, a small constellation located near the Milky Way where it traverses the Swan and the Arrow. Aristotle might well have observed that any specific border point of the Milky Way followed the rising of a neighboring star. The logical inference from this should have been the recognition that the Milky Way was, therefore, a part of the realm of constellations. But Aristotle could not break out of his mental pattern, although his remark about the perennial sameness of the Milky Way implied the death knell of his own theory: "the constitution of the Milky Way remains the same . . . but this should not be so if it were a reflection and not a characteristic of the region."[5]

By "region" Aristotle meant something basically different from what was implied in theories *(a), (b), (c),* and *(d).* In all of them the Milky Way lay in the region of the stellar or planetary heavens. In that region there could not be, at least not according to the Aristotelian interpretation of nature, any material change. Such change to Aristotle was synonymous with corruption and blemish. The irregular contours of the pale band of the Milky Way appeared to him exactly that, and he spared no effort, including some somersaults in logic, to talk the Milky Way out of the realm of stars where it patently belonged. His feat could not have been more farcical. The region which he declared to be the real region of the Milky Way was in the "meteoric" section of the sublunary world; that is, "beyond the air (meta-aer)" and below the orbit of the moon. Everything below the moon's orbit was by definition subject to change, while everything above it was changeless and incorruptible. This dichotomy between the superlunary and sublunary world was the very heart of the Aristotelian world system. If there existed any major phenomenon of nature, which, because of its perfect changelessness, could not be part of the sublunary world, it was the Milky Way. But Aristotle wrote the *Meteorologica* "with a will," as was aptly remarked,[6] and for a zealous systematizer even the impossible could at times become possible. Writing "with a will" easily blinds one to the obvious, and Aristotle's theory of the Milky Way is a classic case in point. For the man, who rejected several theories on the ground that they contradicted the ever-sameness of the Milky Way, made no serious effort to show how that ever-sameness would be satisfied in his own theory.

What Aristotle lacked in this instance certainly was not space or time. He waxed unusually verbose in expounding the details of his own explanation of the Milky Way. If it had any logical implication, it was that the Milky Way should be constantly changing both in shape and in shade. First, there had to be a steady efflux of the dry exhalation from the marshy regions of the earth. Then, this had to rise at a steady rate across the turbulences of the atmosphere many thousands of miles into a specific, ring-shaped volume of space with most intricate contours, located directly under that belt of the sphere of fixed stars where stars were most numerous. Once there, the given volume of dry exhalation had to maintain the same density and spread to secure the same rate of incandescence. This slow burning was in turn caused, according to Aristotle, by frictional heat from the daily revolution of the great number of stars in the belt of the Milky Way (Illustration II). Although Aristotle in another context[7] tried to explain how friction could be generated in the presumably frictionless realm of stars and planets, and how the heat of friction could be transmitted to the sublunary regions, he failed to consider an equally crucial implication of his theory of the Milky Way. Contrary to the basic postulates of Aristotelian cosmology, the Milky Way, although part of the strictly immobile sublunary world, exactly followed the daily rotation of the heavens.

The physical process giving rise to the Milky Way in Aristotle's theory, was the same by which he tried to explain the apparition of comets. For, as he put it:

> *(e)* the Milky Way might be defined as the tail of the greatest circle produced by the foregoing material formation.[8]

Characteristically enough, the theories of comets which Aristotle rejected "on purely logical grounds"[9] placed the comets in the starry region among the planets.[10] His own theory of comets was nothing short of an absurdity and "on purely logical grounds" at that. It certainly contradicted his remark on scientific methodology which introduced his discourse on the topic. "We consider that we have given a sufficiently rational explanation of things inaccessible to observation by our senses if we have produced a theory that is possible."[11] The measure of possibility for cometary formation, in the way Aristotle imagined it, was well-nigh zero. It was based on that mysterious dry exhalation which was as inaccessible as the marshy regions themselves. Aristotle, who claimed, among other things, that wind was *not* air in motion but the movement of dry exhalation,[12] therefore associated the frequency of comets with dry and windy years. He lost no time in noting that a great comet had appeared when the winter was dry and the wind strong and northerly; that again, during the archonship of Nicomachus, a comet appeared in the equinoctial circle for a few days while a storm was blowing in Corinth. Such details were at best diversionary tactics. Its true value was exactly the same as the explanation given by Aristotle for the famous meteor that fell at Aegospotami in 467 B.C., which became the object of worship, awe, puzzlement, and some daring theories during antiquity. Since Aristotle could not allow the extraterrestrial origin of the huge black stone, he had no choice but to explain miracle by magic. The rock, he claimed, was lifted into the air by wind (dry exhalations) at night and fell to the ground during daytime. "Its fall," so Aristotle argued, "coincided with the appearance of a comet in the west."[13]

Such incredible argument from the master of logic had one overriding motivation. The heavens, divine by definition in Aristotle's system, had to be kept free of all transitory phenomena that, in the case of comets, also bore a palpable appearance of impurity or blemish. Thus, Aristotle was forced to state that the stars merely provided the indirect spark to ignite a specific volume of exhalation. The crucial factor in the production of comets was, as Aristotle most revealingly put it, "an exhalation from below of suitable consistency."[14] It was on this "consistency" that the shape, motion, and duration of the comet depended. If the exhalation extended in all directions, then a comet appeared; if it extended only lengthwise, a "bearded star" was the result. Again, if the composition of the exhalation was such as to permit a slow burning in one direction, one had a comet which moved; if the burning was very fast in one direction, the result was a shooting star, whereas, in the case of a stationary burning, the outcome was a sheer exercise in words—a "stationary" star or comet. Even more self-defeating was Aristotle's explanation of the question of why comets were few in number and why they appeared with greater frequency outside the tropics. In the tropics the movement of the stars and of the sun, according to Aristotle, not only sucked up the dry exhalation, but also dissipated it. This was clearly equivalent to eating one's cake and having it too. The same held true, as will be seen shortly, of what Aristotle added in the same breath—the chief reason for the small number of comets in general was that much of the dry exhalation "collects in the area of the Milky Way."[15]

It was with this remark that Aristotle turned to the topic of the Milky Way. In discussing it he tried to remain consistent only with his own "first principle," the dichotomy between the heavenly and the terrestrial realms. His inattention to factual consistency and to the obvious implications of his own theory could hardly be more glaring. Thus, he failed to consider why the stars in the belt of the Milky Way did not dissipate the exhalation in the same way as did the stars in the zodiac. It could have been no secret to him that the Milky Way did not disappear in the area where it intersected the zodiac. There, the sun, the moon, and the planets, if they were indeed chiefly responsible for that dissipation in the belt of the zodiac, should have been equally effective. Obviously, the Aristotelian explanation of the Milky Way was a construct which dragged the famous logician and student of nature into a net of *petitio principii* and patent absurdities. His case illustrates all too well the fatal danger against which, in another connection, he warned scholars of all times: "A small initial deviation from the truth multiplies itself ten-thousandfold as the argument proceeds."[16] He became indeed the very embodiment of the attitude for which he criticized some pre-Socratic physicists: "Out of affection for their fixed ideas these men behave like speakers defending a thesis in debate: they stand on the truth of their premises against all the facts, not admitting that there are premises which ought to be criticized in the light of their consequences, and in particular of the final result of all."[17]

The Milky Way

Every age has a partly forgivable weakness, its pride in its scholarship. Since learning is basically cumulative, true scholarship should reflect both awareness of past attainments and errors, and also the ability to unfold new vistas. As to the topic of the Milky Way, modern Aristotelian scholars have displayed a distinctly lesser awareness of its manhandling by Aristotle than did classical scholars of late Renaissance times. The first modern commentator on the *Meteorologica,* J. I. Ideler, in 1834 acknowledged Aristotle's error, but also praised his modesty for speaking only tentatively on the matter.[18] While Ideler took Aristotle to task for presenting a confused version of Democritus' theory, he failed even to hazard a guess as to why a genius like Aristotle could trap himself in a patently absurd explanation of the Milky Way. In his monograph on the meteorological lore of classical antiquity, Ideler simply ignored that for the *Meteorologica,* the major classical treatise on meteorology, the Milky Way formed an integral part of the subject.[19] The first modern French translator of the *Meteorologica,* J. Barthélemy-Saint-Hilaire, handled the text in too cavalier a fashion to see its deeper aspects and problems.[20] More accurate modern students and translators of the text, like F. C. E. Thurot,[21] F. H. Fobes,[22] E. W. Webster,[23] J. Tricot,[24] and H. Strohm did not suggest to their readers the baffling implications of Aristotle's theory. Strohm, however, presented convincing evidence that there could be no justification in assuming that Aristotle had spoken only tentatively on the Milky Way. [25]

Critical editions of texts may not be the appropriate forum to air some deeper conceptual difficulties. But this cannot be said of translations with commentaries. Certainly, no excuse can be found for studies which were meant precisely to shed a new light on the background and development of Aristotle's thinking about nature, as expressed partly or exclusively in the *Meteorologica.* In the most renowned of such works, W. W. Jaeger insisted on the need to assign the *Meteorologica* to the latest and most mature phase of Aristotle's thought, in which the Stagirite had broken completely with the mythical outlook of Plato and engaged in a systematic collection of observational and experimental data about the external world. [26]

This also meant that Aristotle's explanation of the Milky Way was neither a youthful vagary nor a momentary blunder, but a conviction which he must have been nurturing for a long time. Yet, Jaeger, who referred in general to blunders committed by Aristotle in his works on natural history, failed to see that the classic example was the Milky Way, precisely because Aristotle never wished to extricate himself from the foremost Platonic myth, the absolute perfection and divinity of the heavens. H. Strohm, author of the only major modern analysis of the conceptual structure of the *Meteorologica,* also failed to note the incompatibility of the presence in the heavens of an elongated blur, the Milky Way, with the Aristotelian concept of the perfection of the heavenly regions. Strohm's oversight was all the more curious in that his attention centered on the point that Aristotle, a strong advocate of the transcendence of the heavens, had obvious difficulty in arguing that the motion of heavenly bodies could still govern in minute detail the phenomena of the lower regions.[27] Strohm did not try to fathom the question of why Aristotle felt compelled to turn the Milky Way into a component of the sublunary or essentially perishable and changing part of the universe.

One is, however, truly at a loss to understand why only six lines could be devoted to Aristotle's dicta on the Milky Way in the most detailed modern analysis of the provenance and meaning of Aristotle's account of the physical world. Its author, the outstanding Aristotelian scholar, F. Solmsen, left wholly unexplored his perceptive remark that "strong antecedent convictions had to be at work if celestial phenomena, such as the Milky Way, or the comets, were to be traced to one of the exhalations."[28] Those convictions could only be Aristotle's own, and offered evidence of the blind alley into which any great systematizer could be led if his respect for the system was stronger than his respect for the evidence. The baffling dimensions of Aristotle's explanation of the Milky Way were also completely overlooked in the two modern histories of Greek astronomy. In that by Sir Thomas Heath, Aristotle's discussion is simply paraphrased without any critical comment.[29] In the other by D. R. Dicks, Aristotle receives praise on two counts, neither of which bears critical examination. First, the explanation proposed by Aristotle was at least as fantastic as the "fantastic views" which he battled. Second, close as Aristotle was to the often "fantastic" beginnings of scientific astronomy, he ignored one of its best ideas, the theory suggested by Democritus, that the Milky Way was the fusion of the light of an immensely great number of stars.[30]

Aristotle's discourse on the Milky Way was all the more frustrating, for it could not have been, on occasion, closer to the truth in its phraseology. In one of Aristotle's passages on the Milky Way only a subclause or two need be replaced to bring it in line with reality: "The light of the circle itself is stronger in that half of it in which the Milky Way is double, and in this half the stars are greater in number and density than in the other, which indicates that the cause of the light is none other than the movement of the stars: for if the Milky Way lies on the circle in which are the greatest number of stars, and in that segment of the circle in which the stars appear to be of a greater density and size, it is reasonable to assume that this is the most likely cause of the phenomenon."[31] The emphasis placed by Aristotle on the *movement* of the stars in the Milky Way, rather than on their much greater number there, was irony itself. It prevented recognition on his part both of the true nature of the Milky Way and of the merit of its explanation by Democritus.

Aristotle, most intent on surveying previously proposed solutions to a question before tackling it himself, was not always a paragon of objectivity in this procedure.[32] For one thing, Aristotle should have remembered that his version of Democritus' theory implied that the latter pictured the Milky Way as a mirror formed by stars lying on essentially the same spherical surface. This version sharply contrasted with Democritus' conception of an infinite universe of stars which Aristotle himself acidly criticized in his *On the Heavens*. There is also external evidence that Aristotle failed to report accurately the truth about Democritus' opinion on the Milky Way. The strength of that evidence is not weakened by the fact that it forms part of the first extant thematic discussions of the Milky Way which postdate Aristotle by some five centuries. Their authors, Pseudo-Plutarch[33] (fl. 130 A.D.) and Achilles[34] (fl. 200 A.D.), obtained their information from much older and rather trustworthy sources.[35] According to Pseudo-Plutarch,

> *(f)* Democritus [held that the Milky Way was] the luminescence due to the coalition of many small stars which shine together because of their closeness to one another.

In Achilles' version:

> *(g)* others [stated] the Milky Way is composed of very small, tightly packed stars which seem to us joined together because of

> the distance of the heavens from the earth, just as if many fine grains of salt had been poured out in one place.

In the same context, Pseudo-Plutarch assigned to Anaxagoras the view that the Milky Way was the shadow cast by the earth on the sky. Achilles in turn might have meant both Anaxagoras and Democritus, and it is quite possible that Democritus learned the correct theory from Anaxagoras. But in view of some modern reconstructions of the thought of Anaxagoras,[36] it is difficult to believe that he held the opinion which Aristotle ascribed to him. It is even more difficult to credit Aristotle in the case of Democritus, as the genuineness of many of his utterances is established beyond doubt, and they reveal him as a highly astute interpreter of the physical world. Knut Lundmark, director of the Observatory of Lund University, was obviously mistaken when he followed Aristotle's guidance in a little essay "On Demokritos' Conception of the Milky Way," and spoke of "the primitive stand of the astronomy of Demokritos." According to Lundmark, the "conception of the Milky Way as composed by stars dates back to Anaxagoras or earlier and thus is not peculiar to atomism."[37] Rather curious and arbitrary reasoning! After all, it was an integral part of Democritus' atomism that the number of stars was infinite and that in consequence the stellar realm was infinitely deep. Those stars (worlds) were separated by uneven intervals and subject to collisions, destructions, and re-emergence through the chance combinations of atoms. In keeping with this, there were in Democritus' universe some worlds "devoid of living creatures or plants or any moisture," in others there was no sun, whereas some of his "worlds" had more and larger suns and moons than ours. The formation of fiery bodies, such as the sun and the stars, Democritus attributed to whirls that invariably developed as the chance collision of atoms went on. In these whirls all material substance became gradually dry and finally ignited.

While some features of Democritus' universe could help greatly in forming a correct understanding of the Milky Way, the random distribution of worlds (stars) with respect to size, luminosity, and place, was not at all in agreement with Democritus' conception of the Milky Way. The dense packing of very small stars within a seemingly ring-shaped volume of space represented anything but randomness. There is no evidence that Democritus ever tried to resolve this contradiction, or that any ancient or modern[38] student of

Democritus had seized on it. The singularity of the Milky Way was as much at variance with the idea of an infinite and (even randomly) homogeneous universe of stars as it was with the finite world of Aristotle, enclosed by a crystalline sphere, allegedly free of any blemish or imperfection. A way out of the dilemma offered itself through the analogy of salt-grains. Unfortunately, the analogy, as ascribed to Democritus, did not imply the third dimension of depth. This was all the more frustrating since Democritus' idea of the world possessed "depth." One wonders what paths scientific cosmology might have taken if Democritus had perceived the analogy of a forest, where the visually solid front is caused by the fused images of many individual trees standing behind one another in many rows. Such a coniecture raises the intriguing question of whether the recognition of Olbers' Paradox, or the optical paradox of an infinitely large number of stars, was within the reach of ancient Greek scientific minds. These certainly possessed the acumen to recognize paradoxes,[39] and also the geometrical tools to formulate the paradox in question.

At any rate, the Milky Way was as much a stumbling block to the infinite universe as it was to the closed universe of Aristotle. No intimation of this could, however, make itself felt in ancient Greek science which, for all practical purposes, ignored the Milky Way. The reports of Pseudo-Plutarch and Achilles are in fact more revealing by their omissions than by their almost cryptically short dicta. In addition to mentioning theories *(a), (b),* and *(c),* all of which he ascribed to the Pythagoreans, Pseudo-Plutarch reported the view of Parmenides (fl. 470 B.C.) that the Milky Way was the mixture of a thick and thin substance resembling the color of milk *(h).* This accorded well with Parmenides' explanation of everything in terms of pairs of opposites. Pseudo-Plutarch also reported that, according to Metrodorus (fl. 350 B.C.), the Milky Way was the motion of the sun in its circle. Metrodorus thus seemed to differ from the Pythagoreans, who, as Pseudo-Plutarch also recalled, spoke of a former path of the sun in that connection.

Achilles made little contribution by reporting the myth of the Milky Way in the version of the poet Eratosthenes—that it was Hera's milk *(i).* More valuable was his remark that some saw in the Milky Way the line where the two hemispheres of the celestial sphere were fused together *(j),* presumably by some process of melting or soldering. He ascribed *(b)* to Oenopides of Chios (fl. 450 B.C.), but in recounting *(f)* as the theory that the Milky Way was a formation of clouds and condensed air, he failed to mention Aristotle. Such, at least, was the description of the Milky Way which Achilles seemed to favor, but there is no evidence that it was also the opinion of Aratus, a philosophical writer (fl. 260 B.C.), whom he commented upon.

There is only a short, descriptive reference to the Milky Way in Aratus' *Phaenomena,*[40] a work which had the good fortune to survive while many, far more important works of Greek philosophical and scientific literature were lost. This circumstance explains only a part of the meagerness of ancient Greek speculations on the Milky Way. The extent of the loss can scarcely account for a total absence from the lists of Pseudo-Plutarch and Achilles of any names representing a great astronomical tradition. This spanned six centuries between Eudoxus (fl. 375 B.C.) and Ptolemy (fl. 140 A.D.), and included such figures as Calippus, Aristarchus, Apollonius of Perga, Cleomedes, Hipparchus, Geminus of Rhodes, Heraclides, and Eratosthenes. Yet, among all these, only Geminus (fl. 75 B.C.) is known to have made a reference to the Milky Way. It was as short as it could be. In listing the great circles used by positional astronomy, Geminus noted that of these only the Milky Way was a visual object, and that it consisted of very small particles of cloudy material.[41] He added, however, that because of the unevenness of its breadth, it was not usually marked on celestial globes, a remark which clearly indicated that the Milky Way was of no real use in the type of astronomy where Geminus' interest lay.

In modern phraseology, the astronomy in question was strictly operational. Its motto was the hallowed phrase "to save the phenomena."[42] It meant that the astronomer's task consisted in finding a mathematical (geometrical) method which allowed accurate prediction of celestial phenomena such as eclipses, conjunctions, oppositions, transits, etc. The essence of the procedure was a composition of circles, with no consideration as to whether the system, perhaps in the form of wheels, was physically possible. The chief representative of this astronomical tradition was Ptolemy. Needless to say, he could not completely ignore the objections raised by the realist school, distinct minority though it was. But the little that Ptolemy had to say about problems connected with the physical constitution of the celestial mechanism[43] would hardly include a problem considered second-rate at best: the nature of the Milky Way. As the greatest authority in astrology, ancient or modern, Ptolemy did what all astrologers still do in connection with the Milky Way—he sidestepped it in his lengthy *Tetrabiblos*,[44] which became the bible of astrologers. The presence of so many stars in one narrow belt in no sense assisted the arcane business of casting usually ambiguous horoscopes.

Ptolemy, the operationalist astronomer, interested in "mere descriptions," gave an account of the Milky Way which remained its standard mapping until the advent of celestial photography. This statement on the Milky Way follows his renowned star-catalog, which concludes the Seventh Book of his *Almagest* and opens its Eighth. "It is easily seen," begins Ptolemy, "that the Milky Way is not simply a circle but a zone having quite the color of milk, whence its name; and that it is not regular and ordered, but different in width, color, density, and position; and that in one part it is double. These particulars we find needing careful observation."[45] The meticulous attention which he gave to details of contour and shade is well illustrated in his description of the Milky Way in the area of Cassiopeia: "The Milky Way encloses the whole of Cassiopeia except the star at the tip of the foot. The southern arc is bounded by the star in Cassiopeia's head, and the northern arc by the star in the foot of the throne by the star in Cassiopeia's calf. The other stars about her all lie in the Milky Way. Near the arcs the parts are of a thinner mass, while those in the middle of Cassiopeia show an extended density."[46] For all his attention to these details, Ptolemy said nothing explicit about the real nature of that variation in density, or of its unchanging intricacy.

The reason for this cannot be sought in Ptolemy's conservatism. Bold innovators, like Aristarchus of Samos, were not known to have said anything startling about the Milky Way. The fact that Ptolemy derided Aristarchus' heliocentric theory as madness, proves that uncommon opinions were remembered even if they failed to create a following. Had any of the great Greek astronomers dissented from established opinions on the Milky Way, or discussed them to any considerable extent, the fact would almost certainly have been recorded in one of the many scientific and philosophical writings extant from Hellenistic times. Still, it was not an astronomer, but a man of letters, Plutarch (c. 46-c. 120), who had the boldness to devote a whole book[47] to the claim that the moon was of ordinary matter, that it might have denizens of its own, that it was under the influence of the earth, that the earth was not necessarily the center of the universe—propositions that clashed head-on with accepted views. On the topic of the Milky Way, Plutarch said nothing, prolific though he was. Since he was principally a moralist and a historian, Plutarch can scarcely be blamed for this.

The matter is more puzzling in the case of the Stoics, who represented the largest body of philosophers in the post-Aristotelian era, and were keenly interested in cosmological and physical problems. Thermodynamical considerations, which put a distinctive stamp on their physics, might have found a rewarding object in the Aristotelian explanation of the Milky Way as a low and steady burning of dry exhalation. The only Stoic known to have speculated on the Milky Way was Poseidonius (c. 135 B.C.-c. 50 B.C.). Possibly the most learned man of his time, he was the author of more than twenty works known today only by title and subject matter. He suggested that the Milky Way was a stream of stellar heat, the purpose of which was to temper with its warmth the chill of the universe *(k)*.[48] This, however, merely illustrated the preference of Stoics to explain phenomena as heat processes. The heat in question, frustratingly enough, cannot be taken as an indication that for Poseidonius the stars were hot bodies. He held fast to the Aristotelian distinction between heavenly and terrestrial matter. Of the two, only the latter could grow warm or catch fire.[49]

The Stoics, as natural philosophers, are perhaps best remembered for their staunch vindication of the inherently continuous nature of matter against the claims of the atomistic school. With respect to the Milky Way, the atomists differed not a whit, as far as silence was concerned, from the Stoics. The failure of Lucretius (c. 99 B.C.-c. 51 B.C.), the renowned popularizer of atomism in classical antiquity, to say a word of the Milky Way in his *De rerum natura* appears baffling. His case is all the more strange since he offered marvelous illustrations of the compound effect of a very large number of minute particles or visually small objects. Thus, he noted that a broad, diffused field of light arose when sunlight was reflected from the weapons of thousands of soldiers maneuvering in formation over distant countryside.[50] Herein lay a remark of great potentialities. The fusion of light, in this case, did not arise from the contiguous placement of shiny metal objects, but from the relatively close positioning of rows of soldiers behind one another. Could not the light of the Milky Way be produced by a similar grouping of stars? The idea of Democritus about a randomly homogeneous scattering of stars through immensity was not enough to lead to this bold inference. What was needed instead was the assumption that the scattering could be confined to a specific pattern. Moreover, it had to be recognized that the dense, contiguous packing of stars suggested by Democritus could be a mere optical "packing" or effect. Clearly, all this implied a great leap forward at a very early phase of scientific cosmology. Ancient Greek science made many a bold leap and, in this case, the conceptual elements of the correct solution were not hopelessly out of reach even in the early, or more creative, phase of Hellenistic times. But, as already noted, the Milky Way could not be logically fitted into either the perfect heavens of the closed Aristotelian universe or its rather unsophisticated counterpart, the infinite universe of Democritus. The result was a general apathy on the part of the best investigators of nature in classical antiquity to probe into the riddle of the Milky Way.

The situation had already been criticized in late antiquity. Philo (c. 20 B.C.-c. 50 A.D.), the leading Hellenistic Jewish scholar, was sufficiently explicit when he wrote: "As for the belt of the Milky Way it possesses the same essential qualities as the other stars, and though it is difficult to give a scientific account of it, students of natural phenomena must not shrink from the quest. For, while discovery is the most profitable, research is also a delight to lovers of learning."[51] The passage is from Philo's defense of the presence of providence in a world which is the product of a personal, rational Creator. His antagonist was his nephew, Alexander, who in order to justify his apostasy from Judaism pointed at the well-being of the wicked and the afflictions of those fearing God. Philo then marshalled concepts which illustrated the difference between material and spiritual values, the recondite ways of divine justice, and the fundamental goodness of nature. Natural disasters could be shown to have many beneficial side-effects. At any rate, phenomena of nature, however mysterious, always offered opportunity for research, a most valuable enterprise. This was the gist of his remark on the

Milky Way, which for Alexander merely illustrated the futility of the disputation of philosophers.

Alexander's rejection of philosophy was not without some justification. Late Hellenistic philosophy showed more than one sign of the failure of nerve. It increasingly abandoned the spark of rationality, and permitted itself to be enveloped in a murky atmosphere of pantheistic mysticism which lessened all restraint about astrology and mythical allegorization. The only notable reference in Neoplatonist literature to the Milky Way clearly illustrates this point. It is from *The Homeric Cave of the Nymphs* by Porphyry (233-c. 304), who was prompted to make mention of the Milky Way only because it passed near the sacred gates of the extremities of the sun's passage among the stars. In a world-description filled with nymphs, naiads, and continual references to the marital union between the sky and the earth, Pythagoras was chosen as the authoritative spokesman about the Milky Way: "According to Pythagoras, also, the *people of dreams,* are the souls which are said to be collected in the galaxy, this circle being so called from the milk with which souls are nourished when they fall into generation. Hence, those who evocate departed souls, sacrifice to them by a libation of milk mingled with honey; because, through the allurements of sweetness, they will proceed into generation; with the birth of man, milk being naturally produced."[52]

While Porphyry's own view of the Milky Way had little merit, it was not much inferior to what the best Roman compilers of scientific information could offer on the subject. Seneca (c. 3 B.C.-65 A.D.), in his *Quaestiones naturales,* devoted a full book to comets, but referred only once to the Milky Way. This he did merely to refute the opinion that a conjunction of planets could produce comets, and in particular a huge one, equal in length to one-fourth of the whole span of the Milky Way.[53] On this point Aristotle might have agreed with Seneca, whose work covered much the same topics—thunders, rainbows, comets, winds, earthquakes—that also formed the subject matter of Aristotle's *Meteorologica.* Was Seneca's refusal to discuss the Milky Way as a subclass of comets, in the style of Aristotle, a slighting of the Stagirite's explanation? More likely, Seneca's silence evidenced disinterest and lack of sensitivity to a challenging topic.

The Milky Way

In the massive *Natural History* of Pliny, there is again only one reference to the Milky Way, which reveals that it was imagined to be the source of lactation in the broadest sense. Pliny based this claim on the fact that the Milky Way passed "through the Archer and the Twins, cutting the equinoctial orbit twice at the sun's centre-point, the intersections being marked by the Eagle on one side and the Little Dog on the other."[54] Consequently, the Milky Way, which Pliny cryptically enough called a constellation, could have universal influence because the foregoing constellations represented points "at which the centres of the sun and the earth correspond." From this it followed that if "on the dates of these constellations [i.e., the days of their rising and setting] the atmosphere is clear and mild and transmits this genial milky juice to the earth, the crops grow luxuriantly; but if the moon scatters a dewy cold after the manner previously described, the admixture of bitterness, like sourness in milk, kills off the infant offspring."[55] Pliny merely quoted this fantasy with scientific veneer as he tried to explain why, considering the convex surface of the earth, such a disaster does not strike every part of the earth simultaneously. His dicta carried no more scientific value than the often-quoted lines of Ovid: "There is a high way, easily seen when the sky is clear. 'Tis called the Milky Way, famed for its shining whiteness. By this way the gods fare to the halls and royal dwelling of the mighty Thunderer. On either side the palaces of the gods of higher rank are thronged with guests through folding-doors flung wide. The lesser gods dwell apart from these. Fronting on this way, the illustrious and strong heavenly gods have placed their homes. This is the place which, if I may make bold to say it, I would not fear to call the Palatia of high heaven."[56] The widespread acceptance of such a view of the Milky Way can be seen in Cicero's "The Dream of Scipio"[57] in which the Milky Way is described as the abode of virtuous souls. Part of the baffling history of scientific ideas was that Cicero's moralizing remarks on the Milky Way were to keep alive the meager information on it through dark centuries.

Typically enough, the two Roman compendia of astronomical (and astrological) lore were of no greater help in this regard. They were the *Astronomica* of Manilius (fl. 40 A.D.) and a work by Hyginus (fl. 25 B.C.) with the same title. Manilius, whose work seems to have been written somewhat before the middle of the first century A.D.,[58] devoted more lines to the explanation of the Milky Way by Phaethon's fate, or by Hera's milk, or by the dwelling of blessed souls there, than to the possibility that its light might be due to fusion of the light of a major agglomeration of stars. But at least he advanced this last possibility, though in the form of a question.[59] In speaking of the great and small circles Hyginus merely mentioned that the equinoctial circle was cut in two halves by the Milky Way at Aquila and Procyon.[60]

The only detailed and truly informative Latin discussion of the Milky Way dates from the last, rapidly declining phase of the Roman empire and culture. The discussion is part of the commentary by Macrobius[61] (fl. 400) to the "Dream of Scipio" of Cicero. To Cicero the Milky Way had significance only as the abode of such virtuous souls as Scipio. For Macrobius, Cicero's moralizing offered an excuse to summarize whatever scientific information about the universe still circulated in the Latin part of the empire. His work served as a principal scientific sourcebook in the West prior to the recovery of the Greek scientific corpus during the 12th and 13th centuries. Much of what Macrobius wrote in connection with the Milky Way is rather spiritualistic, but he also gave a list of previous opinions about the physical origin of its appearance. He mentioned the mythical theories, without, however, including any detail. At the same time, he gave Democritus' opinion in a more detailed version, which deserves to be quoted in full: "Democritus' explanation was that countless stars, all of them small, had been compressed into a mass by their narrow confines, so that the scanty spaces lying between them were concealed; being thus close-set, they scattered light in all directions and consequently gave the appearance of a continuous beam of light" *(h)*.[62] From Macrobius we know that not all atomists agreed with their master. Diodorus of Cronus (fl. 340 B.C.), chiefly known as a logician of the Megarian school, took the view *(l)* that the Milky Way "was fire of a condensed and concentrated sort, pressed together into one belt with crooked boundaries, its difference in consistency being responsible for its being massed together, as a result it became visible, whereas the rest of the celestial fire did not present its light to our eyes since it was diffused by the extreme subtlety of its nature."[63] In Macrobius' eyes the prevailing opinion about the Milky Way closely coincided with the definition given by Poseidonius who, in Macrobius' rendering, held *(k)* "that the Milky Way is a stream of stellar heat which crosses the zodiac obliquely; inasmuch as the sun never passes beyond the boundaries of the zodiac and has left the remaining portion of the sky without a share in its heat, the purpose of the Milky Way, lying athwart the path of the sun, is to temper the universe with its warmth. We have already indicated where this intersects with the zodiac."[64]

Rather surprisingly, Macrobius made no mention of Aristotle, nor did he give any information about those who in his time preferred an opinion different from the one championed by Poseidonius. On the other hand, he reported that the two-hemisphere theory of the Milky Way *(j)* was proposed by none other than Theophrastus, Aristotle's successor at the Lyceum.[65] Theophrastus was no slavish admirer. Science in antiquity might have taken a different course had most of his books survived. Such a conjecture is based on much more than the fine reputation enjoyed by his great historical surveys of the various branches of sciences. One of his extant works, the *Metaphysics,* reveals that living in the shadow of Aristotle's genius did not dull his critical sense. Theophrastus could not have been more to the point than when he noted, in the foregoing work, that one should be on guard against carrying teleological considerations too far in scientific explanations.[66] The remark struck at the very heart of Aristotelian science, and had devastating implications for Aristotle's cosmology and physics. Aristotle's explanation was an absurd outgrowth of his preoccupation with showing that every process in nature was "for the best," in accordance with the Socratic dictum.[67] That the Milky Way was the fusion of two hemispheres was certainly a far-fetched conjecture, but at least it was not saddled with the contradictions of Aristotle's theory.

Theophrastus' independence of mind might exemplify the principle that no one is truly great to his assistant. Only as centuries went by did Aristotle begin to appear perfection incarnate in the eyes of latter-day admirers. Distance in time or space can embellish or conceal many a flaw, as may be seen by the manner in which Alexander of Aphrodisias (fl. 190 A.D.), a famed director of the Lyceum, commented[68] on Aristotle's dicta on the Milky Way. Alexander was the first major commentator on Aristotle, but he wanted to do more than make Aristotle's doctrine the center of attention. He started a trend of thought which culminated in the almost fanatical veneration that the Averroists had for Aristotle. What was meritorious in Alexander's comments did not go beyond incidentals. According to him, the reflection theory originated with Hippocrates[69] and suggested that the scattered stars *(sporadai)* to which Aristotle referred, were clusters of stars visible in Orion and Virgo. To buttress this latter point, Alexander recalled a short phrase of Aratus about *sporadai* below Aquarius.[70] In typical Aristotelian fashion, Alexander made no attempt to put into quantitative form the refutation of theory *(d)*. A generation after Ptolemy and four centuries after Aristarchus, nothing could have been more natural than to quote numerical data about the size of the earth and the sun and their distance from one another to show that the earth's shadow fell far short of the sphere of the fixed stars. Alexander merely said that the distance of the earth from the stars was "many times" greater than the earth-sun distance.[71] Here Alexander might have been excused on the ground that perhaps he did not wish to repeat the results obtained by "astrologers" (astronomers) since these results were obviously well known. He should have pointed out that Aristotle's refutation of theory *(d)* implied the claim that multiple reflection was not possible.

On this point Alexander of Aphrodisias was already censured in late antiquity by Olympiodorus[72] (fl. 530), a Neoplatonist, who felt no qualms about laying bare Aristotle's errors. He noted at the outset of his comments on Aristotle's explanation of the Milky Way that all those whom Aristotle criticized offered better ideas than he did.[73] "All of them were eager to have the galaxy in the heavens, except Aristotle who mistakenly insisted upon saying that it was in the upper air."[74] Yet, in the same breath, Olympiodorus began to excuse Aristotle's error. He claimed that the Stagirite put forward his explanation as a theory and not as a fact. True, Aristotle said in an earlier chapter that, of the various meteorological phenomena, some admit explanation while others do not,[75] but the tone of his account of the Milky Way revealed no second thoughts, a fact that should have been obvious to Olympiodorus. Undoubtedly, by trying to exculpate Aristotle, Olympiodorus started a trend, instances of which occurred time and again during the pre-Galilean centuries of the story of the Milky Way.

Olympiodorus couched his rejection of Aristotle's explanation of the Milky Way in arguments already formulated by his master,

Ammonius (fl. 520), the last director of the Lyceum, and a distinguished disciple of Proclus. Whether Ammonius owed his arguments to Proclus cannot be decided, for the extant writings of Proclus contain no reference to the Milky Way.[76] "The great Ammonius," as Olympiodorus referred to him,[77] made four points against Aristotle's theory. First, if the galaxy consisted of warm and dry exhalations, then its size and luminosity must be greater during the summer months when the dry exhalation emerged more abundantly from below the surface of the earth. Second, no particular effect of the air (exhalation) was known to spread all over the earth, whereas the galaxy was a phenomenon encircling the whole globe. Third, the moon had to be below the galaxy since the latter showed no parallax; here Ammonius recalled the reasoning in Ptolemy's *Almagest* that of two celestial objects the one closer to the earth showed the larger parallax. Fourth, Ammonius argued that if Aristotle's theory were true, the Milky Way had to appear whiter at the areas where it intersected the zodiac. There the motion of the planets, of the sun, and of the moon should have contributed heavily to the alleged effect of the stars on the exhalation below.

A rather curious feature of Olympiodorus' commentary was that he gave Democritus' opinion of the Milky Way only in Aristotle's version. The better form of Democritus' opinion was given, but without Democritus' name, in a collection of philosophical opinions composed in the early 6th century by Joannes Stobaeus.[78] The omission of the more appealing form of Democritus' opinion in the commentary by Joannes Philoponus (fl. 535) on the first book of the *Meteorologica* is rather surprising.[79] On the other hand, his discussion of Aristotle's theory shows Philoponus at his best as a scholar and a debater. He was eager to display his erudition from the outset. Regarding the myth of Phaethon, he suggested that it might have been a corrupted form of the legend of Atreus, an astronomer, who first explained to men that the planets, the sun, and the moon revolved in a sense contrary to the motion of the sphere of the fixed stars. His punishment in the hands of the gods then became coupled to the fate of Atreus, who murdered his father, King Pelops. The sun, unable to watch the crime, turned its chariot from its customary path, hurling a great many stars from their places. The route along which they fell became the Milky Way. At any rate, Philoponus continued with his interpretation of the myths about the

Milky Way: comets could readily be looked upon as the disowned children of the sun who tried in vain to follow the exact course of their father. They had to settle for another path, the Milky Way.

All this was merely an introduction to show that Aristotle handled the subject matter too curtly. Thus, in connection with the theory ascribed by Aristotle to Democritus and Anaxagoras, Philoponus marshalled the numerical data which Greek astronomers long before his day, and in part at the time of Aristotle, had derived concerning the relative distances of the sun, the earth, the planets, and the fixed stars. According to these figures the earth's shadow could not reach beyond the orbit of Mercury. Philoponus' lengthy paraphrases of the reflection theory contained nothing noteworthy, except for a reference to Ammonius.[80] The same holds true for Philoponus' rendering of Aristotle's theory which he did not interrupt with critical remarks. These were left to the closing section which opened with the phrase: "In that diction and opinion of the Philosopher there are many troublesome errors and absurdities which imply crass contradictions."[81] These were now laid bare by an array of questions, not all of them original, but never before arranged in such a relentless sequence.[82]

How is it, Philoponus asked, that, unlike the comets, the Milky Way never changed its position and size, neither in whole, nor in parts? How is it that the Milky Way could form a gigantic exception to the invariably short duration of all phenomena that take place in the atmosphere? How is it that the quantity and quality of exhalation, a very volatile substance, never changes as it feeds the Milky Way and only when it does precisely that? How is it that this exhalation always clings to the same position with respect to the stars? Why is it that the Milky Way is not brighter in the area of Orion, if its brightness is a function of the frequency of stars? Why is it that the Milky Way shows no change when the sun, the moon, and the planets cross over it, whereas there is a big change in the seasonal weather according to the position of the sun? How is it that a huge object like the Milky Way can exhibit below the moon's orbit that perpetual changelessness which is the distinctive feature of the celestial regions? How is it that the Milky Way does not dim the light of the planets as they pass behind it, while clouds can completely hide even the brilliance of the sun? Why is it that in the belt of the Milky Way individual stars and planets do not become the origin of individual comets, whereas they do, according to Aristotle, outside it? Why is it that, in connection with the Milky Way, one does not find the phenomenon often observed with clouds which cover the sun with respect to one area on earth, but not at the same time to all other areas? This question of Philoponus concerned the crucial issue about any sublunary phenomenon, its inevitable parallax. The Milky Way could not appear always, and from any point of the earth, against the same stars if it were located below the orbit of the moon (Illustrations IIIa and b).

Philoponus could now wax sarcastic.[83] He ridiculed Aristotle's "physiology," that is, his theory of the Milky Way, as nothing better than the myth which claimed it to be Juno's milk or the path of noble souls to heaven. Was Juno's milk always pouring out in the same quantity, and were those shining souls always ascending at the same rate to ensure the sameness of the Milky Way? That sameness meant incorruptibility, a quality which Aristotle, as Philoponus noted, heatedly denied to anything below the orbit of the moon, while holding that the heavenly world itself was eternal and immutable. Was it not Aristotle who claimed that in a dispute only those conclusions should be admitted which always or almost always were borne out by the evidence? Was not the Milky Way always the same, and had it not, therefore, to be part of the heavens? Would it not fit in very well with a variety of celestial phenomena such as the varying size, color, and brightness of the stars?

The answer to these questions was too clear to be spelled out. Philoponus himself set great store by the last point in his criticism of the eternity of the universe. The variety in question was, for him, evidence that the heavens were not immutable and eternal. As a Christian he could not accept the divinity and eternity of the realm of stars, a tenet generally shared in classical antiquity. His faith in the Creator also led him to the conviction that the ultimate explanations rested with Him alone. It was natural to take "ultimate" not only in a philosophical sense but also in a spatial one. The heavens were clearly the ultimate frontiers, and Philoponus held their fundamental laws and properties to be unfathomable for man. The real reasons for the actual place and number of the orbits of the planets and stars, and for their true nature, were the secrets of the Creator. The same held true, according to Philoponus, of the Milky Way. One could ascertain that it was part of the sphere of the fixed stars, but nothing certain could be known as to why it did occupy a specific place in that sphere.[84]

There was indeed nothing specific in the definition of the Milky Way which Philoponus gave in his *De opificio mundi,* the first major effort to reconcile the six-day creation story of the Bible with the science of the day. Here Philoponus wrote that the Milky Way is a "feature *(pathos)* of the sky and inseparable from it, being co-existent with it and of the same nature."[85] This seemed all the more disappointing since it was in connection with the Milky Way that Philoponus uttered, in his commentary on the *Meteorologica,* the magnificent precept: "physical phenomena should be explained by physical causes."[86] The immediate context was Philoponus' rebuttal of Damascius, a Neoplatonist colleague of Philoponus at the Lyceum, who curiously enough found something good in the view that the Milky Way was the path of souls. In the same breath, Damascius also held, according to Philoponus, that the Milky Way was the "feature *(pathos)* of the incorruptible heavens, rich in stars, shining in the hue of milk with the united light of many small stars."[87] Philoponus, who had had enough trouble in escaping the obscurantism of Neoplatonism, took Damascius to task for mixing physics with mysticism. At the same time, he rejected the starry composition of the Milky Way on the ground that the "position and figure of those many stars could not remain hidden to us."[88]

It was with such remarks that the first phase of man's search for an understanding of the Milky Way came to a frustrating close. The Neoplatonists were too deeply steeped in mystical vagaries, so far as the Milky Way was concerned, to go beyond the silence of Greek astronomers and the scattered hunches of Greek philosophers. Philoponus, who originated several conceptual breakthroughs in physics and cosmology, failed to advance toward a broader synthesis. Still, his criticism of some pivotal points of Aristotelian physics and cosmology might have speeded up the development of physics had his writings not remained unknown in Europe until the late fifteenth century. But Philoponus would have been of meager assistance to the medievals in regard to the Milky Way. On this point it mattered little whether the dicta of ancient Greeks reached the early Middle Ages in their full form or only in Macrobius' short summary. The latter contained in a nutshell all that the Greek genius had produced on the topic. The achievement, insignificant when compared with many other accomplishments of ancient Greek science, nevertheless represented an indispensable starting point. Its historical significance is best seen in contrast with the absence of speculation on the Milky Way in the vast annals of ancient and medieval Chinese science. [89] Those Greek hunches on the Milky Way constituted the kind of immense advance which Bacon must have had in mind when he observed that truth emerges far sooner from saying something than from making no statement at all.

References

[1] The next two paragraphs are largely based on material in *Star Lore of All Ages: A Collection of Myths, Legends and Facts, concerning the Constellations of the Northern Hemisphere* by William T. Olcott (New York; G. P. Putnam's Sons, n.d.), pp. 391–98. See also the study of some Chinese and Japanese tales connected with the Milky Way in *The Romance of the Milky Way and Other Studies and Stories* by Lafcadio Hearn (Leipzig: Bernhard Tauchnitz, 1910), pp. 19–74.

[2] See *Charlemagne* by Jean-Baptiste H. R. Capefigue (Paris: Langlois et Leclerq, 1842), vol. 1, pp. 376–77.

[3] References will be to *Meteorologica* with an English translation by H. D. P. Lee (Loeb Classical Library; Cambridge, Mass.: Harvard University Press, 1952). The Milky Way is the topic of chap. 8, of Book I, or section 345a11–346b15.

[4] Eudoxus (fl. c. 367 B.C.), whose work was familiar to Aristotle, might have anticipated Aristarchus on the distances and sizes of the sun and the moon. See, concerning this, Sir Thomas Heath, *Aristarchus of Samos, the Ancient Copernicus: A History of Greek Astronomy to Aristarchus together with Aristarchus's Treatise on the Sizes and Distances of the Sun and Moon* (Oxford: Clarendon Press, 1913), pp. 331–32.

[5] *Meteorologica,* 345b.

[6] F. Solmsen, *Aristotle's System of the Physical World* (Ithaca, N.Y.: Cornell University Press, 1960), p. 404.

[7] See, for instance, Book II, chap. 7, of *On the Heavens,* where Aristotle "explains" how the divinely "cool" bodies of the sun and stars by their revolution set afire the air below. See the Greek text and English translation by W. K. C. Guthrie (Loeb Classical Library; Cambridge, Mass.: Harvard University Press, 1939), p. 179.

[8] *Meteorologica,* 346b. The translation includes the word "possibly," although neither the Greek text, nor Aristotle's tone of thought seems to warrant this qualification.

[9] *Ibid.*, 343b.

[10] Aristotle recalled that according to Anaxagoras and Democritus the comets were conjunctions of stars, whereas the Pythagoreans considered comets to be planets seen only on rare occasions. The tail of a comet was thought to be the vapor drawn upwards by those "planets" and "stars". See *Meteorologica,* 342b–343a.

[11] *Meteorologica,* 344a.

[12] *Ibid.*, 360a.

[13] *Ibid.*, 344b.

[14] *Ibid.*

[15] *Ibid.*, 345a.

[16] *On the Heavens,* 271b.

[17] *Ibid.*, 306a.

[18] *Aristotelis Meteorologicorum libri IV* (Greek text with Latin translation and commentaries; Leipzig: F. C. G. Vogel, 1834–36), see especially vol. 1, pp. 408–23. Ideler's work was based on the critical edition of the text by I. Bekker, *Aristotelis Meteorologica* (Berlin: G. Reimer, 1829).

[19] *Meteorologia veterum Graecorum et Romanorum: prolegomena ad novam meteorologicorum Aristotelis editionem adornandam* (Berlin: G. C. Nauck, 1832). Ideler also ignored the comets—another "meteorological topic" for Aristotle.

[20] *Météorologie d'Aristote* (Paris: Ladrange, 1863). The work also contained the translation of the spurious *De mundo* and running notes to the text.

[21] "Observations critiques sur les *Meteorologica* d'Aristote," *Revue archéologique* 20 (1869): 415–20; 21 (1871): 87–93, 249–55, 339–46, 396–407.

[22] *Aristotelis Meteorologicorum libri quattuor* (Cambridge, Mass.: Harvard University Press, 1919).

[23] *Meteorologica,* translated by E. W. Webster (Oxford: Clarendon Press, 1923). There are no critical remarks on Aristotle's handling of the Milky Way in the comments accompanying the translation by H. D. P. Lee, quoted above.

[24] *Les Météorologiques: nouvelle traduction et notes* (Paris: J. Vrin, 1941). Tricot even failed to call attention to the misleading character of Democritus' opinion as given by Aristotle, and quoted several times with approval from Aquinas' commentary, perhaps the most slavish of all medieval commentaries on the *Meteorologica.*

[25] In *Aristoteles Werke in Deutscher Übersetzung,* edited by E. Grumach and H. Flashar, Vol. 12, Part I. *Meteorologie* (Berlin: Akademie Verlag, 1970). See p. 149.

[26] *Aristotle: Fundamentals of the History of his Development,* translated with the author's corrections and additions by R. Robinson (Oxford: Clarendon Press, 1934). See especially pp. 294, 307, 331 and 333.

[27] "Untersuchungen zur Entwicklungsgeschichte der Aristotelischen Meteorologie," *Philologus* 28 Suppl. (1935): 25–39.

[28] Solmsen, *Aristotle's System of the Physical World,* pp. 411–12.

[29] See Sir Thomas Heath, *Aristarchus of Samos,* pp. 247–48.

[30] D. R. Dicks, *Early Greek Astronomy to Aristotle* (London: Thames and Hudson, 1970), pp. 209–10. Dicks seems to be unaware of the non-Aristotelian version of Democritus' idea of the Milky Way, although he mentions (p. 82) the fragment (Diels 68 A 91) containing it. For further details on Democritus, see pp. 10–12, 16, and 19.

[31] *Meteorologica,* 346a.

[32] As amply illustrated in *Aristotle's Criticism of Presocratic Philosophy* by H. Cherniss (1935; reprinted, New York: Octagon Books Inc., 1964). But Aristotle's version of Democritus' opinion is reported there (p. 328) without a warning about its divergence from another version.

[33] For passages on the Milky Way, see the edition of the text by G. N. Bernardakis, *Plutarchi Chaeronensis Moralia* (Leipzig: B. G. Teubner, 1893), vol. V, pp. 316–17 (Book III, Section I, in "De placitis philosophorum").

[34] For passages on the Milky Way, see Achilles' *Isagoge in Aratum* in A. Maass, *Commentariorum in Aratum reliquiae* (Berlin: Weidmann, 1898), pp. 55–56.

[35] The sources in question are the *Vetusta Placita* (c. 150 B.C.), and the sixteen books of Theophrastus' "Opinions of the Physicists," (c. 340 B.C.).

[36] Concerning Anaxagoras, see D. E. Gershenson and D. A. Greenberg, *Anaxagoras and the Birth of Physics* (New York: Blaisdell Publishing Co., 1964), pp. 338–39. For a convenient documentation on Democritus, see G. S. Kirk and J. E. Raven, *The Presocratic Philosophers: A Critical History with a Selection of Texts* (Cambridge: Cambridge University Press, 1957), pp. 410–11.

[37] Lundss Universitets Arsskrift, N.F. Avd. 2. Bd. 30. Nr. 15.; (Lund: Hakan Ohlsson, 1935), p. 4.

[38] Attention was directed to this aspect of Democritus' theory of the Milky Way in my book, *The Paradox of Olbers' Paradox* (New York: Herder and Herder, 1969), pp. 11–12. Lundmark who submitted *(op. cit.,* p. 4.) the claim

"that the Milky Way light [sic] is due to scores and scores of stars [theory *f*] is founded upon actual (morphological) studies of the Milky Way," rather wrongly ascribed this theory to Plutarch, to say nothing of the injection of the concept of "morphological studies." Clearly, expertise in observational or theoretical astronomy is not expertise in its history.

[39] See A. Grünbaum, *Modern Science and Zeno's Paradoxes* (Middletown, Conn.: Wesleyan University Press, 1967).

[40] See the Greek text and English translation by G. R. Mair in *Callimachus and Lycophron, Aratus* (Loeb Classical Library; London: Heinemann, 1921), pp. 418–19 (lines 469–79).

[41] *Gemini Elementa astronomiae,* edited by C. Manitius (Leipzig: B. G. Teubner, 1898), pp. 67–68.

[42] It was around this principle that the great controversy revolved in ancient, medieval, and Renaissance astronomy between formalists and realists, as masterfully documented and interpreted by Pierre Duhem in his *To Save the Phenomena: An Essay on the Idea of Physical Theory from Plato to Galileo,* translated from the French (1908) by E. Doland and C. Maschler (Chicago: University of Chicago Press, 1969).

[43] In fact, Ptolemy warned his reader in the opening section of his *Planetary Hypotheses* that mechanical analogies provided only a superficial understanding of the motion of planets. For a real insight one had to assume, according to Ptolemy, that planets moved like living beings, as if by instinct. Their regular pace was a replica of the uniform motion of a group of dancers and of soldiers drilling with their weapons. See *Opera astronomica minora,* edited by J. L. Heiberg (Leipzig; B. G. Teubner, 1907), pp. 71 and 120.

[44] See edition of Greek text and English translation by F. E. Robbins (Loeb Classical Library; Cambridge, Mass.: Harvard University Press, 1940).

[45] Translated by R. Catesby Taliaferro, in *Great Books of the Western World,* vol 16., (Chicago; Encyclopaedia Britannica Inc., 1938), p. 258.

[46] *Ibid.,* p. 260.

[47] *Concerning the Face which Appears in the Orb of the Moon,* vol. 12. in *Plutarch's Moralia,* with an English translation by H. Cherniss and W. C. Helmbold (Loeb Classical Library; London: W. Heinemann, 1957).

[48] The information is derived from Macrobius' commentary (c. 400 A.D.) on Cicero's "Dream of Scipio," which will later be discussed in detail.

[49] See S. Sambursky, *Physics of the Stoics* (New York: Macmillan, 1959), p. 42.

[50] *Lucretius, De rerum natura,* with an English translation by W. H. D. Rouse (Loeb Classical Library; Cambridge, Mass.: Harvard University Press, 1937), p. 109 (Book II, line 325). The silence of Lucretius on the Milky Way is all the more curious since, according to him, the stars were shooting out fiery light, and had their store of fire replenished from "the depths of the universe." *Ibid.,* p. 361 (Book V, line 305).

[51] *De Providentia,* Book II, sec. 51., in *Philo* with an English translation by F. H. Colson (Loeb Classical Library; Cambridge, Mass.: Harvard University Press, 1941), vol. 9, p. 493.

[52] See the translation in *Select Works of Porphyry,* by Thomas Taylor (London: Thomas Rodd, 1823), p. 193.

[53] *Physical Science in the Time of Nero: Being a Translation of the Quaestiones naturales of Seneca,* by John Clarke (London: Macmillan, 1910), pp. 288–89 (Book 7, chap. 15).

[54] *Pliny, Natural History,* with an English translation in ten volumes. Volume V. Libri XVII–XIX, by H. Rackham (Loeb Classical Library; Cambridge, Mass.: Harvard University Press, 1950), p. 367 (Book 18, chap. 69). Pliny's description, in Book 2, of the sky as a principal part of the cosmos, contains no reference to the Milky Way.

[55] *Ibid.*

[56] *Ovid, Metamorphoses,* with an English translation by F. J. Miller (Loeb Classical Library; London: Heinemann, 1916), p. 15 (Book I, lines 169–78).

[57] In Book 6, of his *De re publica.* See *Cicero, De re publica, De legibus,* with an English translation by Clinton W. Keyes (Loeb Classical Library; London: Heinemann, 1928), p. 269.

[58] *M. Manilii Astronomica,* edited by Iacobus van Wageningen (Leipzig: B. G. Teubner, 1915).

[59] *Ibid.,* p. 28 (Book 1, lines 755–58).

[60] *Hygini Astronomica,* Texte du Manuscrit Tironien de Milan, edited by E. Chatelain and P. Legendre (Paris: Libraire Honoré Champion, 1909) p. 6 (Book 1, chap. 2, par. 6).

[61] *Commentary on the Dream of Scipio,* translated with an Introduction and notes by W. H. Stahl (New York: Columbia University Press, 1952). This work also contains the full translation of "Scipio's Dream" by Cicero.

[62] *Ibid.,* p. 149.

[63] *Ibid.*

[64] *Ibid.,* pp. 149–50.

[65] *Ibid.,* p. 149.

[66] *Theophrastus, Metaphysics,* translated by W. D. Ross and F. H. Fobes (Oxford: Clarendon Press, 1929), pp. 31 and 39.

[67] The classic insistence by Socrates on a "new physics" concerned with purposefulness occurs in *Phaedo,* where he rejects the mechanistic physics of some Presocratics as destroying the possibility of vindicating morality and values. It was that insistence and fear of Socrates which ultimately determined the large and small features of Aristotelian physics. See also chapter I, "The World as an Organism," of my *The Relevance of Physics* (2nd printing; Chicago: University of Chicago Press, 1970).

[68] *Alexandri in Meteorologicorum libros commentaria,* edited by P. Wendland (Berlin: G. Reimer, 1901), pp. 37–43.

[69] *Ibid.,* p. 38 (line 31).

[70] *Ibid.,* p. 42 (lines 24–25).

[71] *Ibid.,* p. 38 (line 8). Aristotle used the same vague expression, "many times," *(pollaplasios).*

[72] *Olympiodori in Aristotelis Meteora commentaria,* edited by G. Stüve (Berlin: G. Reimer, 1900), p. 69 (lines 11–12).

[73] *Ibid.,* p. 66 (lines 18–19).

[74] *Ibid.,* p. 66 (lines 19–20).

[75] *Meteorologica,* 339a.

[76] Except perhaps the rejection by Proclus, in his commentary on Plato's *Republic,* of the opinion that Plato referred to the Milky Way by the expression "straight line." See *Procli Diadochi in Platonis Rempublicam commentarii,* edited by W. Kroll (Leipzig: B. G. Teubner, 1899–1901), vol. 2, p. 194. For Plato, see *Republic,* 616b.

[77] *Olympiodori . . . commentaria,* p. 75 (lines 24–25). The summary of the arguments runs from p. 75 (line 25) to p. 76 (line 5).

[78] *Ioannis Stobaei eclogarum physicarum et ethicarum libri duo,* edited by A. Meineke (Leipzig: B. G. Teubner, 1860), vol. 1, pp. 156–57. Stobaeus also reported in another section (p. 133) that Parmenides spoke of the Milky Way and the sun as manifestations of the celestial fire, when he described the heavens as consisting of "wreaths" in a manner similar to Anaximander's loops of fire on which circular holes represented the sun and the moon.

[79] *Joannis Philoponi in Aristotelis meteorologicorum librum primum commentarium,* edited by M. Hayduck (Berlin: G. Reimer, 1901), pp. 100–18. Concerning the date of composition of this work of Philoponus and its place in the development of his thought, much information can be gained from the study of E. Evrard, "Les convictions religieuses de Jean Philopon

et le date de son Commentaire aux 'Météorologiques'," in *Bulletin de la Classe des Lettres et des Sciences Morales et Politiques, Académie Royale de Belgique* 39 (1953): 299–357. A Latin translation of the commentaries of Olympiodorus and Philoponus on the *Meteorologica* was published in 1551 under the title, *Olympiodori . . . in Meteora Aristotelis commentarii. Ioannis Grammatici Philoponi scholia in I. Meteorum Aristotelis, Ioanne Baptista Camotio . . . interprete* (Venice: Aldus), which leaves something to be desired in accuracy.

[80] *Philoponi . . . commentarium,* p. 106 (line 9). It was with reference to Ammonius that Philoponus criticized the failure of Alexander of Aphrodisias to recognize Aristotle's disregard of the fact of multiple reflection in mirrors.

[81] *Ibid.,* p. 113 (lines 34–35).

[82] *Ibid.,* pp. 114–15.

[83] *Ibid.,* p. 115 (lines 17–28).

[84] *Ibid.,* p. 116 (lines 7–8).

[85] *De opificio mundi,* edited by W. Reichardt (Leipzig: B. G. Teubner, 1897), pp. 144–45 (Book 3, chap. 12).

[86] *Ibid.,* p. 117 (lines 30–31).

[87] *Ibid.,* p. 117 (lines 20–21).

[88] *Ibid.,* p. 118 (lines 11–12). This passage, without its context, is among the many representative texts translated by W. Böhm in his *Johannes Philoponus Grammatikos von Alexandrien* (Munich: Schöningh, 1967), p. 326.

[89] Joseph Needham, noted both for his learning and for his eagerness to present ancient Chinese science in the most favorable light, found only one small scrap of evidence of Chinese speculation on the Milky Way. It dates from the 4th century A.D., and ascribes the tides to the "swelling" of the Milky Way, which was pictured as passing beneath the ocean during its daily revolution in the starry sky. See J. Needham, *Science and Civilization in China,* vol. 3, *Mathematics and the Sciences of the Heavens and the Earth* (Cambridge: Cambridge University Press, 1959), p. 489. Needham's two other references to the Milky Way have no connection with Chinese science. The barrenness of ancient Chinese science on this subject is further evident in the absence of references to the Milky Way by ancient and medieval Japanese astronomers, who relied heavily on the Chinese masters. See Shigeru Nakayama, *A History of Japanese Astronomy: Chinese Background and Western Impact* (Cambridge, Mass.: Harvard University Press, 1969), p. 100.

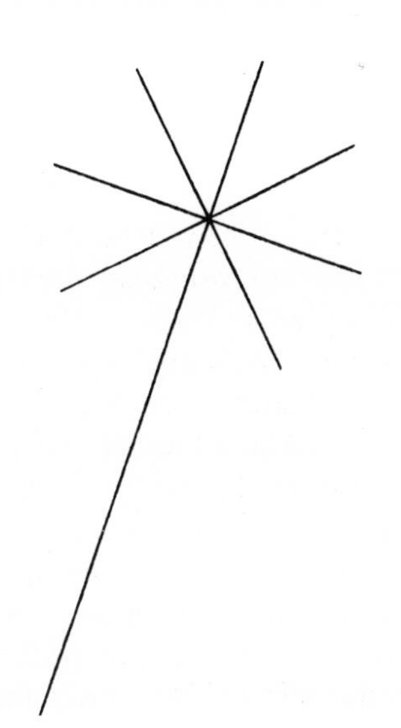

CHAPTER TWO

Medieval Gropings

"THE zone called milky is luminous because all stars pour into it their light." Such was the laconic account of the Milky Way in the *De imagine mundi,*[1] as unpretentious a summary of scientific information about the world as could be found in Western Europe around the middle of the twelfth century. Its author is as uncertain as its exact date of composition, but its popularity is well attested. Its contents were a faithful echo of the oldest medieval scientific tradition which began in seventh-century Spain, with the *Etymologia* of Saint Isidore (560–636). In that work, resembling a topical dictionary, the learned bishop of Seville covered the sundry fields of the seven liberal arts. He certainly owed much to Aristotle, including the latter's *Meteorologica,* but in his summary of astronomical lore he must have followed the elementary works by Geminus, Manilius, and Hyginus. Concerning what he called the "circulus candidus," he wrote: "The milky zone is a way, that can be seen on the [celestial] sphere, and its name is due to its whitish splendor. Some say that it is the path traversed by the sun, and that it receives its light from the luminous passage of the sun."[2]

Compared to Isidore of Seville, the author of *De imagine mundi* showed, in respect to the Milky Way at least, the better judgment. His short dictum about the Milky Way as the fusion of the light of all stars, by which he probably meant the light of the many stars in the belt of the Milky Way, was not very enlightening, but it did lead in the right direction. At any rate, it did not propagate Aristotle's fantasy and willfulness on the subject. Aristotle's idea of the

Milky Way received no consideration in *De mundi coelestis terrestrisque constitutione,* another popular work, composed about 1170 by an unknown author known today as Pseudo-Bede. His omission of Aristotle's opinion and his listing of some other opinions in regard to the Milky Way indicates that Pseudo-Bede had consulted Macrobius before he wrote: "In a serene weather one may also see the 'milky circle'; about it both fabulous and physical explanations are offered . . . Physicists [say] that God made two hemispheres, and that the luminosity is greater at the line where these are joined together. Other [physicists] say that there are small stars there, densely packed; their closeness produces that light and also effaces their distinctness. Again, others say that the stars of the zodiac shine there, and that the sun expends along the tropics the benefits of its light and fire toward both polar regions."[3]

Another major line of medieval scientific tradition, represented by the School of Chartres, produced no detailed comments on the Milky Way. The cosmological speculations of that school were limited to the framework of Plato's *Timaeus* where the Milky Way was not mentioned. On the other hand, there must also have existed in Chartres a keen awareness of the information which Macrobius provided on the Milky Way. William of Conches (c. 1080–c.1154), possibly the most original natural philosopher of that school, referred to Macrobius those who wished to learn about the cause and nature of the Milky Way.[4] Bernardus Sylvester, another member of the School of Chartres, spoke of the Milky Way as the combined light of many closely spaced stars. His two references to the topic appear in *De mundi universitate,*[5] and represent the few relatively enlightened details in a work permeated with gross animism.

The time when Bernardus Sylvester flourished also marked the beginning of the period when copies of most of Aristotle's scientific works spread rapidly across Western Europe. Among these was the *Meteorologica,* first translated into Latin by Gerard of Cremona (1114–1187) and his collaborators in Toledo.[6] It was far from the best of their translations, and was particularly misleading so far as the Milky Way was concerned. The blame for this rests largely on the fact that the translation was based upon an Arabic version of the *Meteorologica* made around 820 by an otherwise eminent Arab translator, Yahya ben al-Bitriq.[7] In his translation—which often degenerated into summaries and paraphrases—the comets were discussed after the Milky Way, an inversion of the original sequence. Al-Bitriq's account of the theories of Pythagoras, and of Anaxagoras and Democritus, including their refutations by Aristotle, represented the original reasonably well, but the reflection theory was only summarized. Aristotle's own exhalation theory of the Milky Way was completely omitted by al-Bitriq—clear evidence of its unreasonableness and of the translator's willfulness. According to al-Bitriq, Aristotle explained the Milky Way as the refraction of starlight in the hot, fiery air penetrating the region of fire: "The air which is near the celestial regions is hot and fiery. At the place where the Milky Way is seen in the sky [sphere], there are many small, densely placed stars and also big ones which are bright and close to one another. Since their light shines from behind that region which is hot and fiery, an elongated patch of light will be seen there. These stars are fixed and some of them touch one another and they receive their light from the sun. Such is the light which fuses into one. Therefore the Milky Way is seen in one and the same place in the sky and does not move from there."[8] To make matters worse, al-Bitriq claimed in the same breath that the Milky Way originated in the same manner as comets. But Aristotle's reasons for claiming this were different from those given by al-Bitriq, who stated that comets and Milky Way occupied much the same position and that their shining was a dispersion of the sun's light.

The emergence of a misleading account at this crucial juncture, the closing decades of the 12th century, befitted a sequence already replete with baffling reversals. True, al-Bitriq's text and that of the *old* translation, as the one by Gerard of Cremona came to be known, emphasized the role of starlight in the formation of the Milky Way, and gave no place to the exhalation theory. Nevertheless, the Milky Way was presented there as a sublunary phenomenon; that is, the refraction of starlight in the fiery air below the orbit of the moon. To make matters even more puzzling, the first medieval commentary on the *Meteorologica* by Alfred of Sareshel in England, written around 1200 in the form of glossae, contained no reference to Aristotle's explanation of the Milky Way as given in the *old* translation. The somewhat disjointed remarks of Alfred of Sareshel suggest rather that to him the Milky Way was of "elementary nature" and consisted of vapors.[9] It was not, of course, altogether impossible for Alfred of Sareshel to learn the true phrasing of Aristotle's explanation. During the closing decades of the 12th century Maimonides, for one, urged a reliable translation of the *Meteorologica,* and a fairly satisfactory Hebrew text was completed in 1210 by Samuel ben Judah ibn Tibbon.[10] By then Averroes' commentary, now a quarter of a century old, remedied the long absence of a good text of the *Meteorologica* among the Arabs.[11]

Whatever the source of Alfred of Sareshel's information on the exhalation theory, it failed to prove influential. It was the version in the *old* translation which was incorporated, around 1230, into the extremely popular encyclopedic work, *De genuinis rerum coelestium, terrestrium et inferarum proprietatibus,*[12] by Bartholomaeus Anglicus. This assured the theory great publicity but no lasting acceptance. In another encyclopedic summary of the various branches of learning, the *Speculum majus,*[13] written during the following decades by Vincent de Beauvais, the Milky Way was treated as one of the two visible great circles. This conjoined discussion of the Milky Way and the zodiac clearly implied a belief that the Milky Way belonged to the realm of the stars. Furthermore, Vincent de Beauvais named William of Conches as his source on the subject, although he failed to report the latter's dictum on the composition of the Milky Way.

The direction which would be followed by medieval opinion on the Milky Way could plainly be seen in the firmness that characterized the statement of Robert Grosseteste (c. 1175–1253), bishop of Lincoln and the first great figure of medieval science. He seems to have been influenced by what he found in Macrobius about Democritus' opinion on the Milky Way, although he also showed great interest in the *Meteorologica.* This may be seen from his treatise on the rainbow and from his discussion of the nature of comets. In the latter work he argued against Aristotle's theory that comets were the incandescence of dry exhalation. Grosseteste clearly wrote with an eye on the *Meteorologica* when he rejected various opinions about comets—one such being that comets were the fusion of the light of some stars (planets). Grosseteste remarked that such an opinion was an unjustified application of the correct explanation of the Milky Way: "Those who considered that many close objects appear continuous when viewed from a distance, know that the galaxy is an agglomeration of closely spaced stars that visually appear contiguous; [they] are inclined to conclude from this analogy that a comet is the grouping of several close stars (planets)."[14] Against such an interpretation Grosseteste noted that most comets were outside the zodiac where the planets moved.

Grosseteste's concise handling of the Milky Way did scant justice to the potentialities of his philosophy on the question of the nature of light. It remained for Grosseteste's foremost disciple, Roger Bacon, to offer a detailed discussion of the Milky Way within the framework of optics and cosmology.[15] At a cursory glance, there is little enlightenment in Bacon's claim that the Milky Way was an optical phenomenon located in the uppermost layer of fire below the orbit of the moon. Again, Bacon inaccurately portrayed Aristotle's idea of the Milky Way when he wrote: "Now the Galaxy is in one sense a circle in the celestial sphere, which is called the Milky Way, consisting of many small stars clustered, and in this part of the heavens [in the zone of fire immediately below the orbit of the moon] produces, according to Aristotle in the first book of the Meteorologics, the impression of continuous light through the meeting of the solar light with the lights of small stars of this kind, and this impression is called likewise galaxy."[16] The passage which introduced this quotation appears even more curious: "In regard to the Galaxy it is strange that it cannot be visible in the celestial sphere, or in the sphere of air, but only in the sphere of fire."[17]

It is, however, to the undying credit of Bacon that he was the first to discuss the problem of the Milky Way in connection with a question which he was probably the first to raise: why is the night sky dark? Today the question is known as Olbers' Paradox, a major challenge to scientific belief in a homogeneous infinite universe. But the paradox of the darkness of the night sky could arise just as logically in a finite world enclosed by a crystalline sphere. In such a universe, according to Bacon, the celestial sphere and the uppermost layers of air should be constantly illumined by the light of stars and especially by the brilliant radiation of the sun. Thus, Bacon argued, the daily dawn should come much earlier than it actually does, for the earth's shadow could hardly block diffusion of sunlight over the sky behind the earth for an entire night.

According to Bacon, "excessive distance together with excessive rarity solves this problem."[18] A very rare medium, such as the outermost sphere, which Bacon believed to consist of very pure water,[19] could not retain light: "The celestial sphere is of such excessive rarity that light cannot be incorporated in it so as to be visible."[20] Bacon, unwittingly enough, spelled out a contradictory aspect of this explanation, since he claimed that light simply passed through the celestial spheres, the last of which carried the stars.[21] But could light really pass beyond the confines of the universe, or was it continually absorbed in the outermost watery layer? Conceptually, the former question could be perceived logically even in the context of Bacon's cosmology and physics. Considering his extensive speculations about the focusing of light rays, to say nothing of his experiments with concave mirrors, he might also have had an inkling of the thermodynamics implied in the second question. At any rate, he certainly knew that layers of water, when exposed to the summer sun, would become warm. The problem no doubt failed to take shape in his mind because he believed that the outermost regions were made of "very pure water."

At the same time, Bacon unerringly sensed that the Milky Way represented a major obstacle against the concept of the universe as being enclosed within a crystalline sphere. A solid sphere must have reflected sunlight everywhere, or, to use Bacon's expression, light would have been "incorporated" into it. In particular, Bacon believed, the light of the stars of the Milky Way would have produced a circular band of starlike brilliance in that sphere. In addition, he illustrated the point by a striking analogy. The light of the Milky Way, he wrote, appears "continuous and lengthened out, although the stars are distinct. But distance causes this, as would be the case if one looked from a distance at a pot perforated in many adjacent parts, in which fire is burning. For owing to the distance, the fire would appear continuous to him because of the proximity of the holes, which the eyes would not distinguish because of the great distance."[22] Bacon was quick to add that the stars of the Milky Way were small, that their light could not be retained in the liquid outer sphere, that the same light was partially lost because of the distance, and that it was too weak to overpower the darkness inherent in the density of ordinary air. The weak light of the stars of the Milky Way could not make itself visible except in a material with a density intermediate between that of the air and that of the outermost waters. That material was, according to Bacon, the layer of fire immediately below the orbit of the moon. In stating this, he clearly endorsed the doctrine of the *old* translation.

Bacon's account of the Milky Way is also noteworthy for its lack of influence from Arabic sources. This is in itself all the more significant since Bacon was an avowed admirer of the Arabic masters, and especially of their astronomer-astrologers. In connection with the Milky Way, he could not learn much from them. The Arabs' disinterest in the topic is evident from their centuries-long neglect in producing a good translation of the *Meteorologica.* What al-Biruni, one of the most enlightened and learned Muslim scholars, wrote in 1029 of the Milky Way, illustrates the minor level of their speculations about it, and their confused ideas of Aristotle's teachings on the subject: "The Milky Way, *kahkashan* [in Persian] is a collection of countless fragments of the nature of nebulous stars. They form a nearly complete great circle which passes between Gemini and Sagittarius, the stars densely packed in some places, more scattered in others, the way sometimes narrow, sometimes broad, and occasionally breaking up into three or four branches. Aristotle considered that it is formed by an enormous assemblage of stars screened by smoky vapours in front of them, and compared it to haloes and nebulae."[23]

So far as the Arabic ambience goes, there is a rather exceptionally fine, though short account of the Milky Way by the Nestorian bishop, Abou'l-Faradj, or Aboulfarag (1126-1286), known also as Bar-Hebraeus, because his father, Aaron, was a Jewish physician. In his "Book of the mind's flight over the form of the heavens and the earth,"[24] an impressive summary of planetary theory and positional astronomy composed in 1279, he discussed the Milky Way in a lengthy chapter devoted to the fixed stars and constellations. Most interestingly, he seemed to consider the Milky Way as a partly resolvable nebula: "There exist in the heavens certain whitish patches, called nebulous stars. Some think that they form part of the Milky Way, because, like the Milky Way, they resemble clouds. Others think that they are a multitude of stars, very small and very close to one another, like that texture similar to the leaves of ivy, which is found above the tail of the constellation Leo. And those who think along these lines also say that the whole Milky Way is formed of stars very small, joined together. It is evident that the Milky Way is not vapor or smoke suspended in the air as the Peripatetics say, for the moon and the five planets do not undergo any change in their luminosity as they pass through it. Rather they modify the [appearance of the] Milky Way. The Pleiades are not [a] nebulous [star], because there the stars are separated from one another, and number exactly six."[25]

Roger Bacon could hardly have been aware of the existence of Aboulfarag's treatise. At any rate, the *Opus majus* was written 13 years earlier. But, in connection with the Milky Way, Bacon seems to have ignored the very extensive Arabic commentary on the *Meteorologica* composed by Averroes[26] (1126–1198). Concerning the Milky Way, Averroes was not "Averroist," that is, a champion of any and all dicta of Aristotle. The text of the *Meteorologica* Averroes commented upon was remarkably reliable, and permitted no doubt about the view Aristotle advocated on the Milky Way. It is almost impossible to imagine a more convincing proof of the illogic inherent in Aristotle's explanation of the Milky Way than to find Averroes himself branding it as a *non sequitur.* According to Averroes, Aristotle's explanation depended on the existence and properties of exhalation for which he offered no independent proof. Less politely, this meant that Aristotle was guilty of *petitio principii,* a rather elementary error. There were other difficulties, as well. If the Milky Way were located below the orbit of the moon, it could not simultaneously be seen from various parts of the earth against the same stars and constellations. In this connection Averroes referred to his own findings on the respective positions of the Milky Way and of the constellation Aquila. Regardless of whether he observed from Cordoba or Morocco, he found one of the stars in Aquila, at the edge of the Milky Way, in exactly the same position. Furthermore, Averroes argued, the Milky Way could not remain the same because the quantity of exhalation must vary according to the weather prevailing in various years.

Averroes remained, however, true to his reputation as a resolute defender of the Aristotelian world view. He rejected the opinion that the Milky Way was part of the celestial sphere, since, in the Aristotelian creed, the heavens had to be eternal and, therefore, absolutely perfect. Nebulous blemishes like the Milky Way could have no part there. So Averroes retained the Milky Way as an image which the rays of many small stars formed in the vicinity of the sphere of fixed stars. The existence of such an image posed more than a few questions which Averroes bravely tried to solve by specious distinctions between active and passive light, and by continual references to the laws of refraction and perspective, without trying to be specific. Unlike Grosseteste, who satisfied himself by stating that objects which were both distant and close appeared continuous, Averroes offered three theories on how the image of the galaxy could be formed. The first explanation started reasonably enough: "The cause of that refracted image is the weakness of vision in its apprehension and sighting because of the smallness of those stars in addition to their distance."[27] The rest was distinctly unclear, beginning with the statement: "refraction already takes place because of the weakness of vision," and concluding with the particularly muddled passage: "it happens that the perception of such an image is weaker in the substance of the galaxy, for the very reason that its kind is to be igneous *per se*."[28] Such complications hardly improved on the mental labyrinths of Aristotle's theory.

The comments of Albert the Great (Albertus Magnus) on the Milky Way[29] show one similarity to those of Averroes. Albert, too, went far beyond a preliminary and partly acceptable statement, although for reasons different from those evident in Averroes' handling of the matter. Averroes' verbosity was caused by the metaphysician's concern to save the perfection of the heavens and the reputation of the "master." Albert became carried away by his attitude as a naturalist, attentive to all details of a broader picture. He began the chapter entitled "De galaxia secundum veritatem," by stating that "nothing else is the galaxy than many small, almost contiguous stars."[30] This was a step in the right direction, and, because of Albert's enormous popularity as a scientist during the fourteenth and fifteenth centuries, it engraved on the minds of many readers the belief that the Milky Way was not a sublunary exhalation, but a configuration of the stars. They could not, however, learn from Albert's comments the more valuable form of Democritus' explanation of the Milky Way. Nor did they receive reliable information from Albert's claim that the Milky Way was also a denser part of the celestial sphere which reflected, more effectively than did other parts of that sphere, the light of the stars, the sun, and the planets.

This might help to explain why Albert defined the location of the Milky Way as "that band of the sphere of stars where the sun's light is diffused." By "diffused" he obviously meant that the denser part of the heavens reflected the sunlight and starlight in such a manner that "one sees there a whitish, smokelike circle."[31] He also set great store by the invisibility of the sublunary zone of fire in order to make more plausible the claim that Aristotle could not have helped but place the galaxy in the firmament of stars. Here one comes face to face with the strangest aspect of Albert's discussion of the Milky Way. He concurred with the *old* translation which glossed over the fact that Aristotle had done his best to prove that the Milky Way was a slowly burning volume of dry exhalation. In fact, Albert insisted that Aristotle could not even have stated that the galaxy was the refraction of starlight in the region of pure fire immediately below the moon's orbit, since that fire was invisible according to Aristotle, Algazel, and Avicenna. He discussed the whole matter without any reference to Averroes. According to Albert, Aristotle held that the galaxy was of the same substance as the stars, a view shared, so Albert claimed, by Ptolemy and Avicenna.

No such misrepresentation of the historical record mars the comments of Aquinas on the *Meteorologica.*[32] Actually, Aquinas was eager to work from the most reliable texts available, a fact amply attested by his close association with William Moerbeke (c. 1215-c.1286), renowned for his translations from the Greek. Among Moerbeke's productions were a translation of Alexander of Aphrodisias' commentary on the *Meteorologica* and the first medieval translation of the *Meteorologica* itself from the Greek, which subsequently became known as the *new* translation.[33] The demand for this translation, completed in 1260, was so great that within a few years Mahieu le Vilain produced a French translation of it.[34] The minor criticisms in Aphrodisias' commentary on Aristotle's theory of the Milky Way failed to alert Aquinas to some of its grave difficulties. Nor did clear presentation of the role of exhalation in the *new* version prompt him to further reflection. Actually, Aquinas offered no comments in the real sense of the word. His dicta on the Milky Way were simply a paraphrase of Aristotle's text, which was too clearly structured to require Aquinas' step-by-step account of its major and minor divisions. Aquinas found nothing more to be desired in Aristotle's explanation of the Milky Way. If he himself contributed anything to the question, it was merely in the area of semantics. He called *vehemens* the effect of the sun, of the planets, and of the stars in the zodiac on the exhalation below. *Efficax* was his word for the allegedly preservative influence which the stars in the belt of the galaxy exerted on that same gaseous substance.[35]

This attention to nomenclature indicated that Aquinas clearly grasped the substance of Aristotle's theory. His acceptance of it accorded with his intention to reject only that part of Aristotle's physical and cosmological theories which patently contradicted tenets of the faith. The Milky Way, as Aristotle explained it, presented no area of conflict.

Whatever the differences between the *old* and *new* translations,[36] there was some petulance in Dante's remark, when he wrote, about 1297, in his *Convivio*[37] that "what Aristotle may have said on this point [the Milky Way] cannot be rightly known, because his opinion does not appear the same in one translation as in the other."[38] In the second treatise of his *Convivio,* where Dante made this complaint, he gave a surprisingly good account of astronomical lore.[39] His principal source was al-Fargani's elementary treatise on astronomy, translated into Latin, in 1134, by John of Seville and later by Gerard of Cremona. Dante was also familiar with more modern works. His information on the Milky Way was gained directly from the *Meteorologica's* two Latin translations. From the *new* translation Dante learned that "Aristotle seems to say that it [the Milky Way] is a congregation of vapours beneath the stars of that region, which ever draw them up."[40] Aristotle was, of course, more emphatic on the matter than just to "seem" to hold the view in question. Dante certainly did not do justice to the *old* translation, for he claimed on the basis of it that Avicenna and Ptolemy shared Aristotle's view that "the milky way is nought else than a multitude of fixed stars, in that region, so small that from here below we may not distinguish them, though they produce the appearance of that glow which we call the milky way; and it may be that the heaven in that region is denser, and therefore arrests and throws back the light."[41] At the same time, Dante scored a good point by noting "that the milky way is an effect of those stars which we may not see, save that we are aware of these things by their effect."[42] In such an explanation of the Milky Way, which he fully accepted, Dante saw a clear illustration of the nature of metaphysics, which in his view "treats of the primal existences, which in like manner we may not understand save by their effects."[43] For if Dante took up astronomical topics in the *Convivio,* it was not for the sake of astronomical knowledge, but merely to provide examples for the analysis of metaphysical and ethical questions. His final remark on the topic was that "the starry heaven hath great similitude to metaphysics," a conclusion which he repeated almost verbatim after discussing the precession of equinoxes.[44]

The *Convivio* represented for a wider audience an explanation of the phenomenon of the Milky Way which at least pointed in the right direction. Its usefulness can best be appreciated when set against the astrological treatises which commanded much broader interest than did the purely astronomical works. In line with a long tradition, the Milky Way was neglected by medieval authors on astrology, lengthy though their compositions often were.[45] Unfortunately, no reference to the Milky Way appeared in the *De sphaera* of Joannes Sacrobosco, composed in 1233, which remained the most popular astronomical textbook well into the middle of the sixteenth century.[46] The same holds true of similar treatises that for a while competed with it. Among these were the *Sphaera* (c. 1225) of Grosseteste[47] and the *Sphaera* (c. 1250) of John Peckham[48] (c. 1225–1292) who ended his career as archbishop of Canterbury. Grosseteste spoke of the Milky Way in another context, as has already been indicated. Peckham considered the topic of the Milky Way in his *Perspectiva* written between 1269 and 1279. There, in the last, or twenty-second proposition, he rebuked those "who were not ashamed to contradict the Philosopher [Aristotle] and state that the galaxy is not produced in the purest region of fire [right below the orbit of the moon], as if visual image *(impressio)* could not be formed in a transparent body. But we can see the rays of sun as they pass through the air in a dark room, although the air is without any tangible density. Clearly, the most vehement radiation of the light cannot remain concealed. The multiple diffusion *(multiplicatio)* of the rays of stars in the highest region of fire may, therefore, appear perceptibly for the same reason."[49]

The passage, a defense of the "Aristotelian" theory as given in the *old* translation is strong evidence of independence of thought on that subject among medieval scholars. Popular as was the *Perspectiva* (it went through some ten printings between 1482 and 1627), this last proposition never achieved its intended impact. John Pena had patently exaggerated in 1557 when stating that Peckham's dictum was taken as an oracle by most physicists.[50] At any rate, no proposition on the Milky Way came to be incorporated in an even more influential medieval treatise on optics composed by

Witelo between 1270 and 1278. His chief source of information was Ibn al-Haitham's (Alhazen's) work, *Kitab al manazir* (The book of vision), and he followed it rather closely.[51] Like Alhazen, Witelo did not discuss optical fusion of the image of distant and relatively closely spaced objects. It must, however, be noted that he had more than one logical opportunity to take up the subject of the Milky Way. He repeatedly analyzed the causes and features of optical illusions,[52] spoke at length of such topics as the much greater separation of stars near the horizon than at the zenith, and argued that if some appropriate cloud were present there, the distance between stars would appear larger.[53]

During the fourteenth century the number of commentators on Sacrobosco's *Sphaera* steadily increased, but the commentaries as a rule avoided discussion of the Milky Way. Among them was the work by Cecco d'Ascoli which became the first of such works to be printed.[54] Far more important was the increase of great commentaries on the cosmological treatises of Aristotle.[55] There ground was broken for some decisive developments in physics and astronomy. The commentaries of Buridan, Oresme, and Albert of Saxony on Aristotle's *Physics* and *On the Heavens* are now recognized as important documents in the history of science. Unfortunately, it is merely a paraphrase of the Aristotelian text on the Milky Way that Buridan wrote in his commentary on the *Meteorologica*.[56] Oresme's exposition of the passage is considerably longer,[57] but he seemed to belabor what had then become obvious; namely, that the galaxy could not consist of "elementary" (terrestrial) substance. Neither Buridan nor Oresme made any effort to rescue Democritus' idea of the Milky Way from the misleading Aristotelian context. For both of them the Milky Way was a denser part of the heavens, which Buridan also stated concisely in his explanation of *On the Heavens*. There, in discussing the nature of the darker spots on the moon's surface, Buridan recalled the opinion of Averroes who identified them as regions of lesser density. Then Buridan added: "And the same holds true of the galaxy; there parts of the stellar orb are in places denser than elsewhere; therefore the Milky Way can retain and reflect the light of the sun although not perfectly; therefore those parts appear whiter than the other sections."[58]

Details of Buridan's teaching on the Milky Way might perhaps be present in the commentary on the *Meteorologica* by Themon

(Themo Judaei). He rose to prominence both as a teacher and administrator at the University of Paris around 1350, when Buridan's long career there came to a close. The importance of Themon's work can be seen both from the impact which it made on Leonardo da Vinci and from the fact that it was reprinted a half dozen times between 1480 and 1522.[59] Themon received most praise from historians of science for his criticism of the Aristotelian doctrine that the center of gravity, the center of the earth, and the center of the universe occupied the same point. The irregular shape of the earth precluded, according to Themon, any such coincidence. This careful attention to the strict exigencies of geometry is further evident in his discussion of the Milky Way. He emphasized that a parallax would be shown by the Milky Way were it located below the orbit of the moon. In that event, observers in Greece and Spain would not see the Milky Way against the same stars. In particular, he singled out as illustrations of the point the position of the constellation Gemini in the Milky Way and the intersection of the Milky Way with the zodiac. To Themon the Milky Way was, as to Buridan, a denser part of the sphere of stars which reflected and refracted the light of stars and of the sun in a special manner. That such a departure from uniformity was possible in the heavens Themon argued with an eye on Aristotle, who had himself claimed that stars were points of very high density in that ethereal orb. In fact, Themon ascribed the Aristotelian theory of the Milky Way to the misguided effort of some translators.[60]

Although Themon hailed from Münster in Westphalia, his administrative positions at the Sorbonne had largely to do with students of the "English nation." It should, therefore, occasion no surprise that his discussion of the Milky Way can readily be recognized in the commentary on the *Meteorologica* written around 1370 by Simon Tunsted,[61] an English Franciscan best remembered for his musical theories. He too tried to exculpate Aristotle, but curiously enough he placed the blame on the translator who produced the *old* version. He waxed rather verbose on the various properties of the darker and whiter parts of the moon's surface. More interesting was his remark that the whiter parts of a piece of alabaster appeared to be more dense than its darker spots. This seemed to agree rather well with the theory that the galaxy was the denser part of the crystalline sphere of the fixed stars.[62]

As the fourteenth century drew to a close, Chaucer had an opportunity to go further than the story of Phaethon which he recounted in *The House of Fame*.[63] But to include a reference to the Milky Way in his famous *Treatise on the Astrolabe*[64] would have meant abandoning rather standard procedures. Traditional forms may have played a part in the omission of the Milky Way from the descriptive astronomy and geography composed by Pierre d'Ailly around 1410, under the title *Ymago mundi*.[65] There was, for instance, no mention of the Milky Way in the "Imago mundi" written in 1245 by Gossouin, which won fame in 1480 when William Caxton translated it as *Mirrour of the World*.[66] A picture of the world could hardly be complete without that magnificent whitish band of the Milky Way, especially when it included sundry details of the zodiac, which d'Ailly called the "cincture of the firmament."[67] The Milky Way would have been a far more realistic choice for that distinction. Again, although d'Ailly described the sublunary region of fire as the place where, according to Aristotle, comets were formed,[68] he failed to mention the Milky Way which, to Aristotle, was a special comet. This strange omission was not due to disagreement on d'Ailly's part with Aristotle's theory of the Milky Way. Not surprisingly, d'Ailly aired his views on the subject in a short commentary on the *Meteorologica*.[69] There he accepted Aristotle's explanation except for one minor detail. According to d'Ailly, the exhalation acted not as an incandescent substance, but rather as a refracting medium for the light of stars which were very numerous in the Milky Way.[70] He put forward his dissent, however, with the deferential remark that Aristotle himself must have had the same explanation in mind.

In the history of science d'Ailly is hardly a figure of progress, nor was the fifteenth century an improvement on the fourteenth in this respect. The dicta on the Milky Way, or rather their absence, are a case in point. This is even true of the most outstanding scientific book of the fifteenth century, the *De docta ignorantia* of Nicolas of Cusa.[71] His bold ideas on an indefinitely large universe, in which the center was everywhere and the circumference nowhere,[72] might have formed a natural framework for a discussion of the optical consequences of an indefinitely large number of stars. Cusanus, perhaps deliberately, chose to leave aside such problems. He pictured each star as having a solid, relatively cool surface which was enveloped in turn by concentric layers of air, water, and fire. By his theory, a star's fiery layer only emitted light and heat in an outward direction. Thus the earth which, together with the moon, was a star, appeared to the denizens of the moon as a bright star; but since the earth was, he declared, within the moon's watery layer, the moon did not appear to us as a fiery disk.

Obviously, Cusanus' mind was firm in its belief in a universe populated everywhere by intelligent beings. His cosmological considerations tried to accommodate those beings, rather than to unfold the physical consequences of some highly interesting premises. The problem of the Milky Way did not profit in such a context from statements that, although enticing in themselves, seemed to close the door on it. Cusanus stated that "the regions of the stars that sparkle are the only ones we see." At the same time, he also referred to the influence on the earth of all those stars "that are invisible to us, since we are outside their region." The influence in question was the light of so many stars: "Each star communicates to another light and influence; but this communication is not the purpose of the stars, for all stars move and sparkle for one sole purpose: the realization of the best possible existence for them; and from that communication follows as a consequence. Likewise, light gives light not that I may see but it is of its very nature to give light; the communication of light takes place as a consequence when I use it for the purpose of seeing."[73]

The taciturnity about the Milky Way which Regiomontanus (Johann Müller), the most notable astronomer of the century, maintained, was probably due to his great admiration for Ptolemy. A major aim of Regiomontanus was to publish a critical translation of the *Almagest,* but he completed only a lengthy commentary on its principal passages[74] by his master, Georg Peurbach. It was only published in 1496, twenty years after Regiomontanus' death. He described the Milky Way "as a belt in the sky of varying color and width, and is called 'milky' because of its hue; it seems to imitate the color of milk for the most part. As it spans the whole firmament, it divides into two branches."[75] This unoriginal piece of information was followed by a list of the principal stars lying within the Milky Way. Finally, Regiomontanus instructed his reader to turn to Ptolemy for further information—not particularly helpful advice, so far as the Milky Way was concerned.

The most voluminous fifteenth-century work on astrology by Giorgio Valla (1447–1500) evinced much the same perfunctory attitude to the Milky Way. Clearly, a broad and steady phenomenon like the Milky Way was an unwelcome detail in a massive opus devoted to "things to be sought and things to be avoided."[76] For purposes of astrology the Milky Way belonged, from Ptolemy on, to the latter class. The few words devoted to it in books on astrology were characterized by the vagueness and contradiction with which astrologers argued their case. Thus, in the *De rebus coelestibus*[77] of Giovanni Pontano (1422–1503), the Milky Way served as an illustration of "hidden things," and the differences in its explanation by Aristotle and by "the mathematicians" were glossed over with the remark that their methods were different. A classic example of the broad but superficial and often inconsistent reasoning of the Humanists of his time, Pontano was best remembered for his amatory poems. His poetical inclinations decided not only the form in which his *Meteororum liber unus* was written, but were also largely responsible for the lack of cogency there. The Milky Way, Pontano claimed, was both the light of many small stars and the incandescence of the upper air.[78] The science of the Milky Way was not helped, either, by the only novel information he offered—that the Milky Way was the path which came into existence when Jupiter routed the giants in a cosmic battle. For all that, the work was quoted with surprising deference several times during the next one and a half centuries.

Pontano's *Meteororum* was an early forerunner of treatises on meteorology which, a century or so later, began to display distinct independence from the framework of Aristotle's *Meteorologica*. However, for the time being, meteorological lore was still discussed in the form of commentaries on it. The list of these during the second half of the fifteenth century begins with the one written by Gaetano de Thiene (1387–1462). His "expositio"[79] is interesting for several reasons. First, it commanded the obvious interest of immediate posterity, for it was printed half a dozen times between 1476 and 1522. Second, Gaetano, a leading Averroist, admitted that Aristotle's theory of the Milky Way was impossible to credit, because that theory could not explain its invariable position in the sky and the perfect steadiness of its shape. Significantly, Gaetano also referred to Albertus Magnus as the exponent of the correct theory. It consisted, in Gaetano's version, of the diffusion of the sun's light by the many, almost contiguously spaced stars of the Milky Way, and included the mutual reflection of their own light. Gaetano also compared the lunar spots with the shade of the Milky Way, arguing that in both cases the physical cause was a higher density of the ethereal material. In this he avowedly followed the lead of the "Commentator," Averroes.

The celestial nature of the galaxy as a denser part of the heavens, like the stars, was firmly voiced around 1500 in Gregor Reisch's extremely popular *Margarita philosophica*[80] and in the commentaries on the *Meteorologica* by Jacob Tymaeus van Amersfort[81] (fl. 1490), by Thomas Bricot[82] (fl. 1480), and by Pietro Tartaret[83] (d. c. 1509). According to them, this was the "opinio communis" and they credited it to Albertus Magnus. Bricot, for one, claimed that the *old* translation had already placed the galaxy among the stars. To Johannes Versor (d. 1485), the *old* translation represented the genuine view of Aristotle on the Milky Way, since he could not imagine that a blatant error, such as the exhalation theory, could have had Aristotle for author. He also praised Ptolemy, Avicenna, and Albertus Magnus for refusing to accept the version of Aristotle's explanation given in the *new* translation.[84] Even by late fifteenth-century standards, these dicta of Versor denoted a rather reprehensible confusion, but they were not as difficult to understand as the position taken by the famed polyhistorian of the age, Jacques Lefèvre, who was also known as Jacques d'Etaples or Jacobus Stapulensis (1455–1537). He had a keen awareness of the need for reliable texts and translations, whether understanding of the Bible or of Aristotle was concerned. He was an eager student of classical authors as well as of astronomy. Nevertheless, in his short commentary on the *Meteorologica* he characterized as "reasonable" the Stagirite's explanation of the Milky Way by exhalations.[85]

No confusion about the intrinsic merits of opinions or about the two translations marred the comments of Hyeronimus de Sancto Marco. He was an Oxford Franciscan, who, in a booklet written during the opening years of the sixteenth century on the universal machinery of the world and on meteorological phenomena, took Albertus Magnus as his guide on the Milky Way, although he expressly noted that he was writing "ad mentem Aristotelis." After reporting the opinions listed by Aristotle and the explanation given by him, the Oxford friar noted: "There is another opinion, namely the one held by Albertus Magnus and by many other philosophers which, I believe, is closer to the truth concerning the 'generation' [physical origin] of the galaxy." The "material cause" of the galaxy is "that part of the heavens which, because of the multitude of stars there, is denser than other parts, and therefore retains the light of the sun and of other stars, and also reflects that light more effectively in those regions of the heavens. Therefore we conclude that the galaxy is of the nature of the heavens and is an integral part of the eighth sphere."[86]

The eighth sphere was not, however, as solid as was usually stated. The sixteenth century witnessed an increasing uneasiness on this point, rooted in part in the skepticism of some leading Humanists. Its best known example is Erasmus' *Praise of Folly* in which the contradictions of philosophers were the target of numerous acid —if not sarcastic—remarks. The same topic was far more systematically and comprehensively treated in the *Examen vanitatis* of Pico della Mirandola.[87] He set great store by the diversity of opinions on the Milky Way, and reproved Albertus Magnus for trying to attribute to Aristotle an explanation of it which was free from the absurdities of the exhalation theory. Interestingly enough, he reported Democritus' opinion in the form that the Milky Way was simply the light of many closely spaced stars. This was perhaps an indication of the validity he attributed to such a conception of the Milky Way. As a polemist, Pico could scarcely be even-handed. He failed to note that the great number of past opinions was only one side of the coin. The other was the indisputable fact of an almost general conviction that the Milky Way was an integral part of the realm constituted by the stars themselves.

Almost contemporary with Pico's work was the commentary on the *Meteorologica* by Francesco da Diacceto (1466–1522), Ficino's successor as leader of Florentine Neoplatonism. Diacceto was a diplomatic advocate of syncretism between Platonist paganism and Christianity. He gave a small but telling proof of his conciliatory attitude when he registered his displeasure with the exhalation theory of the Milky Way "with all the respect due to the greatness of Aristotle."[88] Diacceto also refused to accept Aristotle's claim that Anaxagoras and Democritus failed to see that the shadow of the earth could not reach as far as the sphere of the fixed stars. According to Diacceto, Anaxagoras and Democritus were of the belief that "the Milky Way is the light of very bright stars . . . , which are also very large and so numerous as to seem to touch one another. To our benefit they shine with other luminaries at night, when the earth's shadow covers everything with darkness; this is why the earth's shadow comes into the picture, and this is what Anaxagoras and Democritus meant to say, and this is what stands to reason."[89]

If there was one point on which Aristotelians could agree with Neoplatonists at that time, it was their common disagreement with Aristotle regarding the Milky Way. Even such a resolute Averroist as Agostino Nifo (c. 1473–c. 1545), who found nothing basically wrong with Aristotle's theory, readily admitted that it was not without difficulties which "the Commentator [Averroes] elaborated upon and very extensively at that."[90] To Nifo, the steady shape and hue of the incandescent exhalation was sufficiently explained by the perpetual sameness of stars in the belt of the Milky Way. To non-Averroist Aristotelians, the absurdity of such reasoning could not be so easily glossed over. To exculpate Aristotle one could resort to the old device of blaming some copyist of the original text. This was done by Joannes Dullaert[91] (c. 1470–1513), who also disputed the accuracy of the erroneous "Aristotelian" explanation of the Milky Way in the *old* translation. Here he took his cue from Albertus Magnus, but he showed some independence of thought by arguing against the exhalation theory which turned the Milky Way into a special comet. According to Dullaert, a consequence of that theory would have been the appearance of comets when the Milky Way was visible, because of the greater abundance of exhalations at such times. An even more valuable part of Dullaert's discussion was a diagram, perhaps the first of its kind to be printed, which showed that a sublunary phenomenon (if such were the comets and the Milky Way) should appear against different stars, when viewed from widely distant points on earth (Illustration IIIa).

It was also on the question of comets that Aristotle's theory of the Milky Way received its principal criticism at the hands of Johannes Eck (1486–1543), who earned fame through his debates with Luther. He noted, at the very outset of his comments on the Milky Way, that precisely "because of the objections which he raised against other theories about comets, Aristotle should have spoken rather differently [about the Milky Way] which in all its parts appears the same, whereas exhalations and vapors are not produced at the same rate during the various seasons of the year."[92] Eck sharply censured Lefèvre for his defense of Aristotle's opinion, and expressed doubts about the genuineness of a glossa in some copies of the translation of the *Meteorologica,* which attributed to Aristotle a two-tiered theory of the Milky Way. To Eck the Milky Way was a part of the firmament where its substance was condensed into many small and large stars, and he defended the varying density of the celestial substance by referring to what Themon said concerning this in his commentary on the *Meteorologica.*

Equally incisive were the statements on the Milky Way which Casparo Contarini (1483–1542), the future Cardinal, penned in the late 1510's. After studying in Padua, Contarini spent almost two decades in private pursuit of scholarship before devoting himself entirely to the cause of ecclesiastical reforms. In his *De elementis libri quinque,* which dates from the earlier phase of his life, Contarini singled out the perpetual sameness of the Milky Way as the reason why most Peripatetics failed to follow Aristotle on this point: "It appeared to them incredible that along that way the fiery substance should forever conserve its consistency without ever showing any change whatsoever."[93] According to Contarini, the generally accepted Peripatetic view of the Milky Way was that it "was an *accidens* of the ethereal orb due to the multitude of innumerable stars there: they are so minute that we cannot see them separately: all their lights are seen by us as fused together, and this is what constitutes the brilliance *(candor)* of the Milky Way."[94] To this he added another cause: the possible higher density of the ether in that region of the sky.

In keeping with other baffling detours that had handicapped and were still to hamper progress in understanding the Milky Way, the lectures given in Bologna on the *Meteorologica* by Ludovico Boccadiferro (1482–1545), some time after 1531, were published decades later[95] and failed to receive appropriate notice. This was all the more unfortunate since no one had previously matched his account of the views on the Milky Way of Aristotle and of his major commentators. Boccadiferro was as resolute a critic of Aristotle as he was of his commentators. Of these latter he chided Olympiodorus, Philoponus, Averroes, Albert, and Lefèvre for attempting to present either a modified version of Aristotle's view, or to present it as one submitted only as "probable" by the Stagirite.[96] "But gentlemen," said Boccadiferro to his class in typical lecture style, "I do not think that this is true, because Aristotle does not seem to doubt that he found the real cause of the production of the Milky Way; in fact he believes to have found its very cause!"[97]

No less pointed was Boccadiferro's remark that if Aristotle's claim about comets as causes of dry weather were true, then rain clouds could hardly form under the belt of the Milky Way.[98] Again, Boccadiferro turned the tables on Aristotle in connection with the whiteness of the Milky Way. In the whiteness of stars Aristotle saw proof that their substance was not fiery. He therefore contradicted himself when he attributed the whiteness of the Milky Way to the incandescence of dry exhalation.[99] On the other hand, Boccadiferro defended Aristotle's argument against the reflection theory, but here he seemed guilty of the same error for which he criticized the commentators who tried to put Aristotle's opinion in a better light. Boccadiferro should have remembered what he had said about efforts to credit Aristotle with the two-tiered theory of the Milky Way, one in the heavens, one below the moon: "If you say that this latter circle is a perfect replica of the former then you do what Aristotle did who, by trying to avoid Scylla, crashed into Charybdis, so that he actually made the bigger mistake: one would then have a sublunary body staying forever in the same condition, the very opposite of what Aristotle argued in the first book of *On the Heavens*."[100] Boccadiferro should also have recalled that he had criticized Nifo, a stout defender of Aristotle's view. This fact alone should have kept him from stating without qualification that "all commentators are against Aristotle, the Latins, and the rest, and also the astronomers."[101] No flaw was created for Boccadiferro's generalization by Paracelsus' silence on the Milky Way, perhaps the only saving grace in his writings on "meteors,"[102] which anticipated much of the sickening atmosphere of Fludd's pananimistic lucubrations. However, during Boccadiferro's latter days, two treatises on meteorology were published in Venice, his home city, and both were conspicuous by the absence of any criticism of Aristotle's dicta on the Milky Way.[103] Furthermore, opposition to Aristotle was only half the matter. Boccadiferro himself hesitated to present Democritus' view of the galaxy as the truth. He saw much merit in the position of Philoponus, for whom the Milky Way was a denser part of the firmament. Still, Boccadiferro deserves credit for echoing Philoponus' question of why there were more stars in the Milky Way than elsewhere in the sky.[104]

As for the astronomers, they showed no eagerness to discuss the Milky Way in print. The reason for this was broader than their attachment to the Ptolemaic formalism of the study of planetary motions. Alessandro Achillini (1463–1512) passed over the topic of the Milky Way not only in his *De orbibus* but also in his *De elementis,* although, in the latter, he discussed such topics as the nature of the stellar (planetary) orbs together with several references to comets.[105] Similarly, Oronce Fine (1494–1555), author of a book on the "theory of the eighth sphere and of the seven planets,"[106] and also of a descriptive astronomy,[107] found in the latter no room for the Milky Way when he listed the great circles. One could not, of course, argue with Gianbattista Amico for his omission of the Milky Way. After all, he was an avowed Peripatetic in astronomy, who attempted to give a theory of planetary motions according to "Peripatetic principles," that is, with homocentric spheres and without eccentrics and epicycles.[108] Neglect of the Milky Way seemed most curious in the case of the *Institutiones astronomicae* by Joachim Ringelberg of Antwerp. He discussed the original chaos, some questions of cosmogony, the possible eternity of the universe, and the number of worlds. Next to the fixed stars he listed three other classes of stars: the vagrant, the shooting, and the bearded stars.[109] No details were given by him about the "circulus lacteus."[110] The cause for Peter Apianus' silence on the Milky Way seemed to be astrology. In his *Astronomicum caesareum*[111] the Milky Way was repeatedly marked on magnificently printed celestial maps, but he said nothing about it. After all, the Milky Way was not supposed to have any influence on the birthday of ordinary mortals, let alone on the august rulers of the Holy Roman Empire. Apianus also overlooked the Milky Way in his *Cosmographicus liber.* There he emphasized, after the section on the zodiac, that "all the other circles should be assigned an existence in theory only; they should be imagined without any width or depth."[112] This was aptly said, but it was also true that, to grasp the truth about the Milky Way, one needed an astronomy specifically sensitive to immense celestial depths.

One effective inducement to considering the topic of the Milky Way was an astronomical text to be commented upon, provided this contained a reference to the subject. Peurbach's failure to mention the Milky Way was duly imitated by his commentators, of whom one of the earliest was Francesco Capuano of Manfredonia.[113] But the mention of the galaxy in "Proclus' *Sphaera,*" a late medieval compilation from Geminus' works, served as a lead for Johann Stöffler (1452–1531), author of calendars and ephemerides. His commentary on it[114] included an ample listing of ancient Greek and Latin scientific and literary utterances on the Milky Way. With an eye on Genesis, Stöffler dismissed Theophrastus' two-hemisphere theory of the Milky Way: the heavens, he claimed, were not created in two parts. However, after reviewing the mythological and spiritual interpretations of the Milky Way, he turned to Albertus Magnus for scientific information on the topic. The authority of Albertus Magnus prompted him to reject the theory—largely held by Arab philosophers—that the Milky Way was a phenomenon in the highest layer of fire just below the orbit of the moon. He again referred to Albertus Magnus as he recounted and dismissed Anaxagoras and Democritus for their (alleged) view of the Milky Way as a shadow cast by the earth. Stöffler did not list Aristotle's explanation of the Milky Way, possibly because of its absurdity. At the same time he ascribed the "true explanation" both to Ptolemy and to Albertus Magnus. The latter, according to Stöffler, brought the dispute to a close by stating: "The efficient cause [of the Milky Way] is the many small stars . . . that diffuse their light in that part of the celestial orb. Another part of the efficient cause is the radiation of the sun that falls on those almost contiguous stars. The material cause is . . . that part of the celestial orb which is denser, and therefore retains and reradiates the light of the sun and of the stars, and forms a terminus to one's vision in the same way in which the stars retain light and reradiate it and form a limit to one's vision."[115]

The year 1534 that saw the publication of Stöffler's work also witnessed the printing of a long didactic poem, Marcellus [Stellatus] Palingenius' *Zodiacus vitae.*[116] In its own time, its great popularity derived from its advocacy of a natural morality based on Epicurean moderation. Today its fame rests on its closing, cosmological sections. There Palingenius tried to accommodate, in a truly Humanistic eclecticism, pagan and Christian beliefs about the universe. He advocated its eternity as well as its creation out of nothing. A similar ambivalence characterized Palingenius' doctrine about the extent of the world; to him, it was both infinite and finite. To solve this paradox, he assumed that the light pervading the infinite recesses of the universe was invisible. In such a context the infinite number of stars which he postulated could present no real problem. In the universe of Palingenius, visibility was coextensive with the finite and spherically closed world.[117] As his mind was basically preoccupied with moral and metaphysical questions, he made little headway in directions more scientifically germane to the phenomena of the heavens. Whereas he devoted hundreds of lines to the stars of the zodiac, he allotted to the Milky Way only the following perfunctory lines:

Another circle white there is, whose course by knees doth traine,
Of Gemini, by Scorpius taile, and by the Tropicks twaine,
And through the crooked path of Sunne by midst of Archers string,
And passeth by the Centaurs legges, and by the Eagels wing,
And both the Carter and the Swanne, and Perseus doth it touche.[118]

The translator, Barnabe Googe, although he filled the margins with endless notes, did not advise his readers that the quoted lines referred to the Milky Way. Nor did the extensive index of words and subjects, which he attached to the translation, contain a reference to the galaxy. All this stood in sad contrast to the medieval gropings about the Milky Way, but well matched Palingenius' astrological perspectives. It was these that science had to overcome if a truer picture of the cosmos was eventually to emerge.

References

[1] *Patrologiae cursus completus, Series Latina,* edited by J. P. Migne (Paris: J. P. Migne, 1854), vol. 172, col. 146. The passage concludes a short chapter on the various constellations and is followed by a chapter on comets.

[2] *Isidori Hispalensis episcopi Etymologiarum libri xx,* edited by W. M. Lindsay (Oxford: Clarendon Press, 1911), vol. 1, lib. iii, cap. xlvi. There is no reference to the Milky Way in Isidore's *De natura rerum liber,* edited by G. Becker (Berlin: Weidmann, 1857), although a chapter is devoted both to the light of stars and to the names of constellations; see pp. 46–47 and 48–52. The same is true of the *De universo libri xxii,* a work of Rhabanus

Maurus (c. 780–856), who was mainly interested in the biblical connotations of natural phenomena; see Migne, *Patrologiae . . . Series Latina,* vol. 111 (Paris, 1864), col. 257.

[3] See Migne, *Patrologiae . . . Series Latina,* vol. 90 (Paris, 1862), col. 896.

[4] *Dialogus de substantiis physicis: ante annos ducentos confectus a Vuilhelmo Aneponymo philosopho,* industria Guilielmi Gratalori Medici (Strasbourg: excudebat Iosias Rihelius, 1567; reprinted in Frankfurt: Minerva G. M. B. H., 1967), p. 93.

[5] *De mundi universitate libri duo sive megacosmus et microcosmus,* edited by C. S. Barach and J. Wendel (Innsbruck: Wagner, 1876), pp. 17 and 37.

[6] A critical edition of that translation would be all the more desirable as it is available only in manuscript. Its introductory part alone was printed in Amable Louis Marie Jourdain's *Recherches critiques sur l'âge et l'origine des traductions latines d'Aristote* (Paris: Fantin et Cie, 1819), pp. 461–62. The text of Aristotle's own theory is given in the Vatican Library's *Cod. Urb. Lat. 206* (f. 214r) as follows: "Nos autem dicimus quod esse galaxie est hoc modo quod est quia ignis purus propinquus orbi inflammatus et in locis orbis in quibus videtur galaxia sunt parve multe stelle spisse et magne propinque luminose; cum ergo procedit lumen earum ex illo loco inflammato ignito videtur in eo lumen oblongum, et iste quod stelle fixe quarum quedam tangunt alias et sunt suscipientes splendorem ex sole quare continuatur lumen quarundam earum ac quibusdam; propter ergo illud videtur galaxia in loco uno orbis non recedens ab eo."

[7] *The Arabic version of Aristotle's Meteorology: A Criticial Edition with an Introduction and Greek-Arabic Glossaries,* by Casimir Petraitis (Beyrouth: Dar El-Machreq Editeurs, 1967). I was unable to consult the original form of this work, a doctoral dissertation (Oxford University, 1963), which includes an English translation.

[8] For the translation of this passage, which occurs on pp. 25–26 of the Arabic text, I am indebted to M. Mazzaoui, professor of Persian literature at Princeton University.

[9] My information is based on the doctoral dissertation, "Alfred of Sareshel's Commentary on the Metheora of Aristotle," by James K. Otte (May 1969, University of Southern California).

[10] For details on this translation which remains in codex form, see M. Steinschneider, *Die hebräischen Uebersetzungen des Mittelalters und die Juden als Dolmetscher* (1893; Graz: Akademische Druck und Verlagsanstalt, 1956), pp. 132–35.

[11] As will be discussed shortly.

[12] First printed in Basel around 1470, and followed by John Trevisa's English translation in 1495. The original Latin, reprinted many times, was again

published in 1601 (Frankfurt: Wolfgang Richter; for the Milky Way, see Book VIII, chap. 8, p. 384). The passage on the Milky Way remained unchanged in the famous corrected and enlarged form of the English translation published in 1582 by Stephen Batman under the title, *Batman uppon Bartholome* (London: imprinted by Thomas East, 1582), p. 124.

[13] *Bibliotheca mundi. Vincentii burgundi . . . Speculum quadruplex . . .* (1624; Graz; Akademische Druck und Verlagsanstalt, 1964), vol. 1, col. 173. Classifying the Milky Way among the stars was all the more remarkable of Vincent de Beauvais, as he evidently used the *old* translation. See A. Jourdain, *op. cit*, p. 411.

[14] *De cometis et causis ipsarum,* edited by H. S. Thomson, "The Text of Grosseteste's *De cometis," Isis* 19 (1933): 19–25; see also the chapter on "De cometis" from the *Summa Lincolniensis,* edited by Ludwig Baur, in *Die philosophischen Werke des Robert Grosseteste Bischofs von Lincoln* (Münster i. W.: Aschendorffsche Verlagsbuchhandlung, 1912), pp. 32–41 of Texts. For quotation, see the latter, p. 41. The basic difference between comets and the galaxy, as urged by Grosseteste, is not mentioned in A. C. Crombie's *Robert Grosseteste and the Origins of Experimental Science 1100–1700* (Oxford: Clarendon Press, 1953), pp. 87–90.

[15] *The Opus Majus of Roger Bacon,* a translation by Robert Belle Burke (Philadelphia: University of Pennsylvania Press, 1928), vol. 2, pp. 517–21 (Part Five, Optical Science, Second Part of Perspective, Third Distinction, Chapter I).

[16] *Ibid.,* p. 517.

[17] *Ibid.*

[18] *Ibid.,* p. 519.

[19] See First Part of Perspective, Ninth Distinction, Chapter I, *ibid.,* p. 483.

[20] *Ibid.,* p. 518.

[21] *Ibid.*

[22] *Ibid.,* pp. 517–18.

[23] *The Book of Instruction in the Elements of the Art of Astrology by Abu'l-Rayhan Muhammad Ibn Ahmad Al-Biruni,* with translation facing the text by R. Ramsay Wright (London: Luzac & Co., 1934), p. 87. There is no reference to the Milky Way in Al-Fargani's discussion of the fixed stars, translated into Italian with introduction and notes by Romeo Campani, *Alfragano (Al-Fargani): Il 'Libro dell' Aggregazione delle Stelle' secondo il Codice Medico-Laurenziano Pl. 29.-Cod. 9* (Città di Castello: Casa tipografico-editrice S. Lapi, 1910); see especially pp. 139–41 on the number and magnitude of stars and their distribution in the sky. My evaluation of the Milky Way in Arabic sources is based on material available in translation. I

have not found any reference to the Milky Way in bibliographies on Arabic astronomy. A special study might well repay the effort.

[24] *Le livre de l'ascension de l'esprit sur la forme du ciel et de la terre: Cours d'astronomie rédigé en 1279 par Grégoire Aboulfarag, dit Bar-Hebraeus,* edited and translated by F. Nau (Paris: Emile Bouillon, 1899–1900).

[25] *Ibid.,* pp. 92–93 of the French translation.

[26] *Aristotelis opera cum Averrois commentariis* (Venice: apud Iunctas, 1562–74; reprinted in Frankfurt a. M.: Minerva, 1962), vol. 5, ff. 409v–414r.

[27] *Ibid.,* f. 413v.

[28] *Ibid.,* f. 414r. There is a short reference to the galaxy in Averroes' "Sermo de substantia orbis" *(Opera,* vol. 9, f. 7v), where Averroes tries to prove that the sphere of stars admits variations of density and reflectivity. Alvaro de Toledo (fl. 1290), in his commentary on Averroes' "Sermo," elaborates on the point that the exhalation theory cannot ensure the steady appearance of the Milky Way. See *Comentario al "De substantia orbis" de Averroes por Alvaro de Toledo,* edited with notes by P. Manuel Alonso, S.I. (Madrid: Bolaños y Aguilar, 1941), p. 190.

[29] "Tractatus II. De Galaxia," in *Liber I. Meteororum,* in *B. Alberti . . . Opera omnia,* edited by A. Borgnet (Paris: L. Vivès, 1890), vol. 4, pp. 493–98.

[30] *Ibid.,* p. 495.

[31] *Ibid.* In his *Liber de passionibus aeris, sive de vaporibus, impressionibus (Opera,* vol. 9, p. 682), Albert reports with seeming approval an opinion which he attributes to Aristotle, that the galaxy is the optical effect produced by starlight in the pure region of fire. Such nuances of Albert's dicta on the Milky Way are wholly missing in "S. Alberto Magno e l'astronomia," a contribution by G. Stein, director of the Vatican Observatory, to vol. xxi of *Angelicum* (1944), devoted to the various aspects of Albert's achievements in natural science. According to Stein, Albert accepted Democritus' opinion of the Milky Way (p. 187).

[32] Lectio XI and XII in *Expositio in Lib. I. Meteororum,* in *Sancti Thomae Aquinatis . . . Opera omnia* (Parma, 1853–73; New York: Musurgia Publishers, 1949), vol. 19, pp. 322–26.

[33] It is this translation which is printed in Aquinas' commentary as the *antiqua* in parallel columns with a more modern translation designated as *recens.* The *antiqua* should not therefore be confused with the *old.*

[34] *Mahieu le Vilain, Les metheores d'Aristote. Traduction du xiiie siècle,* edited by Rolf Edgren (Uppsala: Almquist & Wiksells Boktryckeri Aktiebolag, 1945).

[35] *Opera omnia,* vol. 19, p. 325.

[36] An exact evaluation of these differences will have to wait until publication of a critical edition of the text of the *old* translation, a project on which J. D. Schoonheim was reported to have been working in the late 1960's. See Petraitis, *The Arabic Version of Aristotle's Meteorology,* p. 13.

[37] References are to the English translation, *The Convivio of Dante Alighieri* (London: J. M. Dent, 1908).

[38] *Ibid.,* p. 126.

[39] Dante's numerous references to astronomical topics were discussed in a lenghty monograph by Edward Moore, under the title, "The Astronomy of Dante," in his *Studies in Dante, Third Series, Miscellaneous Essays* (Oxford: Clarendon Press, 1903), pp. 1–108. He devoted only a dozen lines to the galaxy, with no reference to the earlier history of the topic. The question is treated with far greater care in *Dante and the Early Astronomers* (London: Gall & Inglis, 1914, pp. 303–306) by Mary Acworth Orr, an admirer of Dante and an amateur astronomer, who in 1906 became the wife of John Evershed, director of the Kodaikanal and Madras Observatories. While she was right in insisting that Dante admired Aristotle too much to charge him with error, the idea of the Milky Way as a dense part of the firmament could still, contrary to her claim, perfectly serve Dante's metaphysical interpretation of astronomy and of the Milky Way in particular.

[40] *The Convivio of Dante Alighieri,* p. 126.

[41] *Ibid.*

[42] *Ibid.*

[43] *Ibid.*

[44] *Ibid.,* p. 127.

[45] A good illustration of this is the massive work by Guido Bonatti (c. 1223–c.1297), *Decem continens tractatus astronomiae* (Augsburg: Erhardi ratdolt imprimendi arte, 1491), at the end of which (ff. EE 1–7) are discussed in detail such meteorological phenomena as shooting stars, circles around the sun and moon, rainbows, and—last but not least—comets, traditionally associated with the Milky Way.

[46] On the history of the influence of Sacrobosco's work, together with the texts and translations of several medieval commentaries on it, see L. Thorndike, *The Sphere of Sacrobosco and its Commentaries* (Chicago: University of Chicago Press, 1949). The work contains the text and translation of the commentaries by Robert Anglicus (c. 1271), by Michael Scot (c. 1230), by Cecco d'Ascoli (c. 1320), and by several anonymous authors, and also excerpts from the *Sphaera* of John Peckham. The absence of reference to the Milky Way in the master text was rarely if ever infringed in the commentaries on it.

[47] Edited by L. Baur, *op. cit.* (see note 14 above), pp. 10–31.

[48] See Thorndike, *The Sphere of Sacrobosco,* pp. 445–50.

[49] My translation was based on *Joannis archiepiscopi cantuarensis perspectivae communis libri tres* (Cologne: in officina Birckmannica, sumptibus Hermanni Mylij, 1626), f. 47r. The work has just become available in a modern critical edition by D. C. Lindberg, *John Pecham and the Science of Optics: Perspectiva communis* (Madison: University of Wisconsin Press, 1970); see pp. 237–39. Little can be learned about the Milky Way from commentaries written at that time in Oxford by Gualterus de Burley (c. 1275), by Adam de Buckfield (c. 1280), and by Guilelmus de Bonkes (c. 1290). In Burley's commentary (MS Oxford, Bodl. Digby 98, XV) there are only four lines on the Milky Way (f. 50r). Buckfield, and probably Bonkes too, depended on the glossae of Alfred of Sareshel to the *Meteorologica.*

[50] In the Preface to his *Euclidis optica et catoptrica nunquam antehaec graece edita eadem latine reddita* (Paris: apud Andream Wechelum, 1557), f. bbiij r.

[51] Printed together with the Latin translation of Alhazen's work under the title, *Opticae thesaurus Alhazeni Arabis libri septem nunc primum editi ejusdem liber de crepusculis et nubium ascensionibus—item Vitellonis Thuringipoloni libri X omnes instaurati . . . a Federico Risnero* (Basel: per Episcopios, 1572).

[52] *Ibid.,* Lib. IV, pp. 117–89.

[53] *Ibid.,* Lib. X, pp. 403–74.

[54] *Opus sphaericum cum commentis Cicchi Esculani Francisci Capuani et Iac. Fabri Stapulensis* (Venice: Simon [de Gabis] Beuilaqua, 1499).

[55] A publication of the full list of medieval commentaries on Aristotle is now in progress thanks to the scholarly and painstaking labors of Charles H. Lohr, "Medieval Latin Aristotle Commentaries," (authors A–F, G–I, Jacobus-Johannes Juff, Johannes de Kanthi–Myngodus) in *Traditio* 23 (1967): 313–413; 24 (1968): 149–245; 26 (1970): 135–216; 27 (1971): 251–351. The list shows that special topics in the *Meteorologica,* such as comets and the rainbow, attracted more attention than the work itself. Relative neglect of the *Meteorologica* was probably due to the patently arbitrary approach of the Philosopher to several conspicuous phenomena of nature.

[56] In reading the microfilm copy of manuscript C. A. 432, Erfurt Stadtbücherei, of Buridan's "Quaestiones eiusdem super meteorologicorum," have received indispensable help from the Rev. Joseph Brown, of the Institute of Advanced Study, Princeton, N.J. For the section on the galaxy, see ff. 35r–36r.

[57] Through the courtesy of Prof. Marshal Clagett of the Institute of Advanced Study, Princeton, N.J., I had access to the photocopy of the text in

Codex 839, St. Gallen. For the section on the Milky Way, see ff. 39v–42r. Here again, I wish to acknowledge the kind help of the Rev. Joseph Brown.

[58] *Johannis Buridani Quaestiones super libris quattuor de caelo et mundo,* edited by Ernest A. Moody (Cambridge, Mass.: The Medieval Academy of America, 1942), p. 217.

[59] Its printed text is in *Quaestiones et decisiones physicales insignium virorum,* edited by G. Lokert (Paris: in aedibus Iodoci Badii Ascensii & Conradi Resch, 1518), which contains, in addition to Themon's commentary on the *Meteorologica* and *De anima,* commentaries on other works of Aristotle by Albert of Saxony and Buridan.

[60] For Themon's discussion of the Milky Way, see Fol. CLXVI.

[61] His work was often ascribed to Duns Scotus and was reprinted in vol. 4, of *Joannis Duns Scoti Opera omnia* (new ed.; Paris: L. Vivès, 1891). On the galaxy, see pp. 97–100.

[62] *Ibid.,* p. 100.

[63] In *The Complete Works of Geoffrey Chaucer,* edited by the Rev. Walter W. Skeat (Oxford: Clarendon Press, 1894), vol. 3, pp. 28–29.

[64] *Ibid.,* pp. 175–232.

[65] See the edition of the original Latin with a French translation by Edmond Buron, *Ymago mundi de Pierre d'Ailly* (Paris: Maisonneuve Frères, 1930). The omission of the Milky Way was all the more conspicuous, since the fixed stars, their sphere, and the zodiac were discussed in detail. See pp. 167–99.

[66] *Caxton's Mirrour of the World* [1480], edited by Oliver H. Prior (London: Kegan Paul, Trench, Trübner, 1913). In the work there would have been several logical opportunities to mention of the Milky Way, especially in the closing sections of the Third Part, where the distance and number of stars, and their crystalline sphere is discussed. See pp. 171–74.

[67] *Ymago mundi,* p. 169.

[68] *Ibid.,* p. 185.

[69] *Tractatus breuis . . . venerabilis Episcopi Petri Cameracensis . . . que in Prima Secunda atque Tercia regionibus aeris fiunt . . .* [with the subtitle on fol. 2r] *De impressionibus aeris . . . Libell[s] sup libros Metheorou Arestotelis* [Leipzig: Conrad Kachelofen, 1495?].

[70] *Ibid.,* Fol. A [Vv].

[71] References are to the English translation by G. Heron, *Of Learned Ignorance* (London: Routledge & Kegan Paul, 1954).

[72] *Ibid.,* p. 111.

[73] *Ibid.,* p. 113.

[74] *Epytoma Joannis de monte regio in almagestum Ptolomei* (Venice: arte . . . Johannis Haman de Landoia, dictus Hertzog, 1496).

[75] *Ibid.,* p. iiv.

[76] *De expetendis ac fugiendis rebus opus* . . . (Venice: in aedibus Aldi Romani, 1501). Valla mentioned only the position of the Milky Way; see ff. bbii r and bbiii r.

[77] *Ioannis Ioviani Pontani De rebus coelestibus libri xiiii, eiusdem De luna fragmentum* (Basel, 1530), p. 101. Titles of earlier editions of the work contain the more correct version of *libri xv.*

[78] In *Pontani opera* (Venice: in aedibus Aldi et Andreae Asulani Soceri, 1513), ff. 132r and v.

[79] *Gaetani de thienis Vincentini philosophi . . . in metheororum Aristotelis libros expositio* (Padua: per Petrum Mauser, 1476); on the Milky Way, see [p. 24, col. 1]. Far less important than Gaetano de Thiene were his contemporaries, Henricus de Gorkum and Henricus Parker, authors of as yet unedited commentaries on the *Meteorologica.* See Lohr, *art. cit.,* (1968), pp. 225 and 227.

[80] The original title reads: *Aenumeranturque in hoc opere dicto Margarita philosophica: contineantur MARGARITA PHILOSOPHICA totius Philosophiae rationalis* etc., in twelve books, of which the seventh deals with astronomical and cosmographical topics. The colophon reads: Chalcographatum primiciali hac pressura Friburgi per Johannem Schottum Argen. citra festum Margarethae anno gratiae MCCCCCIII [1503]. For the section on the Milky Way, see ff. miiij [+2r–3r]. It remained unchanged not only in the second edition of 1504 (also printed in Freiburg), but in the edition of Basel, S. Henricpetri, 1583, p. 482, and in the Italian translation, *Margarita filosofica,* by Giovan Paolo Gallucci (Venice: appresso Iacomo Antonio Somascho, 1599), pp. 355–57. There were many other editions as well. Reisch was a Carthusian and confessor of Emperor Maximilian.

[81] Its colophon reads: *Commentaria trium librorum metheororum Arestotelis per . . . Iacobum de Amsfordia . . . ex diversis praecipue tamen doctoris magni Venerabilis Alberti dictis . . . collecta* (Cologne: per H. Quentell, 1497). On the Milky Way, see ff. xix v–xx r.

[82] *Incipit textus abbreviatus Aristotelis super octo libris physicorum et tota naturali philosophia nuper a magistro Thoma Bricot copilatus . . .* (Paris: per Wolfgangum hopyl, 1494), ff. LXX r and v.

[83] *Clarissima singularisque totius philosophiae necnon metaphysicae Aristotelis . . . expositio* (Lyons, 1498), f. xciiij r.

[84] *Quaestiones Versoris super, De coelo et mundo, De generatione et corruptione, Metheororum, Parva naturalia* [Cologne: H. Quentell, 1488]. See f. vii r of *Metheororum.*

[85] *Littere librorum. Liber Physicorum . . . Liber meteororum . . . Liber de longitudine et brevitate vitae.* The colophon reads: Impressum Parisii Anno domini millesimo quadringentesimo nonagesimo secundo [1492]. See ff. *v,* iiii r and v. There is no reference to the Milky Way in his *Opus novum astronomicum Iacobi Fabri Stapulensis cum lucidissima expositione Christiani Sculpini Hangeltensis . . .* (its colophon reads: Impressum Coloniae in domo Quentell Anno MCCCCCXVI,) although its last three chapters are on the eighth, ninth, and tenth spheres.

[86] *Opusculum de universali mundi machina ac de meteoricis impressionibus a Fratre Jeronimo de scto Marcho . . .* (Oxford: Richard Pynson, 1505), ff. xvii r and v.

[87] *Ioannis Francisci Pici Mirandulae Domini, et concordiae comitis, Examen vanitatis doctrinae gentium, et veritatis christianae disciplinae, distinctum in libros sex* [Mirandulae, 1520]. The Milky Way is discussed in Lib. I. cap. xii, ff. xviii–xxi.

[88] *In Aristotelis Meteorologica Paraphrasis,* in *Opera omnia Francisci Catanei Diacetii . . . nunc primum in lucem edita* (Basel, 1563), p. 307. The Milky Way is discussed on pp. 305–08. The whole *Paraphrasis* is less than thirty pages long. It was composed in the 1510's according to P. O. Kristeller, "Francesco de Diacceto and Florentine Platonism in the Sixteenth Century," in *Miscellanea Giovanni Mercati,* vol. IV, *Letteratura classica e umanistica* (Città del Vaticano: Bibliotheca Apostolica Vaticana, 1946), p. 287.

[89] *Ibid.,* p. 307.

[90] *Suessanus in libros Metheororum. Augustini Niphi . . . in libris Aristotelis meteorologicis commentaria . . .* (Venice: impensis . . . Octaviani Scoti . . . et sociorum, 1531), f. 35v.

[91]The title is *Habes humanissime lector librorum metheororum Aristotelis facilem expositionem et quaestiones super eosdem Magistri Joannis Dullaert de Gadavo,* while the colophon reads: Impresse vero Parisius a Thoma Rees . . . anno Domini 1512. 22. Aprilis. For the Milky Way, see ff. 29v–32r. See f. 31v for the diagram illustrating the parallax of sublunary objects. His definition of the galaxy expresses his reasoning very well: "Unde galaxia est pars coeli magis densa in qua etiam sunt valde multe parue stelle et propterea est lucidior aliis partibus. Et nego quod illud includat aliquam absurditatem. Stella est densior pars sui orbis secundum Aristotelem." (See f. 32r.)

[92] *Aristotelis Stagyritae Libri . . . Meteororum iiii Boetio interprete adiectis Eckij commentariis* (Augsburg: excusa in officina Sigismundi Grim . . . et

Marci Buyrsung, 1519), f. XC v. Aristotle's opinion was stated in a few lines in Celio Calcagnini's (1479–1541) "Paraphrasis trium librorum meteororum Aristotelis," in *Caelii Calcagnini Ferrarensis. . . . Opera aliquot* (Basel: Froben, 1543), p. 427.

[93] In *Gasparis Contarini Cardinalis Opera* (Paris: apud Sebastianum Niuellium, 1571), p. 24.

[94] *Ibid.*, p. 25.

[95] *Ludovici Buccaferrei Bononensis lectiones super primum librum Meteorologicorum nunc recens in lucem aeditae* (Venice: ex officina Joan. Baptistae Somaschi, 1565). The Milky Way is discussed on ff. 52v–66r.

[96] *Ibid.*, f. 57v.

[97] *Ibid.*, f. 58v.

[98] *Ibid.*, f. 63v.

[99] *Ibid.*, f. 64r.

[100] *Ibid.*, f. 65r.

[101] *Ibid.*, f. 64r.

[102] Two works of Paracelsus deserve to be mentioned in this connection, the *Philosophia de generationibus et fructibus quatuor elementorum* and *De Meteoris,* in Theophrast von Hohenheim gen. Paracelsus, *Sämtliche Werke, I. Abteilung, Medizinische, naturwissenschaftliche und philosophische Schriften,* edited by Karl Sudhoff, vol. 13 (München: Druck und Verlag von R. Oldenbourg, 1931), pp. 7–123 and 125–206, respectively. In the former Paracelsus discussed comets and new stars; in the latter he passed from the stars directly to the topic of winds.

[103] They were Ermolao Barbaro (1454–1493), author of a *Compendium scientiae naturalis ex Aristotele,* edited by Daniello Barbaro (Venice: apud Cominum de Tridino Montisferrati, 1545), f. F3r, and Sebastiano Fausto (fl. 1540), author of *Meteorologia, cioè discorso de le impressioni humide e secche* (Venice, 1542), ff. 39r–40r.

[104] *Ludovici Buccaferrei . . . lectiones,* f. 65r.

[105] *Alexandri Achillini Bononiensis philosophi celeberrimi opera omnia in unum collecta* (Venice: apud Jeronimum Scotum, 1545), f. 27r.

[106] *La théorique des cielz, mouvemens et termes practiques des sept planètes . . . rédigée en français* [par O. Finé] (Paris: J. Pierre, 1528).

[107] *Orontii Finei . . . De mundi sphaera, sive Cosmographia, primave astronomiae parte libri V.* (Paris: apud Simonem Colinaeum, 1542), see especially ff. 10r–26v on the various circles.

[108] *Opusculum de motibus corporum coelestium iuxta principia peripatetica sine eccentricis et epicyclis, denuo aeditum* (Venice: ab Augusto de Bindonis, 1537).

[109] Venice: J. Antonius de Nicolinis de Sabio, 1535, ff. 16v and 17r.

[110] *Ibid.*, f. 33r.

[111] Ingolstadt: in aedibus P. Apiani, 1540.

[112] Edited by Gemma Phrysius; Antwerp: in aedibus Rolandus Bollaert, 1529, f. Vr.

[113] *Theorice nove planetarum Georgii Purbachii . . . ac in eas . . . Domini Fracisci Capuani de Manfredonia . . . sublimis expositio . . .* (Venice: per Simonem Bevilaqua, 1495).

[114] *Joannis Stoefleri Iustingensis . . . in Procli Diadochi . . . Sphaeram mundi commentarius* (Tübingen: ex aedibus Morhardinis nostris, 1534).

[115] *Ibid.*, f. 72v.

[116] *Stellati Zodiacus vitae sive de hominis vita libri xii,* edited by C. H. Weise (Leipzig: C. Tauchnitz, 1832).

[117] *Ibid.*, p. 327 line 76.

[118] *The Zodiake of Life by Marcellus Palingenius,* translated by Barnabe Googe (1576), with an Introduction by Rosemund Tuve (New York: Scholars' Facsimiles and Reprints, 1947), p. 205.

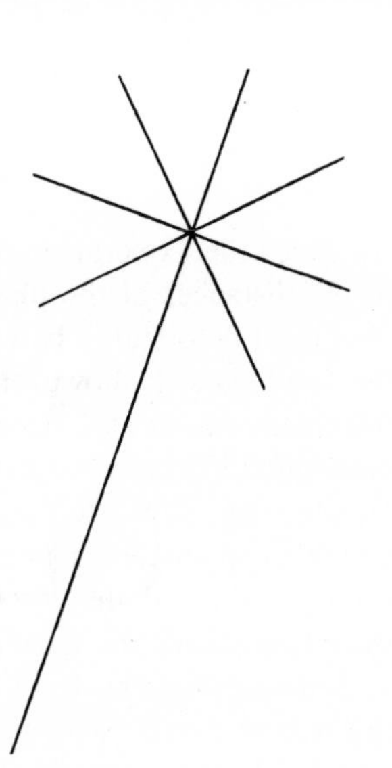

CHAPTER THREE

Copernican Silence

WHEN Palingenius' *Zodiacus vitae* was published in 1534, a long treatise on planetary motions, remarkably free of astrological lore, was being drafted near the misty shores of the Baltic Sea. But the really revolutionary aspect of the *De revolutionibus orbium coelestium*[1] of Copernicus lay elsewhere. The rotation of the earth on its axis and its revolution around the sun implied not only startling consequences for man's understanding of the motion of terrestrial and planetary bodies. The orbital motion of the earth also made it necessary to abandon the belief that the distance to the sphere of the fixed stars was only some 20,000 times the radius of the earth. The basic reason for this was stellar parallax, or change in the apparent relative positions of stars, a logical consequence of the earth's annual journey around the sun. From the minimum of angular resolution that could then be achieved by astronomical instruments, and from the accepted value of the earth-sun distance, one could readily deduce a minimum distance for the stars to explain the apparent absence of stellar parallax. Copernicus himself made no such calculations. He knew, of course, that the motion of the earth entailed a drastic increase in estimates of the distance of stars: "The heavens are immense in comparison with the earth," he wrote, and compared their respective sizes to that of a mere point in relation to a large body and to that of a finite to an infinite magnitude.[2] He insisted that even the radius of the earth—not a negligible quantity—had to be viewed "as nothing in comparison with the sphere of the fixed stars."[3]

Since the Copernican doctrine required no such drastic increase in the distances of the planets, it followed that an immense emptiness must be pictured between the orbit of Saturn and the sphere of the fixed stars. If, however, such new and "empty" depths could be tolerated, was it still reasonable to assume that all the stars were embedded like jewels, so to speak, in the wall of one single sphere? Would it not have been far more reasonable to assume that their location in space also spanned immense depths? Before long, developments showed that such were indeed logical questions. Within their framework the question about the Milky Way would have assumed a new aspect. But Copernicus' main interest lay with the sun, not with the stars. With the worshipful attitude of a Pythagorean he wrote: "In the center of all rests the sun. For who would place this lamp of a very beautiful temple in another or better place than this wherefrom it can illuminate everything at the same time. As a matter of fact, not unhappily do some call it the lantern; others, the mind and still others, the pilot of the world. Trismegistus calls it a 'visible god'; Sophocles' Electra, 'that which gazes upon all things'."[4] Beside the brilliance of the sun shedding light and warmth on everything around it, the starry realm paled into insignificance—and so did the Milky Way. It did not appear in any form in the world of Copernicus.

Equally uninterested in the topic were Copernicus' early defenders. Fascination with the marvelous possibilities and paradoxical implications of the new ordering of planets could but distract attention from more remote consequences of the Copernican theory. Such was the case in the writings of Rheticus, the first champion of Copernicus. In this respect a brief reference to the Milky Way in Robert Recorde's *The Castle of Knowledge,*[5] the first English book to mention Copernicus' theory, was highly revealing. The work, a dialogue between the "Master" and the "Scholar" (student), evidenced Recorde's sympathy when, as the "Master," he did his best to defend Copernicus, "this man of greate learninge of much experience and of wonderfull diligence in observation," against the derogatory remarks of the "Scholar."[6] The Fourth Treatise of the book opened with a list of 25 topics and the claim that in the book "are the Proofes of all that is taught before and other divers notable conclusions annexed therto, but nothing in a maneer without demonstration and good proofe."[7] The twenty-third of these topics read:

"The description of the Mylke way in the skye, whiche is commonly called Watlynge streete, and what is the cause of that coulour in it."[8] The preceding hundred and the succeeding two hundred pages contained no explanation of that whiteness. From the "Master" one could learn that many in England called the Milky Way "Watlyinge streete," after "one of the greate highe waies," and that it was a way "in the skye it selfe, as all men hath confessed and their eyes do testifye." The situation was not quite so simple, but Recorde had no concern on this score. For him the "Mylkie way" proved, "if it served for none other purpose,"[9] that the heavens moved.

The question of the silence about the Milky Way becomes particularly pressing when one examines that famous diagram in Thomas Digges' small essay of 1576 on the Copernican system.[10] There, for the first time, stars were placed outside their circle, which was further separated from the orbit of Saturn by a wide gap. "This orbe of starres fixed," reads the first line of the circular inscription, "infinitely up extendeth hit self in altitude spherically."[11] In the essay itself, Digges added further emphasis to Copernicus' statement that the orbit of the earth was but a point with respect to the distance of the sphere of fixed stars. To see how small a portion of an immense universe the gods permitted man to admire, one only had to think "of that fixed Orbe garnished with lightes innumerable and reachinge up in *Sphaericall altitude* without ende."[12] This was certainly a bold leap in the right direction. It must also have been exhilarating. In such a state of mind, Digges could hardly be expected to think of the fact that his diagram, where with obvious care he placed the stars at equal distances from one another, resembled only remotely the true situation and not in the least the baffling frequency of stars in the Milky Way. Actually, Digges could easily ignore the obvious. Three years earlier, when he had published a booklet[13] on the famous nova of 1572, nothing would have been more natural than to ask if the new star had something to do with the Milky Way. But Digges failed to do what others did, and excitement over Copernicus was then no excuse.

The most vehement Copernican of the period was, of course, Giordano Bruno. Rarely is it remembered that he censured Copernicus for not seeing far enough; that is, beyond the sphere of the fixed stars into infinity.[14] Bruno also ridiculed Palingenius for wrapping the infinite realm of stars in an invisible "Platonic" light.[15] Bruno could therefore have been expected to delve into the problems of an infinitely extended realm of stars, all the more so in that he constantly referred to denizens prospering on planets around each and every star. He dealt with those problems very superficially, attributing the invisibility of those planets, infinite in number, to the low reflectivity of their watery and dry surfaces. That only relatively few stars could be seen of an infinite number, Bruno ascribed to the immense distance of most stars.[16] He had more difficulty in demonstrating that so many opaque planets did not eclipse at least a large portion of the stars. The problem savored of the rigor of geometry, to which Bruno paid some lip service without wishing to introduce it into his universe.

No less geometrical was the problem set by the strangely close grouping of stars in some constellations. For explanation of this, he referred to trees in a forest, an analogy that should have set him on the threshold of glimpsing the optical paradox (Olbers' Paradox) of an infinite number of stars. "Stupid man," he addressed himself to his hypothetical antagonist, "look around in a forest in which the trees are planted everywhere in the same density. Would you believe that those which are closer, are separated from one another by a greater distance, and those which are farther are more closely grouped, so that in the far background the agglomeration of an innumerable crowd [of trees] would fuse into sameness, and make one out of thousands?"[17]

Clearly, Bruno must have shut his eyes, as if by instinct, to the true implication of all this. To him, the infinity of the universe was the sacred cause, the great proof of his pantheism, which demanded a divine and unfathomable uniformity on the most pervasive and fundamental level of existence. Thinking through the quantitative or geometrical consequences of such a world view was another matter. So was mental alertness to evidences observable in the sky that could be marshalled for or against that view. Thus, as he mentioned the area around the eye of Taurus, the heart of Leo, and the Pleiades, he refused to admit that the various magnitudes of stars were an indication of their distances. A spacing of stars according to their apparent luminosity was not to Bruno's liking. Vitalism, not quantitative correlations, ruled supreme in his universe, which held no room for major patterns of ordering.

This seems to be the reason why Bruno remained silent about

the Milky Way although the grouping of stars there almost cried out for a discussion in the same context. His silence was all the more conspicuous since it was almost immediately preceded by his reference to Democritus' universe.[18] Nonetheless, he ignored Democritus' idea of the Milky Way with which he must have been fully familiar. But Bruno discussed at length, with biting sarcasm, the Aristotelian theory of comets, without so much as an oblique reference to its connection with Aristotle's theory of the Milky Way.[19] Again, the Milky Way must have appeared on his mental horizon when he referred to the nova of 1572 and to its interpretation by Cornelius Gemma and Christopher Rothmann.[20] For both of these, as will be seen, the nova raised the question of whether or not it was part of the Milky Way. Finally, there are Bruno's references, in his *La Cena de le Ceneri,* to Aristotle's theory of the clouds and of long-term changes on the earth's surface.[21] Aristotle discussed these matters in the *Meteorologica* only a few chapters removed from his theory of the Milky Way.

Bruno was a Copernican who wanted no pattern in the universe apart from the pattern of planets surrounding each star. His pantheism, steeped in magic, was well-nigh impossible to reconcile with a universe of reason manifested in an all-pervading pattern. However, Bruno should not be judged too harshly. Kepler, an equally outspoken Copernican, failed to see the pattern where it really existed—in the Milky Way. In his *Mysterium cosmographicum,* published in 1597, the universe was essentially the co-ordination of planets in terms of the perfect solids. The nova of 1604 which appeared in Serpentarius forced him to look seriously at the starry realm,[22] and, with an eye on Bruno, he anxiously forged proofs in support of a finite universe. His crucial point was that Bruno's infinite universe had a faulty presupposition: namely, its homogeneity. According to Kepler, who read too much logic into Bruno's passionate dicta, astronomy did not bear out the contention that the universe had to appear the same from any point. His argument[23] rested on a belief, generally shared by astronomers of his time, that all stars were at the same distance, and that their apparent diameter was of the order of a few minutes of arc. To these points Kepler added his estimate of the distance of the stars, which he assumed to be located within one, not overly wide spherical shell. The estimate was based on the minimum distance required that there should appear no stellar parallax through the earth's orbiting around the sun. On this basis, Kepler could reasonably argue that the three closely spaced, second-magnitude stars in Orion should appear, when viewed from one another, as suns five to six times larger in diameter than our own sun.

The argument was not without cogency, though its foundations were to be swept away before long, but a most ironical weakness in Kepler's proof resulted from what he said about the Milky Way. Had he pictured it as the agglomeration of many small stars, he could have conjured up an image of the sky in the Milky Way blazing day and night with the brilliance of innumerable large disks of suns. This would have certainly represented a striking argument against Bruno's homogeneous, infinite universe. The few words Kepler uttered about the Milky Way[24] described it as the place and the substance from which the new star of 1604 and the nova of 1572 had emerged. To prove his point he not only quoted parts of Ptolemy's meticulous mapping of the Milky Way, but he also submitted that both Aristotle and Pliny might have meant novae when they referred to comets appearing in the vicinity of the Milky Way. Most important, he emphasized that this explanation of the origin of novae is simply the "exceedingly beautiful idea" of Tycho Brahe. Until 1610 this was Kepler's only reference to the Milky Way, and he owed it, typically enough, to his anti-Copernican master.

This is not to suggest that Kepler found much concerning the Milky Way in Tycho's works, although he prepared for publication the major one, *Astronomiae instauratae progymnasmata.*[25] It was a motley collection of Tycho's observations, smaller works, and astronomical letters. Clearly, systematic thinking about the astronomical universe did not prompt Tycho's short utterances on the Milky Way. The "culprit" was the nova which had appeared in 1572 near Cassiopeia. When he proved in his famous booklet, *De nova et nullius aevi memoria prius visa stella,*[26] hastily written in 1573, that the new star was located above the lunar and planetary regions, Tycho passed over the question of its physical origin. Only after a lapse of years did he contend that the new star originated from the highly refined liquid which constituted, he thought, the primary substance of the spherical heavens. It was that substance which in its vaporous state gave rise to the Milky Way, and, in its highly condensed form, to stars.[27] Tycho's failure to see in the Milky Way an agglomeration of stars was not due to his patently Aristotelian and heavily animistic interpretation of nature.[28] With the same firmness which characterized his placement of the Milky Way among the stars and his assertion of the identity of its substance with that of the stars, Tycho also turned against Aristotle's theory of the Milky Way. By joining the comets and the Milky Way as products of the terrestrial exhalation, Aristotle, so Tycho argued, made two errors out of one; Tycho did much the same when he claimed that the Milky Way and the comets were of the same celestial substance. His proof consisted of the somewhat dubious generalization that most comets originated alongside the Milky Way, or in its very belt.[29] Aristotle generalized about the location of comets with similar zeal since his theory of the Milky Way demanded that no comet could possibly appear in its vicinity.

Tycho's case also illustrates that preoccupation with planetary theories rather effectively drew attention from the problems of the stellar realm, among them the Milky Way. The pattern is plain in the case of those who earned their reputation as astronomers by their discussion of the motion of planets. This is all the more revealing because several also published works in which the topic of the Milky Way could have been considered quite naturally. In the "De meteorologia," part of a work by Caspar Peucer on prognostication,[30] the topic of the Milky Way was passed over because of its old nemesis, astrology. From the invariability of the Milky Way one could scarcely collect clues to a highly variable future. In the case of Alessandro Piccolomini, author of a philosophy of nature,[31] the reason for the omission is more difficult to fathom, for he devoted some hundred pages to questions connected with the nature of stars, among them this: why is it that there should be so many stars on the sphere of the fixed stars, whereas only one star is carried on each of the planetary spheres?[32] Such a question could not, of course, be resolved on the basis of a purely formalistic astronomy guided by the hallowed principle of "saving the phenomena," which influenced the thought even of those who tended toward realism in astronomical method. Thus, Michael Maestlin, Kepler's teacher, and very sympathetic to Copernican realism about the new ordering of planets, restricted the domain of astronomy to questions that could be handled by geometry. In his *Epitome astronomiae*[33] Maestlin dealt only with phenomena caused by the revolution of the heavenly sphere and with the system of planetary motions. These two subject matters formed, according to him, spherical and theoretical astronomy, or that which might properly be called astronomy. Questions about the material cause of heavenly phenomena formed the domain of physical astronomy, which Maestlin left to the physicists. On the nature of stars he merely said that they were the denser parts of the heavenly sphere.[34]

The same generally accepted distinction and adherence to an established text would explain the silence concerning the Milky Way in the commentary on Peurbach's *Theoricae novae planetarum* by Erasmus Reinhold,[35] Christian Würsteisen,[36] and Erasmus Oswald Schreckenfuchs.[37] The latter broke his silence on the Milky Way only when he was forced to do so by the words of some traditional text. Thus, Schreckenfuchs noted in commenting on Proclus' (Geminus') *Sphaera* that "without question this circle [of the Milky Way] is part of the starry sky and advances with the fixed stars one degree of arc in a hundred years."[38] He attested to the overwhelming acceptance of this view when he added that "the opinions of the philosophers are not at variance with the truth."[39] Slavish conformity to established texts was further evident in Schreckenfuchs' silence on the Milky Way as he offered his comments on the *Sphaera* of Sacrobosco.[40]

The continued printings of the two *Sphaeras*[41] perpetuated the twofold tradition of complete silence or, at most, of a few words. Of the latter sort, a good example was the translation of Proclus' *Sphaera* by Thomas Linacer [Linacre], printed with Jacob Tusanus' comments.[41] Ariel Bicard's commentary on Sacrobosco's text illustrates the former,[42] but clearly, it became more and more incongruous to observe complete silence on the Milky Way, even for Sacrobosco's commentators. Thus, in Francesco Giuntini's commentary on Sacrobosco from 1577, the Milky Way occupied, of a total of 476 small octavo pages, a little more than half.[43] As to the physical explanation of the Milky Way, Giuntini fell back on the mysterious "perspicuum purum" advocated almost three centuries earlier by Peckham in his *Perspectiva communis.* A considerable part of Giuntini's short treatment of the topic was taken up by a poetical account of Juno's milk as the origin of the Milky Way. Nothing more on the Milky Way can be found in the massive two volumes of Giuntini's collected works on astronomy although he dealt with the topic of comets and exhalations in his account of various observations of the fixed stars.[44]

The commentary by Christopher Clavius on Sacrobosco's *Sphaera* deserves special consideration since it served as the major textbook on astronomy for almost half a century. To Clavius, Sacrobosco's short text was merely a traditional stepping-stone to astronomical topics, and his commentary on it clearly constituted an original and lengthy treatise. Yet, what Clavius said on the Milky Way in the first edition of the work in 1570[45] remained unchanged throughout the subsequent five revisions, between 1581 and 1608, which he saw through print.[46] This curious insensitivity to current discussion seems all the more striking since Clavius' treatment of the Milky Way was frustratingly short. He rejected both Aristotle's theory and the one which pictured the Milky Way as a congeries of stars, without, however, mentioning either Democritus or Albertus Magnus. "I consider with others," whom he did not name, "the opinion to be more likely that the Milky Way is a continuous part of the firmament and more dense than other parts of the heavens so that it can absorb the light of the sun, unlike other stars, which are far denser parts of the firmament and more distant from one another; poets may tell any fable they wish about the milk of Juno and the combustion done by the sun."[47] Clavius threw consistency to the winds when, in the same breath, he quoted twenty lines from the astronomical poem of Manilius and four lines from Ovid. His final remark was that those who wished to know more about the Milky Way should consult Ptolemy and, typically enough, Stöffler's commentary on the *Sphaera* of Proclus.

The short shrift Clavius gave to the Milky Way was part of his rigidly traditional treatment of the starry heavens in general. He firmly upheld the traditional count of stars set by Ptolemy at 1,022, as may be seen from what he added on this subject to the second revised edition of his work. There he faced two objections to the traditional count.[48] One was based upon the extremely large number of stars visible on clear wintry nights, especially in the direction of the North Pole. According to Clavius, the visual appearance was either due to a greater clarity of the air which permitted very minute stars to be seen, or to the "hallucination" of one's vision. This in turn was caused, so Clavius claimed, by the more intense scintillation of stars on clear nights and by the inability of the human eye to keep its line of sight steadily fixed on a single star. He made no mention of the fact that the number of stars visible in the belt of the Milky Way was patently larger on similarly clear nights. The other objection Clavius faced was of no scientific merit, but afforded a glimpse of some of his deeper preoccupations. The traditional count sharply contrasted with the words of Genesis comparing the number of Abraham's children to the multitude of stars, even if one assumed that there were 10,000 stars in each of the 48 constellations. The new total, about half a million, still seemed to fall far short of the number of progeny in question. Clavius knew, of course, what figurative speech was, and he felt no further qualms in retaining the hallowed figure of 1,022.

The paucity of words on the Milky Way in professional astronomical treatises in this period appears even more perplexing when compared with the poetical cosmography of the renowned Scottish humanist, George Buchanan. He bested many an astronomer with his beautiful lines on the Milky Way in *De sphaera,*[49] a poem on which he worked intermittently for twenty-seven years, from 1555 on.[50] An acid critic of Copernicus, Buchanan was at heart not a man of science but a philosophical poet. As a poet, he gave a stylish account of the myths, especially that of Phaethon, a favorite figure for Buchanan's soaring imagination. As a philosopher, intrigued by the cause of phenomena, he concluded with words that should have come from astronomers in the first place: "But those who cared to investigate the causes of hidden things, came to believe that many small, not sufficiently bright stars produce, by fusing their light, a luminescence resembling the evening crepuscle, or the starlight dimmed by the rising sun."[51]

The search for causes of physical explanation was a strong factor which readily prompted discussion of the Milky Way. An excellent illustration of this was provided by Bernardino Telesio, a celebrated thinker of his time, who tried to reduce all physical causality to the interaction of two principles, hot and cold. In the system which he elaborated at great length in his *De rerum natura,* the principal embodiment of cold was the earth, while heat resided primarily in the stellar skies. On such basis the Milky Way was somewhat intractable, and Telesio could only remark that the Milky Way was a denser part of the stellar sphere which reflected the light of the sun as the moon did.[52] There was not much enlightenment in Telesio's thematic discussion of the Milky Way, a booklet entitled, *De cometis et lacteo circulo.*[53] Of its nine chapters, only two dealt with the Milky Way, and half of that space was taken up by a rather needless refutation of Aristotle's theory which, as Telesio himself stated, had no adherents either before or during his time.[54] The remarks on heat and cold which Telesio added to the "correct" explanation of the Milky Way, namely, that it was a denser part of the firmament, were mere verbalization. Clearly, philosophical interest was one thing, the usefulness of one's philosophical principles another.

The same disparity between speculative interest and soundness of principles marred the dicta on the Milky Way by that erratic genius, Girolamo Cardano (1501–1576). In the corrected and revised edition of his *De subtilitate,*[55] Cardano mentioned the Milky Way as an example of the fusion of light from many closely spaced sources. In the same breath Cardano also attributed the appearance of the Milky Way to the differences of density in the substance of the celestial sphere. According to him, the stars were the less dense parts in that sphere, and so were the comets which he located among the stars because of their "invariability." Here, Cardano obviously sensed some difficulties, for he mentioned that the variations of the air also might be helpful in producing the appearance of those "invariable" comets which were visible only occasionally. The extent to which Cardano's thinking was influenced by speculations on comets may be seen in his *De rerum varietate.*[56] There the Milky Way once more became a subclass of comets, but Cardano no longer emphasized the role of many individual, closely spaced stars. He spoke of the Milky Way as a reflection of light, in close analogy to comets whose durability he obviously overstated.

Rather revealingly, Cardano's description of the universe started with a discussion of comets, which he considered not only the most spectacular but also the most fundamental feature of the universe. Spectacular and magic phenomena received much attention in J. Bodin's ambitious work, *Universae naturae theatrum.*[57] It contained no reference to the Milky Way, a rather curious omission in view of the subtitle of the book, which promised a discussion of the efficient and final causes of all things. The work gave ample evidence of the basically obscurantist proclivities of its author who in his time stirred much controversy by his rationalist approach to problems of political science, ethics, and theology.

No threat to orthodoxy was posed by Cornelius Gemma (1535–1579), professor of medicine at Louvain, whose discussion of the Milky Way was included in a book[58] which had much in common with the weird atmosphere of Cardano's and Bodin's accounts of nature. The two pages which Gemma offered on the Milky Way were part of a chapter in which he reviewed the diverse prognostications to be derived from monsters, strange worms, oddly shaped stones, bizarre dreams, and from earthquakes and comets. Obviously, it was the topic of comets which prompted him to discuss the Milky Way, concerning which, he claimed, the number of opinions was infinite. He himself defined its material as a transition between ordinary air and heavenly ether, and placed it below the orbit of the moon.[59] Gemma's attention to the Aristotelian connection between comets and the Milky Way was not entirely without some merit. In 1577, he did what none of the notable astronomers of his time cared to do when he made at least one short reference to the Milky Way in connection with the big comet which appeared in that year. "The milky circle," he wrote, "is of its nature between the corruptible substances and the ones that appear eternal."[60] True, even an indirect connection between comets and the Milky Way might have savored of Peripatetic backwardness, but truth would emerge the sooner from error than from silence. At any rate, unlike Tycho and Digges, Gemma did not forget the Milky Way in the pamphlet which he published upon the appearance of the famous nova of 1572 at the edge of the Milky Way. He felt that the new star was too bright to be composed of exhalation. In the diagram attached to his pamphlet, the nova was located well inside the Milky Way, but Gemma was unsure of the latter's nature: "It is the perennial foam of exhalations; physicists discuss it at length though the problem has not been settled as yet."[61]

The study of physics still remained a heavily philosophical enterprise, and philosophy was not made more attractive by the meticulous distinctions through which Francesco Patrizi (1523–1597) tried to probe into the "great and arduous question" of the nature of the Milky Way in his massive *Pancosmia,* or universal natural philosophy.[62] Pedantically, he eliminated at the outset the empyrean heavens, or the immovable sphere enveloping the sphere of the fixed stars, as a possible location of the Milky Way. His reasoning proceeded in this manner: If the galaxy moves with the stars, it is in the ether and has the nature of stars. Its light comes either from it [the ether] or from a separate cause. This latter source of light is either the sun or the stars. If the light of the Milky Way comes from the stars, then from which of the stars? If from all the stars, why is the whole sphere not illuminated? Patrizi saw in this last question the crux of the problem. He sought a solution in the flamelike nature of the stars: "The stars are flame and they illumine because they are both light and fire."[63] The corollary was equally arcane —if not stupefying. "Therefore the Milky Way shines of necessity with a luster, because in all its length and width the light is more dense and the flame is burning; either both of these factors play a part or neither of them. But the latter case is not possible because nothing except fire shines on the earth as well as in the air."[64]

All that could be gained from this was the inconclusive declaration that "the Milky Way shines either because it is light or flame or neither; but light is not flame; flame is light."[65] Patrizi did not profess to know why an extra amount of flame and luster was present in the sky only in the belt of the Milky Way. He felt that to try to fathom this was to probe into the mind of the Almighty. "The Creator of all who wanted the heavens to declare his glory, decreed the flame to be more dense there; . . . no philosophy can rise to those heights; . . . His counsels are inscrutable."[66] Equally inscrutable for Patrizi was the fact that fire on earth produced warmth. It should not, therefore, be surprising that he suspended judgment on the question of whether the stars in the Milky Way were located in, above, or below it. He entertained no doubt on one point. The ether in the Milky Way was more dense than it was elsewhere in the sky, but less dense than it was in the stars. Patrizi's lucubrations on the Milky Way were all the more frustrating in that he prefaced them with a well-informed and witty survey of past opinions on the subject. He carefully distinguished the version of Democritus' opinion as given by Aristotle from the one preserved by [Pseudo] Plutarch. Averroes' "desertion" of Aristotle on the question of the Milky Way was specially welcome grist for Patrizi's mill. He was clearly no friend of the Stagirite. "What should we do with that man," Patrizi exclaimed, "who forever repeats the same points in his philosophy with the same industry and negligence?"[67]

Patrizi in his time was not alone in his forceful denunciation of Aristotle's theory of the Milky Way. A generation earlier, John Pena (1528–1558), professor of mathematics at the Sorbonne, was at a loss for words to castigate Aristotle sufficiently. His opinion on the Milky Way, Pena wrote, "is wholly condemned by the science of optics," and he singled out the fact that the Milky Way showed no parallax.[68] About the same time, Ortensio Landi (1512–1553), a Humanist with a particularly sharp pen, made Aristotle his target in the *Paradossi*.[69] There he tried to prove thirty paradoxical propositions. Some of these were strictly satiric in intent, such as the claims that it is better to be foolish than wise; that it is better to be in war than at peace; that it is better to be sick than healthy; that women are more talented than men; that Boccaccio's writings are not worth reading and the like. He seemed to be rather serious about the claim in proposition 29: "Aristotle was not only ignorant but he was also the most conceited person of all times."[70] As proof of this, Landi referred (among other things) to Aristotle's explanation of the Milky Way and exclaimed: "We fools worship him as an idol and respect all his conclusions as replies from an oracle, although there is hardly a statement in the works of that repository of wisdom which could not be disproved by mathematics."[71] Landi, as a Humanist, was not the man to provide those mathematical proofs, and he showed his true colors when he tried to prove in another paradox that Aristotle was not the author of the works ascribed to him.

Landi's remark on the usefulness of mathematics to refute Aristotle would certainly have delighted John Pena. The principal criticism he leveled against Aristotle's explanation of "meteorological" phenomena, such as comets, rainbows, halos and the like, related to exactly those quantitative aspects largely ignored by Aristotle.[72] Pena took to task for the same reason the "physicists" who, he declared, espoused almost to a man Aristotle's theory of the Milky Way. By physicists he seemed to mean the "philosophers of nature," going about their business in Aristotelian fashion. Pena did not live to see that, for the next half century, it was just those "physicist-philosophers" who kept alive the topic of the Milky Way in their discussions of "meteorological" topics and in their commentaries on Aristotle's *Meteorologica.* In Pena's own profession of mathematical astronomy, only those of the "lower rank," busy with popularization and navigational instruction, were to find space in their now extremely rare works for the Milky Way.

That they were of the "lower rank" is borne out by the fact that their names nowhere occur in general histories of astronomy. Francesco Barozzi, author of a *Cosmographia,*[73] was a Venetian nobleman who also had an interest in astrology, but his short discussion of the Milky Way was not tainted by it. His interest in practicality can be seen in his reluctance to commit himself to any of the "modern" theories of the Milky Way. He also gave undeserved credit to Ptolemy for demonstrating the falsity of Aristotle's opinion. Even more obscure than Barozzi was the "mathematicall lecturer in the Citie of London, sometime of Trinitie Colledge in Cambridge," as Thomas Hood described himself in 1592, on the title page of his *The Use of both the Globes, celestiall, and terrestriall.*[74] He described the Milky Way as part of the firmament, not as thick as were the parts formed by the stars and not as thin as elsewhere. In the latter case, the light of the sun would have passed through the Milky Way (he did not consider the question of what lay beyond the firmament and how anything could move beyond it); in the former case, the Milky Way "would glitter, and shine as the Starres themselves doe." But "beyng neither so thinne as the one, nor so thicke as the other, it becommeth of that whiteness which we see." That such was the "best opinion" on the Milky Way had already been asserted by Hood two years earlier in his *The Use of the Celestial Globe in Plano.*[75] His account of various opinions on the Milky Way should appear more entertaining than informative. He not only dismissed Aristotle, but also failed to endorse the opinion of "the Philosophers (and chiefly Democritus)," who "affirme the cause of the thing to be the exceeding great number of starres in that parte of the heauen, whose beames meeting together so confusedly, and not comming distinctly unto the eye, causeth us to imagine such a whitenes as is seene." Hood's fancy was mostly attracted to Juno's milk as the cause of the Milky Way. He reported the myth in four different versions, and found it important to remind his reader that of that milk "you must suppose some sufficient quantitie." From Hood one could also learn that in England "some in sporting manner doe call it Watling streete," a detail which somewhat perplexed him: "why they call it so, I cannot tell, except it be in regard of the narrowness that it seemeth to have, or else in respect of that great high way that lieth betwene Douer and S. Albons, which is called by our men Watling streete." Hood was, however, supremely confident of his knowledge about the purpose of the Milky Way in the firmament. It was there to show us that "the starres mooue not in their spheres, as fishes in the sea, or as birdes in the ayre." Had such been the case, it is most certain, Hood argued, "that the starres which are in this circle at this present, woulde by and by shift the same, and passe out of it into some other place of heauen, which never falleth out to be so."[76]

On the other hand, Paul Merula, professor at Leiden, mentioned the Milky Way in his *Cosmographia* in connection with circles, for which "there is no use."[77] Equally evasive, but in a different sense, was the treatment accorded the Milky Way in S. Girault's *Globe du monde,*[78] a dialogue between Charles and Marguerite. The latter, after being informed that the Milky Way was the fusion of the light of many small, closely spaced stars, then wanted to know whether the apparent differences of stars were caused by variety in size, or by the fact that some of them were fixed in regions of the firmament closer to us. The question implied not only an extreme width for the starry sphere, but also a view in depth of the Milky Way. To these enticing inquiries, the answer was that enough had already been said about the firmament.

Among the treatises on meteorology, which were not strict commentaries on Aristotle's *Meteorologica,* chronological priority belongs to Antoine Mizauld's treatise on weather forecasting, published in 1547.[79] He offered a grand total of 376 weather forecasting aphorisms based on various atmospheric phenomena among which he listed comets. The Milky Way was, according to Mizauld, a sign of serene weather, scarcely a major discovery. The meteorological work of Marcus Fritsche in 1555 had greater importance not only because of its special chapter on the Milky Way,[80] but also because of its impact on a group of treatises on meteorology, authored by teachers and graduates of the University of Wittenberg. The first to mention it is Michael Stanhufius, who in his *De meteoris*[81] contrasted the grave dissensions on the Milky Way in ancient times with the almost unanimous agreement among the "more recent philosophers" on the opinion that it was a congeries of many small stars. He reported the correct form of Democritus' opinion from Macrobius, and quoted Pontano as often as he could, though the gain was merely stylistic.

The *Meteorologia* of Johann Gartze (Garcaeus)[82] was, unlike the work of Stanhufius, filled with endless distinctions and classifications. Still, he was at one with Stanhufius in stating that "it is commonly held today among scholars that the galaxy is of celestial nature and that the many, closely spaced, small and hither-tither scattered stars, called sporads, form a whitish band in the firmament because their lights are seen as fused into one; due to their smallness and great distance they cannot be seen separately." This opinion, he concluded, "seems to be most consonant with reason."[83] The main arguments in support of this were, according to Gartze, the following: (1) all exhalation disappears and is subject to corruption, whereas the galaxy is a perennial circle which always appears on clear nights; (2) the galaxy always shows the same extent (quantity); were Aristotle's theory true, this quantity would change when the two hot planets, Jupiter and Mars, crossed the Milky Way, thereby adding their influence to that of the fixed stars on the exhalation below; (3) if the galaxy consisted of exhalation, it would be dissipated by the sun when its path crossed the Milky Way in Sagittarius and Gemini, but the galaxy shows no diminution.

Gartze's attentive recall of Pliny's opinion on the influence of the Milky Way was a point emphasized also by Wolfgang Meurer, who served from 1543 to 1570 as professor of Aristotelian philosophy at Wittenberg. But his lecture notes on meteorology were not published until 1587, two years after his death.[84] Through his mother he was a relative of Melanchthon, the founder of Protestant scholasticism and an astronomer of sorts, who himself avoided the question of the Milky Way.[85] Meurer might have been a teacher of both Stanhufius and Gartze, though with respect to the Milky Way neither followed Meurer's efforts to shore up Aristotle's theory. To Meurer, who rejected the explanation given by Albertus Magnus, the steadiness of the Milky Way constituted no proof against Aristotle's theory, for the simple reason that the Milky Way was very often invisible on clear nights. He disagreed with Aristotle only on the matter of the clouds of exhalation: he saw in them a medium of refraction for the light of the stars in the Milky Way, whereas Aristotle had presented them as incandescent. An addict of astrological lore, Meurer quoted with full approval Pliny's opinion that the Milky Way was the cause of all milk in mammals and of all juiciness in fruit. While he did not avoid the problem of why the clouds of exhalation always gathered in the same area to form the Milky Way, his explanation of this was that clouds formed in any region where nothing opposed them. For him, the case was analogous to the multiplication of frogs where storks were not present, and to the proliferation of rabbits and thieves in any area full of hiding places.

As was the case with the authors from Wittenberg, the name of William Fulke has never been included in any roster of scientific men. For all that, his *Goodly Gallery to behold the naturall causes of all kynde of Meteors*[86] is far more informative in respect to the Milky Way than the silence of the great astronomers of those times. A believer in lactation on earth as the "finall cause" of the Milky Way, Fulke rejected Democritus' theory, which he carefully recalled in both versions, on the ground that whitish patches should then appear everywhere in the sky where stars were closely spaced. He thought that the Milky Way was a denser part of the sphere of the fixed stars which, he asserted, had no light of their own. Equally indefensible, though in a different sense, was his report of an ancient explanation of the Milky Way by the myth of Hebe, Jupiter's cupbearer, who "on a time stumbled at a straw, and shed the Wine or Milk that was in the Cup, which coloured that part of Heaven to this day: wherefore she was put out of her office." After all, did the king of Olympus ever request a cup of milk?

No such details enliven two other treatises on meteorology published in Italy. In his *Meteoria*[87] Bartolomeo Arnigio claimed to provide prognostication on all kinds of weather and for every land. Although he admitted that most astronomers held the Milky Way to be a congeries of stars, he nevertheless chose to believe that it was the reflection (dispersion) of starlight in the dry exhalation. This was all the more baffling, as his short chapter on the Milky Way closed with a listing of Philoponus' arguments against Aristotle's theory. The best aspect of the discussion of the Milky Way in Cesare Rao's *I meteori*[88] related to its history. Rao knew the difference between the *old* and *new* translations of the *Meteorologica* and was also familiar with the dicta of Albertus Magnus, Peckham, and the ancient commentators, Olympiodorus, Ammonius, and Philoponus. According to Rao, the Milky Way was caused by variations in the density of the sphere of the fixed stars, and he held the resulting effect to be strictly analogous to the differences in luminosity of various parts of the moon's surface.

The number of systematic commentaries on Aristotle's *Meteorologica* from the period under discussion is comparatively small. Obviously, to the extent that a more genuine scientific method was gaining ground, Aristotle's *Meteorologica* was falling into disrepute. In Jakob Degen's commentary of 1550[89] the four lines devoted to the nature of the Milky Way implied an acceptance of the Aristotelian theory. Quite the opposite in more than one sense was the approach to the topic by Francesco Vicomercato (c. 1505–1565). Unlike Degan, Vicomercato was an accomplished scholar. He not only published a critical edition of Aquinas' commentary on the *Meteorologica,* but also of two earlier Latin translations. To these he added his own, based on the best available Greek text.[90] His lengthy commentary[91] shows him in full command of the tradition up to Albertus Magnus. Vicomercato pointedly noted (possibly the first to do so) Ptolemy's silence on the nature of the Milky Way, and rebuked Albertus Magnus for crediting Ptolemy with the opinion that the Milky Way was the fusion of the light of many small stars. Such was, of course, the view which Vicomercato himself defended against Aristotle and against Alexander and Averroes. He had some critical remarks even for Olympiodorus, whom he reproached for trying to put Aristotle's explanation in a better light. As for Aristotle, Vicomercato took the view that the Stagirite went astray in much the same way as did the Pythagoreans. These were criticized by Aristotle in Book II of *On the Heavens* for explaining everything by the number ten, and as a result came up with "many false and dreamy assertions," as Vicomercato recalled the matter. What trapped Aristotle, whom Vicomercato hastened to describe as "a man endowed with the highest perspicacity in investigating the nature of things," was his systematic reliance on the two kinds of exhalations. It was to these that he reduced the cause of many sublime things, among them the Milky Way, in spite of the fact that its perpetual sameness demanded that it should consist of some unchanging, that is, heavenly substance.[92]

The freedom with which Aristotle was criticized by Vicomercato, himself a Peripatetic, was also much in evidence in the commentary of Francesco de Vieri on the *Meteorologica.*[93] He censured Aristotle for not following his own dicta on the love of truth as set forth in the *Nichomachean Ethics*. De Vieri's remark, that only God knew why the Milky Way existed, echoed an already old, rudimentary probing into cosmological causes.[94] Well acquainted with the arguments of Ammonius and Philoponus, de Vieri could wax disputative. He heaped criticism on the efforts of those who either claimed that Aristotle expressed only a probable opinion, or who assigned Aristotle's theory to the mischief of a copyist.

Such efforts were the work of those who venerated Aristotle as the infallible oracle of truth. No criticism of the Stagirite, however well argued, could be left by them without a reply. Vicomercato's discussion of the Milky Way earned some vituperative remarks from the Venetian polemist, Antonio Polo (d. 1582), author of a small treatise on the Milky Way. Its title claimed to defend Aristotle against all Peripatetics.[95] The most valuable parts of Polo's often hairsplitting discourse were borrowed from Boccadiferro, especially the interpretation of Aristotle's somewhat obscure dicta on the reflection theory. But Polo had no patience with any critic of Aristotle. Boccadiferro and Vicomercato were among those whom Polo dismissed with the declaration: Juniors should not be listened to! He decried as insane and stupid those commentators of Aristotle who tried to shore up the Stagirite's theory by deviating from the obvious sense of his words. Aristotle, so Polo insisted, could not become oblivious to his prinicples. He was the "semper bonus scientificus," the "secretary of nature," who perceived everything that could be grasped by the light of intellect.[96] However, the view of Aristotle did need some defense. Polo based this on the minuteness of the changes which might occur in the Milky Way, and which could not be seen because of the great distance. As an analogy, he referred to the absence of observable change in the volume of the sea although some part of it—especially in the northern regions—changed constantly into mist.

Quite different was the tone of the compendia and commentaries on Aristotle's philosophy by Johann Ludwig Havenreuter (1548–1618). His works were largely lecture notes, carelessly published against his will by his students at Strasbourg. Havenreuter, who combined professorship with active medical practice, could only complain about the piracy and correct the faulty texts, at least in part, in later editions. He seems to have been a good teacher, attentive to essentials, but breaking no original ground. In his compendium of Aristotle's teachings about the physical world, he completed his survey of the topic of the Milky Way with the remark that opinions in his time were divided between Democritus' theory and that of Philoponus.[97] His commentary on the *Meteorologica*[98] emphasized careful explanation with constant reference to the Greek text. Criticism played a minor role and little was said about previous authors, although Havenreuter clearly stood in debt to Boccadiferro, to say nothing of Philoponus, whose opinion he held to be the most probable.

What Havenreuter offered on the question appears a trifle unimpressive when set against the commentary composed by the faculty of the Jesuit College at Coimbra around 1600.[99] Although Clavius had strong connections there, his view of the Milky Way did not prevail with the Coimbrenses. They supported the theory of many small stars, adding, however, the possibility that the Milky Way was a denser part of the sphere of the fixed stars. Their commentary also noted that Aristotle's opinion was rejected by the "general consensus of philosophers and astronomers."[100] As for the dissenters among Peripateticians, "some from Paris, some from Louvain, others from the family of Saint Thomas," the commentators recalled their efforts to exculpate Aristotle on the ground that the text of the *Meteorologica* had become corrupted. It is better, they remarked, to recognize that "for all his ingenuity, Aristotle was a mere human, to whom no basic human characteristic was foreign; and very human is indeed to fail and err on occasion."[101]

The possibility of error is especially great when one moves out of his own field, great scholar though he may otherwise be. Joseph Justus Scaliger (1540–1609) was in the eyes of many the finest scholar of his day, yet he followed the wrong theory of the Milky Way because of his admiration for classical poetry. In his edition of Manilius' astronomical poem, he praised Manilius for having followed Aristotle in speaking of the Milky Way in connection with the comets and not as a part of the stellar heavens.[102] It was not, however, on that basis that a blow was aimed at Scaliger's undisputed prominence in the scholarly world. What startled his overawed contemporaries was that the acid champion of historical criticism should, on patently weak evidence, claim to be a descendant of the prestigious La Scala family.

What was said of the Milky Way during the dozen or so years previous to 1610 did not suggest that a spectacular event was in the making. Only a year earlier Georg Heinisch, a mathematician and physician in Vienna, offered a vivid example, in his commentary on Proclus' *Sphaera,*[103] of the extent to which Aristotelian tradition on the Milky Way could still trap a less critical mind. He found confirmation for what he proudly called the "Aristotelian dogma" on the Milky Way in the play on words by which one could turn *Hêraklês* into *Hêras kléas,* or Hera's (Juno's) radiance. Heinisch bolstered this with the physician's remark that the blood changed into milk as it rose from the uterus into the breasts. Juno was therefore the air, the milk of which was the dry exhalation, or the glow of the air, *aéros kléas.* No less naive was the manner in which Heinisch argued, on the one hand, that the Milky Way could not be the fusion of starlight because in that case it should always be visible; on the other hand, he saw Aristotle's theory vindicated by the fact that the Milky Way could not be observed on every cloudless night. Because of the problem of parallax he admitted that the Milky Way, though consisting of dry exhalations, was in the superlunary regions. The "comets" (novae) of 1572 and 1604 were to Heinisch proof that the dry exhalations could rise even above the orbit of the moon. Curiously enough, the appearance of a large comet in 1607 prompted no reflection on the Milky Way, as can be seen, for instance, in the pamphlet on that comet by the Lübeck physician, David Herlicius von Zeitz.[104]

The Tychonian view of the Milky Way transpired in 1603 in the *Uranometria* of the Augsburg lawyer and astrologer, Johann Bayr (Bayer), whose only enduring contribution to astronomy was purely technical and incidental; namely, the designation by Greek letters of the stars in each constellation. Special mention should, however, be made of Bayr's careful recording of the nebulous star, Praesepe, in Cancer (Illustration IVa). He took pains to note that "astronomers compared it to the Milky Way."[105] A few years later, such a comparison was made, most strikingly, by no less an astronomer than Galileo. The Tychonian view was not given unqualified endorsement by Thomas Lydiat when, in 1605, he discussed the motion of the sphere of the fixed stars. On one hand, he claimed that the Milky Way was a fine vapor or smoke near the stars. On the other, he wanted to agree with Aristotle as well as with those who claimed that the Milky Way was a congeries of a great many small stars.[106]

It was the motion of the heavens which also prompted Joseph de Acosta, a Jesuit missionary, to mention the Milky Way in his famous account of Central and South America.[107] For the same purpose, he also found useful the daily revolution of some exceedingly dark spots in the heavens "which I remember not to haue seene at any time in Europe, but at Peru, and in this other Hemisphere I haue often seene them very apparent. These spots are in colour and forme like vnto the Eclips of the Moone, and are like vnto it in blacknes and darknes; they march, fixed to the same starres, alwaies of one forme and bignes, as we haue noted by infallible observation." As to the physical nature of those black areas, he tied it to the nature of the Milky Way about which, so Acosta claimed, philosophers state that it "is compounded of thickest parts of the heauen and for this cause it receiues the greater light." Therefore, he pictured the black areas as being "very thinne and transparent, the which receiuing lesse light seeme more blacke and obscure."[108] But he did not exclude the possibility that such an explanation might not be the true one.

Similar was the opinion of the Milky Way held by William Gilbert (c. 1540–1603), author of the *De magnete,* but the manuscript, entitled "De Mundo,"[109] in which he stated this, gathered dust during the first half of the seventeenth century. Thomas Harriot seemed to think that it was about to be published when he sent word of it on July 13, 1608, to Kepler, who expressed his interest in having a copy in his reply of September 1, 1609.[110] Gilbert's manuscript was collated from his notes by his half-brother, William Gilbert, some time after 1603, and certainly included several details which would have prompted some lively reflections on Kepler's part. One of these details was the diagram of the seemingly unlimited distribution of stars beyond the orbit of Saturn.[111] Another was Gilbert's suggestion that since the stars in the belt of the Milky Way maintained the same appearance everywhere, the Milky Way was probably situated beyond most of the stars.[112] Clearly, Gilbert's idea was that the Milky Way differed from the vast empty space of the ethereal realm which, he believed, did not diminish the starlight. Gilbert associated the Milky Way with nebulous patches in the sky and defined it as the "corporescence of the superior ether,"[113] and "as a real, permanent matter, diffused across the inane of the universe, but distinct from the inane."[114] Nevertheless, Gilbert's discussion of the Milky Way which began with a long and acid refutation of Aristotle's theory, concluded with a restatement of Democritus' idea of the Milky Way: "Democritus thinks it to be the compound light of stars, due to the crowding of many small and contiguous stars shining together." To this Gilbert added a phrase which clearly expressed his own reflection: "The Milky Way may therefore be a visual impression, as very small sources of light have only a blurred effect on the eye and not a distinct one as do the stars."[115]

With that explanation of the Milky Way, Kepler would have disagreed. At the same time, he might have been pleased with the title, "Nova Meteorologica contra Aristotelem," which introduced the third book of Gilbert's manuscript. There Gilbert discussed the most important topics covered in Aristotle's work, including that of the Milky Way. But in September, 1609, when Kepler requested a copy of the manuscript, the most provocative detail in it was proved to be neither the diagram of the stellar realm nor the title of its third book, although both unerringly expressed the irreversible turn of ideas. The true hidden gem of the manuscript was the last sentence of Gilbert's discussion of the Milky Way, which read: "Adspice lacteum etc cum specillis," or "look at the Milky Way etc., with lenses."[116]

Written before 1603, this phrase is, of course, a most intriguing one for the history of the telescope. Dutch lensmakers had worked for several years on a model with a magnifying power of about three before they sought a patent for their invention in 1608. Did Gilbert get early word of their success, and did it occur to him that an instrument which seemed to be useful only to bring closer distant objects on earth, might be turned with profit toward the sky? Such questions may never be answered. What history clearly reveals is that the dramatic implications of Gilbert's phrase for the Milky Way, for astronomy, and for man's view of the universe were to unfold with dramatic suddenness. Some time in the fall of 1609, Galileo, who had just demonstrated his "perspicillum" to the Venetian

Senate with resounding success on August 25, succeeded in increasing the magnifying power of his instrument from about three to thirty. Most important, in a moment of happy inspiration, he looked through it at the canopy of the sky. The immense multitude of stars he saw in the Milky Way merely confirmed, in regard to its true composition, a long-standing consensus which had, until now, been buried under an avalanche of stereotyped claims that Aristotle's "ideas about the non-celestial character of . . . the Milky Way held sway until the revival of astronomy in the sixteenth century."[117] The great initiator of that revival, Copernicus, contributed only silence to that most valuable consensus.

References

[1] See the English translation by Charles G. Wallis, in *Great Books of the Western World,* vol. 16. (Chicago: Encyclopaedia Britannica, Inc., 1938).

[2] *Ibid.,* Book I, chap. 6, p. 516.

[3] *Ibid.,* p. 517.

[4] *Ibid.,* Book I, chap. 10, pp. 526–27.

[5] Imprinted at London by Reginalde Wolfe, Anno Domini 1556.

[6] *Ibid.,* p. 165.

[7] *Ibid.,* p. 97.

[8] *Ibid.,* p. 100.

[9] *Ibid.,* pp. 105–06.

[10] "A Perfit Description of the Caelestiall Orbes according to the most aunciente doctrine of the Pythagoreans, latelye reuiued by Copernicus and by Geometricall Demonstrations approued," ff. N1–O3 in Leonard Digges, *A Prognostication euerlastinge* [etc.] (Imprinted at London by Thomas Marsh, 1576). The full title of the book tells us that Leonard Digges aimed at weather forecasting based on the position of the sun, moon, planets, and comets. More sensible was his attention to the rainbow, clouds, and thunders. Needless to say, his meteorological list did not include the Milky Way, probably because of its unalterable form and position.

[11] The diagram faces the first page of Thomas Digges' essay and is marked, erroneously, Folio 43. The full text of the essay is reprinted, with the diagram and an Introduction by F. R. Johnson and S. V. Larkey, under the title, "Thomas Digges, the Copernican System, and the Idea of the Infinity of the Universe in 1576," *The Huntington Library Bulletin,* April, 1934; 69–117.

[12] *Ibid.*, f. N4r.

[13] *Alae seu scalae mathematicae* [etc.] (London, 1573). The "mathematical scales" stood for the method by which Digges wanted to establish the distances of the planets together with the distance and position of the new star.

[14] See *La Cena de le Ceneri* [1584], edited by G. Aquilecchia (Roma: G. Einaudi, 1955), pp. 90 and 92 (Dial. I).

[15] *De immenso et innumerabilibus* [1591], in *Jordani Bruni Nolani opera latine conscripta,* vol. I, part 2, edited by F. Fiorentino (Naples: Dom Morano, 1884), pp. 295–96 (Lib. VIII, cap. 4).

[16] *Ibid.*, p. 44 (Lib. IV, cap. 8).

[17] *Ibid.*, pp. 127–28 (Lib. V, cap. 4).

[18] *Ibid.*, p. 126.

[19] *Ibid.*, pp. 230–31 (Lib. VI, cap. 19).

[20] *Ibid.*, pp. 227 and 229.

[21] Bruno, *La Cena de le Ceneri,* pp. 173 and 218 (Dial. III, and Dial. V). This work of Bruno is more expressive of his thought than its sequel, the better known *On the Infinite Universe and Worlds* [*De l'infinito universo et mondi,* 1584], translated by D. W. Singer. The translation forms the second part of her *Giordano Bruno: His Life and Thought* (New York: Henry Schuman, 1950). On the physical problems of an infinite universe, see the opening section of Dial. III, pp. 304–05.

[22] *De stella nova in pede Serpentarii* [1606], in Johannes Kepler, *Gesammelte Werke,* edited by M. Caspar, vol. I (Munich: C. H. Beck, 1938), pp. 151–292.

[23] *Ibid.*, chap. 21, pp. 251–57.

[24] *Ibid.*, chap. 22, pp. 257–58.

[25] In *Tychonis Brahe Dani opera omnia,* edited by I. L. E. Dreyer, vol. III (Copenhagen: Libraria Gyldendaliana, 1916).

[26] First published in 1573 and reprinted in a facsimile edition by the Royal Danish Society of Sciences (Copenhagen, 1901) to commemorate the 400th anniversary of Tycho's death.

[27] Brahe, *Astronomiae instauratae progymnasmata,* p. 305.

[28] See his Inaugural Lecture given in 1574; *Opera,* vol. I, pp. 147–73.

[29] Brahe, *Astronomiae instauratae progymnasmata,* p. 306. In 1590, Tycho defended his explanation of the Milky Way in a letter to Caspar Peucer; see *Opera,* vol. VII, pp. 235–37.

[30] *Commentarius de praecipuis divinationum generibus. . .* (Wittenberg: Ioannes Crato, 1553), ff. 235r-257v. Peucer, who accepted Aristotle's explanation of comets, eagerly discussed comets as portents of future events, and noted that even those who claimed them to be of the same nature with the stars admitted their influence on terrestrial processes. In the same year there appeared the revised edition of Peucer's major astronomical work, *Elementa doctrinae de circulis coelestibus, et primo motu, recognita et correcta* (Wittenberg: Ioannes Crato, 1553), but its section on major and minor circles contained no reference to the Milky Way.

[31] Published in two parts, of which the second dealt with particular phenomena of nature, *La seconda parte de la filosofia naturale* (Vinetia: presso Giorgio de' Canalli, 1565).

[32] *Ibid.*, pp. 392–96. There is no reference to the Milky Way in Piccolomini's *De le stelle fisse* (Venice: G. Varisco & Compagni, 1579), a work devoted to the description of forty-seven constellations.

[33] Originally published in 1598. The full title of the third and revised edition reads: *Epitome astronomiae, qua brevi explicatione omnia tam ad Sphaericam quam Theoricam ejus partem pertinentia ex ipsius scientiae fontibus deducta, perspicue per quaestiones traduntur* (Tübingen: Philippus Gruppenbachius, 1610).

[34] *Ibid.*, p. 36. On his definition of the subject matter of astronomy, see pp. 39 and 30.

[35] *Theoricae novae planetarum Georgii Purbachii Germani ab Erasmo Reinholdo Salueldensi pluribus figuris auctae, et illustratae scholiis* (Paris: apud Carolum Perier, 1553). Reinhold's silence on the Milky Way contrasts sharply with the length of his comments and with the fact that on occasion he discussed questions of "physical astronomy" such as the cause of the moon's light. See ff. 103r–105v.

[36] *Quaestiones novae in theoricas novas planetarum* (Basel: ex officina Henricpetrina, 1568).

[37] *Commentaria in novas theoricas planetarum Georgii Purbachii* (Basel: per Henricum Petri, 1556).

[38] *Sphaera Procli cum annotationibus Eras. Osvaldi Schreckenfuchsii* (Basel: per Henricum Petri, 1561), p. 66 note.

[39] *Ibid.*

[40] *Erasmi Oswaldi Schreckenfuchsii commentaria in Sphaeram Ioannis de Sacrobusto* [etc.] (Basel: ex officina Henricpetrina, 1569).

[41] *Procli sphaera Thoma Linacro Britanno interprete, figuris et demonstrationibus illustrata cum annotatiunculis Iacobi Tusani Regij Graecarum literarum professoris* (Paris: G. Cavellat, 1562), ff. 21v–22r, where much atten-

tion is also given to references to the Milky Way by classical poets. The reference in the beginning of the tenth book of the *Aeneid* to Jupiter's starry seat is also taken as an allusion to the galaxy.

[42] *Quaestiones novae in libellum de Sphaera Joannis de Sacro Bosco* [etc.] (rev. ed.; Paris apud Gulielmum Cauellat, 1569).

[43] *Commentaria in Sphaeram Ioannis de Sacro Bosco accuratissima* (Lyons: apud Philippum Tinghium, 1577), pp. 62–72.

[44] See his "Annotationes in cometis," in "Compendium de stellarum fixarum observationibus," in *Speculum astrologiae universam mathematicam scientiam in certas classes digestam complectens* (Lyons: in officina Q. Phil. Tinghi, 1583), vol. II, p. 1124. This volume also contains Giuntini's commentary to Peurbach.

[45] *In Sphaeram Ioannis de Sacro Bosco commentarius* (Rome: apud Victorium Helianum, 1570), pp. 376–77.

[46] There were in addition several reprints of these revised editions.

[47] *Ibid.,* p. 376.

[48] Rome: ex officina Dominici Basae, 1581, pp. 149–50.

[49] In *Georgii Buchanani opera omnia* (Leiden: apud J. A. Langerak, 1725), vol. II, p. 476 (Lib. II, lines 383–412).

[50] For further details, *See George Buchanan: A Memorial 1506–1906,* edited by D. A. Millar (St. Andrews: W. C. Henderson, 1907), pp. 150–65.

[51] Lines 407–12. *Les semaines* of Guillaume de Salluste, Sieur du Bartas, a cosmographical poem of 1578 based on the six days of Creation, was as anti-Copernican as Buchanan's *De sphaera.* See the English translation (1605) by Joshua Sylvester, *Bartas: His Devine Weekes and Works,* reprinted with an introduction by Francis C. Haber (Gainesville, Fla.: Scholars' Facsimiles and Reprints, 1965), p. 120. Du Bartas discussed comets and winds as parts of the First Day's work (pp. 47–48), he spoke at length of the zodiac's signs as part of the Fourth Day, and waxed verbose on such topics as the elephant's way of fighting the rhinocerus (p. 193). He singled out astronomy (p. 222) as special evidence of man's superior capabilities, but ignored the Milky Way.

[52] Originally published in 1565 and again in a revised form in 1570. The Milky Way is discussed in Lib. I, cap. 3. See the modern critical edition prepared by V. Spampanato (Modena: A. F. Formiggini, 1910), vol. I, p. 15.

[53] Venice: apud Felicem Valgrisium, 1590.

[54] *Ibid.,* f. 13r. The two chapters in question are 4 and 9. Since the Milky Way was, according to Telesio, part of the sphere of the fixed stars, there was no reference to it in his booklet on meteorological topics, *De his quae*

in aere fiunt et de terraemotibus liber unicus (Naples: apud Josephum Cacchium, 1570).

[55] Lyons: apud Guliel. Rouillium, 1554, "Unaquaque igitur stella proprium habet lumen cum propriam habet lucem, quo sit ut ex frequentibus sideribus lumen miscentibus lacteus circulus ab oculo aestimetur, adiuvat hoc coeli substantia densa et syderum rara, velut in cometis, qui caudam aut crines habet, nam cum numquam mutetur, constat ipsum in coelo esse non infra" (p. 155). See also pp. 162 and 164.

[56] Avignon: per M. Vincentium, 1558. See Lib. I, cap. 1, pp. 2–3. Cardano rejected the motion of the earth as something "fabulosum". See Lib. II, cap. 11, p. 77.

[57] Frankfurt: apud heredes A. Wecheli, C. Marnium & I. Aubr., 1597.

[58] *De naturae divinis characterismis seu raris et admirandis spectaculis causis, indiciis proprietatibus rerum in partibus singulis universi libri II* (Antwerp: ex officina Christophori Plantini 1575).

[59] *Ibid.*, vol. I, pp. 116–17 (Lib. I, cap. vi).

[60] *De prodigiosa specie, naturaq. cometae qui nobis effulsit altior lunae sedibus insolita prorsus figura, ac magnitudine, anno 1577* [etc.] (antwerp: ex officina Christophori Plantini, 1578) p. 40. In other treatises prompted by the appearance of that comet, there are only a few short references to the Milky Way, as may be seen in the lengthy study by Clarisse D. Hellman, *The Comet of 1577: Its place in the History of Astronomy* (New York: Columbia University Press, 1944).

[61] *De peregrina stella quae superiore anno primum apparere coepit* (Antwerp, 1573), f. A2r.

[62] *Francisci Patricii Pancosmiae de aethere ac rebus coelestibus libri XIII* [comprising Books 9–22 of the whole work] (Ferrara: ex typographia Benedicti Mammarelli, 1591). The topic of the Milky Way is the subject of the whole "Book" 16, of three folio pages (pp. 100–02). Patrizi was better known as a political scientist and a Hermetic philosopher.

[63] *Ibid.*, p. 101.

[64] *Ibid.*

[65] *Ibid.*

[66] *Ibid.*

[67] *Ibid.*

[68] In the Preface to his *Euclidis optica et catoptrica numquam antehaec graece aedita* [sic] *eadem Latine reddita* (Paris: apud Andream Wechelum, 1557), f. bb ii v.

[69] *Paradossi cioè sententie fuori del comun parere: novellamente venute in luce* (Venice, 1545). The work was first published in 1543, and reprinted in 1554, 1564, 1594, and 1602. Landi, a master of hyperbolic sophisms, also penned a refutation of the *Paradossi,* in the same fashion that he praised and deplored Cicero in another series of essays.

[70] *Ibid.,* f. 76r.

[71] *Ibid.,* f. 77v.

[72] *Euclidis optica,* f. bb iij v.

[73] *Cosmographia in quatuor libros distributa . . . ad magnam Ptolemaei mathematicam constructionem ad universamque astrologiam instituens* (Venice: ex officina Gratiosi Perchacini, 1585). On the Milky Way, see pp. 83–85. The section is given without any revision in the Italian translation, *Cosmografia in quattro libri divisa* (Venice: presso Gratioso Perchacino, 1607), ff. 100r–101v.

[74] London: imprinted by T. Dawson, 1592. On the Milky Way, see ff. C3r–C3v. No reference to the Milky Way was made by Hood's English forerunner, William Cunningham, author of *The Cosmographical Glasse conteyning the pleasant Principles of Cosmographie, Geographie, Hydrographie, or Navigation* (Excussum Londini in officina Ioan. Daij Typographi, Anno 1559). This was all the more curious as Cunningham followed Proclus' *Sphaera* in describing the great and small circles. See ff. 18r–40v.

[75] London: imprinted [by I VVindet] for T. Cooke, 1590. On the Milky Way, see ff. 40r–42r.

[76] *Ibid.,* f. 42r.

[77] *Cosmographiae generalis libri tres, item geographiae particularis libri quattuor* (Raphelengij: ex officina Plantiniana, 1605). On the Milky Way, see pp. 61–62.

[78] *Globe du monde, contenant un bref traité du ciel et de la terre* ([Langres]: J. des Preyz, 1592). On the Milky Way, see ff. 38r–38v.

[79] *Phaenomena sive Aeriae ephemerides* (Paris: ex officina Reginaldi Calderij et Claudij eius filij, 1546), f. 64v. No mention of the Milky Way can be found in Mizauld's discussion of the constellations, *Antonii Mizaldi Montluciani asterismi: sive stellatarum octavi coeli imaginum officina* (Paris: apud Carolum Cuillard, 1553).

[80] *Meteororum, hoc est, impressionum aerearum et mirabilium naturae operum loci fere omnes* (Nürnberg: in officina Ioannis Montani et Ulrici Neuber, 1555). On the Milky Way, see pp. 104–11. According to Fritsche, it was generally admitted that the Milky Way heated those parts of the globe over which the sun did not pass directly (p. 109). Fritsche was unaware of the two forms in which Democritus' opinion of the Milky Way survived, but

listed three arguments against Aristotle's theory. His work was reprinted in 1581.

[81] Originally published in 1562. The full title of the second edition is *De meteoris libri duo. Quorum prior tradit de aethere et elementis. Posterior complectitur omnium fere meteororum prolixam explicationem. Recitantur etiam passim Aristotelis, Plinii et aliorum Philosophorum indicia et opiniones, conscripti et editi a M. Michaele Stanhufio Franco* (Wittenberg, 1573). On the Milky Way, see ff. L3v–L4r.

[82] Wittenberg, 1568. On the Milky Way, see ff. 75r–78r.

[83] *Ibid.*, ff. 76v–77r. He described this opinion as that of the "experts" (artifices) in his *Secundus tractatus de tempore sive de ortu et occasu stellarum fixarum ad quodlibet temporis momentum* (Wittenberg [Joannes Crato], 1565), p. 47. There he also described stars as denser parts of the firmament (pp. 17–19).

[84] The work was re-edited by his son, Christopher Meurer (Leipzig: impensis Henningi Grossii, 1606). On the Milky Way, see pp. 102–14.

[85] See Melanchthon's *Initia doctrinae physicae,* cols. 181–412, in vol. 13, of *Opera quae supersunt omnia,* edited by C. G. Brettschneider (Halle: C. A. Schwetschke, 1846), especially cols. 223–29, "Quot sunt sphaerae coelestes?"

[86] London, 1563, On the Milky Way, see ff. 38r–40r. It was reprinted in London in 1670 (for William Leake) under the title, *Meteors; or, A plain Description of all kind of Meteors* [etc.]; see pp. 81–86; quotation is from p. 83.

[87] *Meteoria over discorso intorno alle impressioni imperfette, humide, secche et miste* [etc.] (Brescia: appresso Francesco et Pietro Maria fratelli de Marchetti, 1568). On the Milky Way, see ff. 89v–91r.

[88] *I meteori di Ceasare Rao di Alessano citta di terra d'otranto i quali contengono quanto intorno a tal materia si puo desiderare* (Venice: apresso Giovanni Varisco Compagni, 1582). On the Milky Way, see ff. 35v–38v.

[89] *Iacobi Scheggij Schorndorffensis in reliquos naturalium Aristotelis libros commentaria . . . videlicet . . . meteoron lib IIII et al.* (Basel: per J. Hervagium, 1550). On the Milky Way, see p. 348. His popular name was Schegk (1511–1587). There are no commentaries on the Milky Way in the translation of Aristotle's *Meteorologica* published by Simon Porzio (Portius) in his *Aristotelis Stagiritae tripartitae philosophiae opera omnia absolutissima, ex optimis quibusque, maxime novis interpretibus collecta . . . multis scholijs illustrata* (Basel: per Joannem Hernagium, 1563), see pp. 267–70.

[90] *D. Tho. Aquinatis in Meteora Aristotelis commentaria* [etc.] (Venice: apud Iuntas, 1565). Once a personal physician to the French king, Francis I, Vicomercato went on to teach philosophy in Turin. His basically Aristotelian natural philosophy can best be seen in the posthumous *De principiis rerum naturalium libri tres* (Venice: apud Franciscum Bolzetam, 1596).

[91] *In quattuor libros Aristotelis Meteorologicorum commentarij* [etc.] (Venice: Io. Batistae fratrum, 1565). See especially ff. 38r–47r.

[92] *Ibid.*, f. 46r.

[93] *Trattato di M. Francesco de Vieri . . . nel quale si contengono i tre primi libri delle meteore nuovamente ristampati et da lui ricorretti con l'aggiunta del quarto libro* (Florence: appresso Giorgio Marescotti, 1582). On the Milky Way, see pp. 30–44. The book was first published in 1573.

[94] *Ibid.*, p. 31.

[95] *Digressio de circulo lacteo in defensionem Aristotelis adversus omnes peripateticos* (Venice: apud Simonem Galignanum de Karera, 1578), an essay of 30 octavo pages.

[96] *Ibid.*, pp. 22 and 15. More restrained was the defense of Aristotle's theory in F. Accoromboni's massive list of Aristotle's "more obscure" passages that "needed explanation." He admitted that there was a "mathematical demonstration" (the parallax problem) against Aristotle's theory. See *Felicis Accoromboni . . . interpretatio obscuriorum locorum et sententiarum omnium operum Aristotelis . . .* (Rome: apud Sanctium et Soc., 1590), pp. 475–76.

[97] *Compendium librorum physicorum Aristotelis* (Strasbourg: per Iosiam Rihelium, 1593), p. 422.

[98] *Commentarii Ioannis Ludovici Hevenreuteri . . . in Aristotelis philosophorum principis meteorologicorum libros quatuor* (Frankfurt: e Collegio Musarum Paltheniano, 1605), pp. 89–96.

[99] *Commentarii Collegii Conimbrensis in libros Meteororum Aristotelis Stagiritae* (Lyons: ex officina Iuntarum, 1597).

[100] *Ibid.*, p. 37.

[101] *Ibid.*

[102] *M. Manili Astronomicon a Iosepho Scaligero ex vetusto codice Gemblacensi infinitis mendis repurgatum. Eiusdem Iosephi Scaligeri notae* [etc.] (Leiden: ex officina Plantiniana, 1600), p. 99.

[103] *Commentarius in Sphaeram Procli Diadochi* (Stuttgart: typis Davidis Franci, 1609), pp. 130–34.

[104] *Kurtze aber Trewhertzige Erklerung des geschwäntzten newen Sterns oder Cometen so sich im September dieses 1607. Jahrs hat sehen lassen* [etc.] (Lübeck: durch Johann Witten [1608]). There is no reference to the Milky Way in the voluminous discussion of astronomical (planetary) theories by Nicolas Rymer Baer, mathematician and astrologer of Rudolph II, in his *Nicolai Raimari Ursi Dithmarsi . . . de astronomicis hypothesibus seu systemate mundano: additur astronomicarum hypothesium tractatus astron-*

omicus et cosmographicus [etc.] (Prague: apud autorem, 1597), although he discussed the sublunary location of comets (f. G iij).

[105] *Ioannis Bayeri Rhainani Uranometria, omnium asterismorum continens schemata, nova methodo delineata aereis laminis expressa* (A[ugsburg]: M[angus], 1603). See Tab. XXV, note I ε. For the passage on the Milky Way, see Tab. IV, note D. On Bayr and his star atlas, see Basil Brown, *Astronomical Atlases, Maps and Charts: An Historical Guide* (London: Search Publishing Company, 1932), pp. 19–30.

[106] *Praelectio astronomica de natura coeli et conditionibus elementorum* (London: Joannes Bill, 1605), pp. 60–61.

[107] It was originally published in Latin in 1588. Ten years later it took Europe by storm when a French translation was published and reprinted four times during the next six years. The English translation by E. G. [Edward Grimston] in 1604 appeared immediately in four printings. A German translation followed the next year. References are to the English translation, *The Natural and Moral History of the Indies* by Father Joseph de Acosta, with notes and introduction by Clements R. Markham (London: printed for the Hakluyt Society, 1880).

[108] *Ibid.,* vol. I, p. 7.

[109] It was printed in 1651 in Amsterdam and recently reprinted by Menno Hertzberger & Co., Ltd. [n.d.]. For an informative study on Gilbert's *De mundo,* see *The De mundo of William Gilbert* by Sister Suzanne Kelly, OSB (Amsterdam: Menno Hertzberger, 1965).

[110] For the text of Harriot's letter and of Kepler's reply, see Kepler's *Werke,* vol. 16, *Briefe 1607–1611* (1954), pp. 172–73 and 250–51. Harriot described the *De mundo* to Kepler as a work "contra Aristotelem."

[111] Gilbert, *De mundo,* p. 202.

[112] *Ibid.,* pp. 249–50.

[113] *Ibid.,* p. 219.

[114] *Ibid.,* p. 52.

[115] *Ibid.,* p. 250.

[116] *Ibid.*

[117] This statement, which could be matched by countless similar ones from modern textbooks and monographs on astronomy, was made by the noted astronomer and historian of astronomy, J. L. E. Dreyer, in his *History of the Planetary Systems from Thales to Kepler* (1905), reprinted with some corrections by W. H. Stahl under the title, *A History of Astronomy from Thales to Kepler* (New York: Dover, 1952), p. 122.

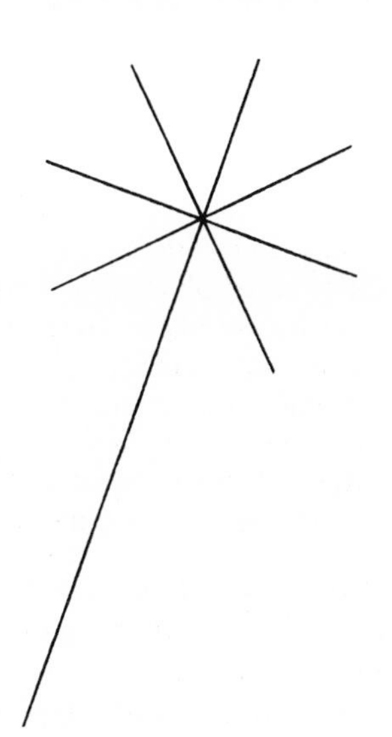

CHAPTER FOUR

Galilean Myopia

On January 30th, 1610, Galileo journeyed from Padua to Venice with the partially completed manuscript of a small book. When the full text saw print on March 12, the last entry was only ten days old. The date of the first entry was January 7, the night on which Galileo with his telescope first observed four moons around Jupiter. He left no indication of the exact time when he first observed the roughness of the moon's surface, or the incredible multitude of stars in the Milky Way. The little book carried the proud title of *Sidereus nuncius,* or *Starry Messenger.*[1] However, the new vistas of the heavens strewn with innumerable stars, particularly in the belt of the Milky Way, commanded distinctly less of his attention than the moon's surface or the satellites of Jupiter. The Milky Way, although prominently featured on the title page as one of the book's five attractions, was actually used as a mere third example of Galileo's discovery with the telescope of many more stars than those visible to the naked eye.

His first example of "the almost inconceivable number of the fixed stars" was the new appearance of the Belt and Sword of Orion. His diagram showed eighty stars in addition to the previously known three in the Belt and six in the Sword. His second example was the Pleiades where thirty-six stars now clustered around the original six. The third example, the Milky Way, received just twelve lines in a book of sixty pages:

> Third, I have observed the nature and the material of the Milky Way. With the aide of the telescope this has been scrutinized so di-

> rectly and with such ocular certainty that all the disputes which have vexed philosophers through so many ages have been resolved, and we are at last freed from wordy debates about it. The galaxy is, in fact, nothing but a congeries of innumerable stars grouped together in clusters. Upon whatever part of it the telescope is directed, a vast crowd of stars is immediately presented to view. Many of them are rather large and quite bright, while the number of smaller ones is quite beyond calculation.[2]

Whitish patches lying outside the Milky Way and shining "with faint light here and there throughout the aether" were also shown by the telescope to be "a tight mass of stars." Even more remarkable appeared to him the fact that "the stars which have been called 'nebulous' by every astronomer up to this time, turn out to be groups of very small stars arranged in a wonderful manner." He did not elaborate on this last point. His next remark was that "although each star separately escapes our sight on account of its smallness or the immense distance from us, the mingling of their rays gives rise to that gleam which was formerly believed to be some denser part of the aether that was capable of reflecting rays from stars or from the sun."[3] Of his observations of several such nebulous stars, two were illustrated with diagrams. One was the nebula in the head of Orion; the other, Praesepe (Manger) between the twin stars Aselli (Ass-colts) in Cancer. In the former, his diagram showed twenty-one stars, while the latter (Illustration IVb) was resolved by his telescope into "more than forty starlets." In all likelihood, Galileo's choice of the nebulous star, Praesepe, was not without consideration for Bayr's description of it in his *Uranometria*. But it was not Galileo's forte to acknowledge the merits of his colleagues, let alone of his medieval forerunners. His short description of previous theories on the Milky Way was far removed from the historical record, whose substance was certainly available to him.

Worse yet, Galileo failed to elaborate further on the Milky Way. This was in sharp contrast to his emphasis on the need for adding the vision of the mind to that of the senses. After all, he was already preoccupied with a new account of the framework and mechanism of the cosmos that must be largely derived from a reliance on the vision of the mind. Twice in the *Starry Messenger* he mentioned his plan to write a "System of the World," but in both cases the references to his cosmological project were made in connection with the moon. In his great *Dialogue concerning the Two Chief World Systems,* the stars played a minor role. He seems to have espoused the belief that the stars were confined within two imaginary spherical shells, forming an enclosure for the universe which to him was finite, since he could only conceive of it as a perfect body, that is, a sphere.[4]

Clearly, in connection with the Milky Way, Galileo looked backward, not forward. He saw his observation of the Milky Way as a feat which ended wordy debates and age-old disputes. He should have seen that these had become almost obsolete by the time he directed his telescope at the sky. More important, he failed to realize that his description of the Milky Way raised more questions than it allegedly solved. Those questions Galileo ignored, and partly under his influence, so did a galaxy of five generations of scientists after him. Scientists who preceded Galileo had said relatively little about the Milky Way, which they could observe only with the naked eye. Nevertheless, they achieved a partially correct view of it, because they trusted in the vision of the mind. For a century and a half after Galileo, scientists said little more of the Milky Way. To their contemplation of that magnificent phenomenon of the night sky—particularly when viewed through the steadily improving telescopes—they failed to add the all-important mental vision.[5]

Considerations about the cosmological implication of this new view of the Milky Way were almost non-existent in the first outburst of enthusiasm and envious criticism that greeted the appearance of perhaps the most important booklet in the history of science. In this respect, even Kepler was no exception. In his lengthy comment[6] on the *Starry Messenger,* he devoted to the Milky Way only three phrases of less than a dozen lines. Still, each of them cast an interesting light on the situation. The first phrase shows a Kepler who could praise his contemporaries generously, a trait not evident in Galileo: "You have conferred a blessing on astronomers and physicists by revealing the true character of the Milky Way, the nebulae, and the nebulous spirals." The second also sheds a more flattering light on Kepler than on Galileo, who did scant justice to previous speculations about the Milky Way: "You have upheld those writers who long ago reached the same conclusion as you: they are nothing but a mass of stars, whose luminosities blend on account of the dullness of our eyes." The third reveals, however, Kepler's reluctance to recall that Tycho Brahe's error was also his own: "Accordingly, scientists will henceforth cease to create comets and new stars out of the Milky Way, after the manner of Brahe, lest they irrationally assert the passing away of perfect and eternal celestial bodies."[7]

Interesting as these remarks are, none of them bears on cosmology. This is a somewhat paradoxical fact, for the newly perceived "overcrowding" of stars in the Milky Way could have been most naturally exploited by Kepler to add a new twist to the argument which he formulated six years earlier with an eye on Bruno against the infinity of the universe. He could have now pointed out that to an observer located amidst the stars of the Milky Way, the appearance of the sky must have been one of practically unbroken brilliance because of the proximity of so many stars, small though most of them might be. But Kepler referred only in general to the larger number of fixed stars visible through the telescope when he elaborated on his assertion of the finiteness of the universe. In comparison to the total amount of starlight, it became even more obvious, Kepler argued, that "the body of our sun is brighter beyond measure than all the fixed stars together, and therefore this world of ours does not belong to an undifferentiated swarm of countless others."[8] This argument's flaw was the mistaken belief that the apparent diameter of stars was of the order of one minute of arc. A thousand of them placed together would have covered an area larger than the sun's surface, yet their total light was incomparably weaker. To the objection that most of those stars were very far away, Kepler replied that, in that case, their bodies should have been much larger than the sun, since they appeared to show a diameter of one minute of arc at much larger distances. The tantalizing part of Kepler's reasoning was that he explicitly faced and dismissed the possibility that starlight might be diminished by the interstellar ether: "But maybe the intervening ether obscures them? Not in the least. For we see them with their sparkling, with their various shapes and colors. This could not happen if the density of the ether offered any obstacle."[9]

Kepler, first to propose that the intensity of light diminished from a point source according to the inverse-square law,[10] now had all the conceptual elements to advance his argument against the idea of an infinite universe in a form equivalent to Olbers' Paradox.

His denial of the absorption of starlight in the ether was, in fact, less faulty in its own context than the contrary position of Chéseaux in 1744 and, independent of him, of Olbers in 1823.[11] Both, but especially the latter, should have considered the thermal consequences of such absorption. Kepler sought to demolish Bruno's reasoning about infinity by exploiting what appeared to be a basic singularity in the universe. Instead of comparing the light of stars with that of the sun, he should have concentrated on the Milky Way. After starting his argument with a reference to the new, countless host of stars, Kepler added: "the more there are, and the more crowded they are, the stronger becomes my argument against the infinity of the universe."[12] Were not those stars by far most crowded in the belt of the Milky Way?

If anyone, then Kepler should have been sensitive to the far greater number of stars in the Milky Way revealed by the telescope, since he promptly claimed to himself the discovery of its theory. It was with some haste that he published in 1611 his *Dioptrice* on the various combination of convex and concave lenses. In the Preface of that work Kepler did his best to emphasize the superiority of telescopic astronomy over naked-eye observations, strengthened as these could be by mathematics. He clearly exaggerated in stating that prior to the telescope the location of the Milky Way among the stars could only be established through cumbersome mathematical demonstrations. Again, he grossly misrepresented the case in claiming that the nature of nebulous stars had previously been wholly unknown. His estimate, that the telescope revealed "ten or perhaps twenty times" as many stars as listed by Ptolemy, was also off the mark. The greatly increased number of stars all over the sky seemed once more to him clear evidence of the finiteness of the cosmos. Around the "movable world," that is the world of planets, the host of stars formed, so Kepler believed, a spherical enclosure, or "concameratio."[13] He assigned no explicit role in this respect to the myriad stars in the Milky Way.

Similarly, Kepler made no reference to the greater density of stars in the Milky Way when, seven years later, he used the Milky Way as an argument against the infinity of the universe in his *Epitome astronomiae copernicanae*. He argued that the "Way, 'Milky' for the Greeks, and 'St. James' path for us,' singled out specifically the position of the earth and of the movable world [planetary system] as compared with any other point in the realm of the fixed stars."[14] That realm was a spherical shell which the Milky Way encircled, dividing it into two hemispheres. Kepler then asked his reader to consider the earth as though it were far out of the plane determined by the Milky Way. In that case "the Milky Way should appear in one single look as a very small circle or an ellipse visible in its entirety, whereas now only one half of it can be seen at any moment." Or place the earth near the Milky Way, Kepler continued, "then that part of the Milky Way shall appear immense [very broad], whereas its opposite segment very narrow."[15] Behind this reasoning obviously lay a concept of the Milky Way as a ring of stars—a somewhat new idea. The cross section of that ring was extremely small as compared to the size of the circular plane along whose rim it encircled the whole finite world.

This frustrating near-miss about the true shape and structure of the Milky Way was the only positive point in the comments on its new picture in the *Starry Messenger.* The first, and possibly the worst, of these comments both with respect to science and to style, was the *Brevissima peregrinatio contra nuncium sidereum* ("A very brief excursion against the Starry Messenger") by Martin Horky.[16] He probably would not have written or published his vituperative libel had he not been urged on by Giovanni Antonio Magini, an old rival of Galileo. Horky's principal target was the new system of satellites around Jupiter, which he dismissed as an optical illusion. In regard to the Milky Way, his aim was to discredit the telescope, whose value he admitted only for terrestrial observations: "The Messenger offered nothing new about the Milky Way. We all know, that it has been the longstanding consensus of all philosophers and mathematicians that the Milky Way was a congeries of an infinite number of small stars."[17]

Horky, a native of Bohemia, not surprisingly tried to bring himself to Kepler's attention in Prague. In a letter to Kepler, he described as a total fiasco Galileo's demonstration of the telescope at Magini's home in Bologna on April 24 and 25.[18] When Kepler again took up, in August, 1610, the defense of the *Starry Messenger,* it was only to communicate his own observations of Jupiter's satellites.[19] But it was by quoting Kepler's *Conversations* that John Wodderborn, a Scottish assistant of Kepler, gave the lie to Horky in respect to the Milky Way. As Wodderborn noted, no less a figure in astronomy than Tycho Brahe had disagreed with the consensus that the Milky Way was a congeries of stars. Politely enough, Kepler was not mentioned as a former champion of Tycho's explanation of the Milky Way.[20]

However, there was not even a short reference to the Milky Way in the attack on Horky by Giovanni A. Roffeni, in early 1611.[21] The epigrams composed by a Briton, Thomas Seggeth, in 1610, contained the famous phrase, "Vicisti Galilaee." but no mention of the Milky Way,[22] although of the 58 lines of nine epigrams one could easily have been spared for it. The Milky Way must have certainly been considered a topic of negligible importance when it was omitted from a lengthy oration. The orator was Father Odo van Maelcote,[23] the place the famed Collegio Romano, the occasion the celebrations given there in honor of the author of the *Starry Messenger,* who in the spring of 1611 made a triumphal visit to Rome. Since he himself had begun by disregarding the significance of the new vistas of the Milky Way, he could hardly be disturbed when, in his very presence, Maelcote's festive address made no reference to it. The Russian aristocrat, Christopher, Duke of Zbaraz, thus acted in style in more than one sense. Having failed to find Galileo in Padua, he sent a letter to him from Bologna on March 8, 1612, in which he mentioned that the fame of Jupiter's moons had spread as far as cold Muscovy. He said not a word about the Milky Way. A stylish "I kiss your hands," rather an obsequious gesture from a prince, were his concluding words to the astronomer-employe of the princes Medici.[24]

Precisely the opposite tone was taken by Jacob Christmann, professor of logic at Heidelberg, who claimed in 1611, in the Appendix to his *Nodus Gordius,*[25] that the telescope—several of which he constructed for himself—multiplied small sources of light, especially when they were scintillating. The telescope, he declared, multiplied but did not magnify the scintillations: "No such scintillation can be seen in the Milky Way or in the spaces between stars, which is an indication that if there existed no star [in a given place] no scintillation or coruscation could emanate from it."[26] Whatever the merit of such reasoning, it ill befitted the Milky Way, and, certainly Jupiter, which, like other planets, does not scintillate. Whether he, with his best telescope (to which he ascribed a magnification of ten), observed some protrusions from Saturn (its rings), is rather doubtful. He was given more to preconceived ideas than to facts, and showed no reluctance to heap abuse on Galileo and especially on those who saw, in the greatly increased number of stars, proof of an infinite universe. He contended that there was no need for so many stars in the universe unless it was to become a chaos. In the same breath, he also noted that "equally useless if not impossible is the construction of telescopes with a thousandfold magnification."[27] Neither logic nor prophecy were his strongest qualities.

Christmann's lucubrations, hapless as they were still kept alive the topic of the Milky Way in the first heat of the debates over the *Starry Messenger*. It was attacked, but without reference to the Milky Way, in two pamphlets published at about the same time. One was written by Francisco Sitio, who concentrated on Jupiter's moons.[28] The other, a work of Giulio Caesare La Galla,[29] concentrated on the new, earthlike features of the moon. Their real concern, however, was the possibility of many, if not infinitely, numerous worlds. Clearly, if the moons of Jupiter, and the roughness of the moon could be rationalized away, the many new stars in the Milky Way and elsewhere could more readily be regarded as bright parts of the crystalline sphere. This was, in fact, the context in which La Galla referred to the telescope's resolution of nebulous "impurities" into swarms of distinct stars. With the notion of the ethereal sphere of the fixed stars defended, the specter of immensely numerous worlds seemed to recede into the background.

The "silent treatment," as a means of warding off an ominous prospect, was used by Cesare Cremonini, an ultra-Averroist of Padua, in his special treatise on the Milky Way, published in 1613.[30] One is indeed at a loss to find any other explanation for the fact that not even the shortest reference to Galileo occurs, nor to the telescope, nor to the new picture of the Milky Way—in a treatise of forty-one quarto pages ostensibly devoted to it. Incredibly enough, Cremonini professed to discredit Dante, whom he described in the Preface of the essay as a mere poet. He reproached even Averroes for abandoning the Philosopher. Cremonini did not recall what Averroes had stated about parallax; on the contrary, he offered a diagram for the "solution" of that problem on the basis of Aristotle's theory. Equally astonishing was Cremonini's reference to "mathematicians" who, he declared, distinguished between nebulosity and albedo, in the sense that the former could be produced by many small stars, but that the latter was the unique feature of the Milky Way.[31] Among those whom Cremonini reproached was Vicomercato nearly the only author whom he mentioned among the "moderns." Neither Brahe nor Albertus Magnus were recalled by Cremonini, whose chief target was Philoponus, the most powerful critic in antiquity of the eternity and unchangeability of the heavens. In an essay full of stultifying discourse, the only truly instructive detail appeared on the last page where Cremonini's true motivation (and Aristotle's) was unabashedly laid bare. Were the Milky Way a congeries of stars in the sphere of the fixed stars, wrote Cremonini, it would represent a "condensation in the heavens, which would be an elongated circle, thick here, thin there, and double in places, and in that case everything would appear contrary to nature." Clearly, in the "perfect nature" of the Aristotelian heavens there could be no place for the Milky Way.

Unlike the remark in Cremonini's essay, reference to nebulae was a truly valuable part in the only major challenge of Galileo's fame as discoverer of the marvels in the *Starry Messenger*. On the title page of the *Mundus Iovialis,*[32] published in 1614, Simon Marius, a minor court mathematician in Güntzenhausen, Germany, claimed that he had already observed the satellites of Jupiter in August of 1609. He also claimed priority in resolving the Milky Way into stars, as well as other nebulous patches such as those in the Pleiades, the Hyades, and Orion. His other claims to priority concerned the sighting of sunspots on August 3, 1611, and the nebula in Andromeda on December 15, 1612. Concerning the latter he remarked that, with his telescope, which he called "perspicillum belgicum" (indicating its provenance from the Low Countries), he could see in it "many distinct stars."[33] This hardly improved his credibility with historians, and the same might be said of his claim that Galileo should have seen that stars were round and even the planets scintillated. Whatever the merit of the *Mundus Iovialis,* a booklet in which the fool's gold outweighed the gems, it certainly came too late to make the world believe that Galileo was "the first only in Italy."

National pride at the Dutch origins of the telescope might have played a part in the omission of Galileo's name when the Milky Way was described at that time in textbooks on astronomy and physics by Dutch scholars. Nicholas Mulerius (Müller), professor at Groningen,[34] Adrian Metius, professor at Alkmaar,[35] Gilbert Iacchaeus, professor at Leiden,[36] and Willem Janszoon Blaeu, the famed mapmaker of Amsterdam,[37] mentioned without any sign of astonishment that the telescope showed a great many small stars in the Milky Way. Galileo fared no better, in this respect, in Denmark. His name was not recorded by Caspar Bartholin, the renowned professor of anatomy at Copenhagen, when he described the Milky Way as seen through the telescope in his *Uranologia.*[38]

On the other hand, telescopic evidence seemed to be rather effective in severing once and for all the speculative connection between comets and the Milky Way. In the year 1618, three comets, one particularly large, stole the celestial show—which led to a celebrated controversy.[39] Neither the prominent participants, Kepler and Galileo, nor the minor ones, Grassi and Guiducci, cared to bring the Milky Way into a dispute which ultimately centered not so much on the nature and place of comets, as on the proper nature of scientific method. The gem of the controversy was Galileo's *Assayer* in which one could read the intriguing but undeveloped remark: "We know positively that a nebula is nothing but an aggregate of many minute stars which are invisible to us."[40]

Those who continued to talk about the Milky Way in connection with the comets of 1618 were insignificant authors. Now that the Milky Way could not be considered, without patent obstinacy, the breeding place of comets, Nicholas Mulerius struck a note of agnosticism concerning their nature in his *Hemelsche trompet.*[41] Isaac Habrecht, a "physician and mathematician," tried however, to keep alive Brahe's theory of comets, admitting at the same time that the telescope ("Brillenrohr") forced him to abandon the Aristotelian theory of comets *and* Milky Way. As to the latter, he noted that the closeness of stars there was only a visual effect, and suggested greatly varying distances for them.[42] Somewhat better known than Mulerius and Habrecht was Libertus Fromondus of Louvain, who appended a lengthy series of comments to a booklet on the big comet of 1618 by his colleague, Thomas Fieno.[43] Fromondus, obviously perplexed by the new picture of the Milky Way, also tried to retain it as a breeding place for comets. He took the view that the Milky Way "should not be pictured as part of the firmament, but as a ring studded with very small stars, as if with as many gems, floating in the ether between the sphere of Saturn and the sphere of the fixed stars."[44] To this he added that some of those small gems could dissolve into a rarified liquid under the impact of the sun, and it was the resulting substance which was then launched as a comet on its "orbital dance."

Such efforts did not reverse the separation of comets from the Milky Way. While this was a welcome clarification, it also contributed to a puzzling neglect of the very topic of the Milky Way. Aside from Cremonini's obscurantist essay, the few paragraphs, which Kepler wrote in the wake of the *Starry Messenger,* were not matched until 1622 by a contribution from a reputable astronomer. But Christian Severin, known as Longomontanus after his birthplace, Longberg, in Jutland, had an ax to grind. From 1589 to 1597 he was employed by Tycho Brahe whom he felt bound to defend, even when he served as astronomer in charge of the construction of the Copenhagen Observatory from 1632 until his death in 1647. To Longomontanus, Brahe was more than a revered teacher. He was the symbol of Danish excellence in astronomy, and there was clearly a touch of chauvinism in the title of Longomontanus' major work, *Astronomia danica,*[45] first published in 1622. In an appendix to the work on the "new phenomena of the sky," that is, the novae and comets which had appeared since 1572, Longomontanus took up the question of the nature of the Milky Way.[46] Clearly, it was Brahe whom Longomontanus wished to defend when he submitted three reasons why he could not accept the conclusion of Galileo's observation that the Milky Way was but a congeries of stars. In addition to stars, there was also some other substance in the Milky Way, first, because otherwise, Longomontanus argued, the light of those very small stars could not be stopped in the celestial region to produce the whitish band. Second, Longomontanus here recalled Brahe, only the "vaporous" substance of the galaxy could give rise to new stars, as the number of existing stars was not to be reduced to provide material for some new ones. Third, the spatial closeness of novae and comets to the Milky Way was, Longomontanus wrote, analogous to the case of herbs and plants which grew only when nutritive substances were nearby.[47] With such reasoning, any opinion could be made to seem plausible.

Longomontanus clearly felt that his explanation of the Milky Way represented a losing fight. He had to admit that even Caspar Bartholin, at that time the intellectual glory of Denmark, had part-ed company with Brahe over the Milky Way. Longomontanus also mentioned Isaac Habrecht as one who accepted the galaxy as a congeries of stars.[48] There were others, too, rather minor figures, who, for one reason or another, registered the truth of Galileo's observations. The names of Jean Lereuchon, writing under the pseudonym, Van Etten,[49] and of Joseph Blancano[50] would sound unfamiliar even to historians of astronomy. Both were Jesuits, but the question of the Milky Way seemed charmingly free of deeper issues. Even to Galileo it consisted of very minute stars, and these could be readily accommodated on the traditional firmament, especially when it was pictured as sufficiently thick. Blancano nevertheless deserves special mention, for he also spoke of the nebulae reported by navigators in the southern sky and insisted that these too were agglomerations of stars as was the Milky Way.[51]

Few if any such gems can be found in the analysis through eighty chapters of twenty-one theories on the nature of comets and novae by Fortunio Liceto.[52] He is better remembered as the bitter critic of Harvey's theory of the circulation of the blood, which he tried to displace by some bizarre speculations. It was indeed through his obscurantism, supported by a most meticulous documentation, that he earned fame in his time. There was nothing new in the two-tiered theory of the Milky Way, which he proposed as he discussed the dicta of Brahe and Fromondus concerning it. According to Liceto's explanation, which he claimed was that of Aristotle, the Milky Way existed both as a ring of stars in the firmament and as a band in the purest region of fire below the orbit of the moon.[53] But in 1622 it was well-nigh impossible to overlook the question of the parallax of a sublunary phenomenon. Liceto sought refuge in sheer verbosity, as he claimed that the lower part of the galaxy was "invisible or hardly visible."[54]

It was about this evasion that Fromondus took him to task nine years later in his *Meteorologicorum libri sex*.[55] If the lower galaxy was not visible, there could be no reason for postulating it. If it was visible, however slightly, it had to show a parallax. Fromondus no longer stressed the fluid aspect of the galaxy, but he firmly retained the distinction between it and the firmament. The galaxy could not be part of the firmament which, by definition, was uniform throughout. His definition of the galaxy as "a shining circle, composed of the firmament's smallest stars, and dividing it into two hemispheres at Gemini and Sagittarius" was, therefore, not without some contradiction. More instructive was the analogy by which he illustrated the fusion of starlight into one luminous band: "a hundred night torches when closely spaced without touching one another would produce from a distance the image of one single zone of fire."[56]

In the 1620's, Raffaele Aversa's discussion of the Milky Way in his compendium of philosophy stood alone by reason of its clarity and its substantial references.[57] He chided Cremonini for his obscurantism, praised the author of the *Starry Messenger* and his telescopes, took note of Longomontanus, but claimed also that all astronomers and most Peripatetics agreed on the starry constitution of the Milky Way. Certainly, apart from Longomontanus, no notable astronomer recorded a dissent from Galileo with respect to the Milky Way. As for the Peripatetics, they were far more prolific in writing than in observing. In 1626, Bartolomeo d'Amici, a Jesuit philosopher in Naples, tried to shore up Aristotle's theory of the Milky Way in his explanation of *On the Heavens*.[58] It would be pointless to review the wordiness with which he tried to eliminate the major difficulties of Aristotle's theory, the unchangeability of the Milky Way, and the question of parallax connected with its position. The only interesting point in d'Amici's discourse was the remark that Aristotle placed the Milky Way in the sky according to the *old* translation. He had to admit that he was unable to consult it. His indirect reference to it was "Georg. I met, tr. I in fine". The obscurity of the reference[59] matched the already highly elusive *old* translation which had not been found worth printing in that rush for medieval manuscripts shortly before and after 1500.

The sheer obstinacy of some Peripatetics in the matter of the Milky Way came to the fore in two notable cases. One was the publication, in 1627, of some opuscules of Federico Buonaventura, of which one concerned the Milky Way.[60] It must have been written shortly before the author's death in 1602, since it contained Acosta's remarks, though with the contention that they confirmed Aristotle's theory![61] Yet, this was only one of the dizzying distortions that Buonaventura bestowed on the subject. He dismissed eleven arguments against Aristotle's theory of the Milky Way. In connection with its perennial sameness, Buonaventura replied that rivers too were always the same. He claimed, in order to shore up Aristotle's theory, that when planets passed through the Milky Way, it was momentarily destroyed by them.[62] The only saving grace in his fifty-page discourse concerned Ptolemy. As Buonaventura rightly noted, Ptolemy could not be considered an advocate of the starry nature of the Milky Way. That Buonaventura remained adamant even when he surveyed the arguments of Ammonius and Philoponus against Aristotle, is as indicative of his mental blindness as his claim that the parallax problem was irrelevant to the case of the Milky Way because of its vastness.[63] Still, much the same was true of the thinking of those who chose to have his opuscule printed seventeen years after the *Starry Messenger*.

The other manifestation of that obstinacy is the commentary on the *Meteorologica* by Giovanni Cottunio,[64] who, in 1631, brazenly ignored the question of parallax. He also ignored Galileo as he tried to turn to his own advantage the immense number of stars in the Milky Way. Cottunio's problem was to show why, if the Milky Way was indeed the slow incandescence of dry exhalation, groups of stars outside the Milky Way did not produce a similar phenomenon. The telescope revealed, Cottunio noted with satisfaction, that the difference in the relative frequency of stars in and outside of the Milky Way was far greater than appeared to the naked eye. For the same reason, Cottunio added, the less lucid areas of the Milky Way were precisely those where the telescope did not reveal many more stars.[65] The rest of what Cottunio wrote in some fourteen folio pages on the Milky Way, proved only the sad extent to which veneration of the Stagirite could blind one's mind to the obvious. Cottunio even claimed, in his misguided defense of the Aristotelian theory of the Milky Way, that it implied a three-tiered structure. The topmost part of it was located in the firmament, in the form of many densely packed small stars. The lowest part was below the orbit of the moon, in the form of exhalation. The middle layer consisted of "aggregation of the other two," but Cottunio failed to assign it to any specific location. More instructive was Cottunio's defense of the large quantity and invariability of exhalation demanded by Aristotle's theory.[66] He pointedly noted that Aristotle posited an eternal universe, in which the substance of the lower Milky Way was forever replenished. In a universe created in time, the same problem could be solved, according to Cottunio, by the perfection of the Creator, whose works could only be perfect, and who created enough dry exhalation at the very outset to secure the invariability of the Milky Way.

The obscurantism of such rear-guard defenders of each and every dictum of Aristotle is alleviated only by the absence of astrology and magic for which, especially in Liceto's book, the new stars and comets might have provided more than enough ammunition. Astrology and magic as was formerly so often the case, kept one's attention from the Milky Way. In Rudolph Goclenius' *Urania*[67] there were, for instance, fifty "important" pointers on the various houses of the zodiac,[68] in addition to chapters on the fixed stars, the zones, and the major circles, but not a word on the Milky Way. The entire second half of the work was about "judicial" astronomy. Preoccupation with astrology and magic seems the reason for the neglect of the Milky Way by Thomas Campanella, the most spirited defender of Galileo. In his *Apologia pro Galileo,*[69] he mentioned the galaxy only when he argued that so many stars, different in size, would hardly keep the same perpetual order if they were to rotate at incredible speed. He felt that the perpetual sameness of the heavens could better be salvaged if the earth moved.[70] This is not to suggest that Campanella maintained a belief in the eternity and unchangeability of the heavens. The real features of his world view were best revealed in his *De sensu rerum et magia,*[71] where he gave free vent to his animistic conception of the world. He not only populated all stars with denizens, but attributed sensory perception to each star, as though they were huge animals. In keeping with this, he pictured the nebulosity of various parts of the sky (he must obviously have had in mind the Milky Way) as vapor and dust arising from battles between stars. Because of that vapor, their infinitely large number could not be perceived.[72]

The author who provided the most startling connection between addiction to astrology and silence on the Milky Way was Robert Fludd (1574–1637), the epitome of obscurantism in his time. A physician by profession, he did not completely neglect observation and experiments. Some of these, neatly discussed and illustrated, are appended to his *magnum opus* of 1626, the *Meteorologica cosmica,*[73] from which rhyme and reason are distinctly absent. Mersenne lacked no serious reason to denounce him as an "evil magician," and Gassendi, at Mersenne's request, in 1628 exposed his aberrations in a long essay.[74] Gassendi, therefore, could not help thinking of Fludd, when, years later, he rebuked astrologers for their inconsistent neglect of the Milky Way which the telescope revealed as an immense number of stars.[75] Psychoanalysis may someday shed light on Fludd's mind which attempted to derive everything, including the origin of the world, from winds. In his big folio the Milky Way appeared only on a turgid diagram of the upper air, but was not mentioned either in connection with comets or with stars. Angels and demons constituted the topic of the last fourth of the book, which offered bewilderingly detailed correlations among the positions of planets, meteorological phenomena, and every conceivable sickness and mental disposition. Perhaps the only consistency in the book was its silence on the Milky Way, reflecting a long-standing tradition in circles astrological.

Fludd is a shocking author. His writings bring one directly in contact with some almost unfathomable aberrations of the human mind. One's perplexity is hardly lessened when one notices the almost complete unawareness of the topic of the Milky Way on the part of Bacon, Descartes, and Pascal, all renowned for incisive, curious, and creative minds. Bacon's case is all the more tantalizing since the short discourse on the Milky Way in his *Descriptio globi intellectualis et thema coeli*[76] is only one chapter removed from Chapter VI, which is entitled: *"That philosophical questions concerning the* Celestial Bodies, *even such as are contrary to opinion, and somewhat harsh, should be received. Five questions are propounded concerning the system itself; namely,* is there a system? *if there be,* what is the centre of it, what the depth, what the connexion, and what the position of the parts?"[77] Could any phenomenon be more relevant to these questions than the newly discovered crowding of stars in a narrow belt encircling the sky? But Bacon took up the question of the Milky Way in Chapter VII when he considered the problem of the substance of the heavenly bodies. Curiously enough, around 1615 he still had some slight reservation about Galileo's discovery. To the question, "Is the milky way a collection of small stars, or a continuous body; and part of the ether, of a middle nature between the ethereal and the starry?" his reply was that it "seems on the point of being settled, if we believe the report of Galileo, who had resolved this confused appearance of light into stars numbered and placed."[78] Such a perplexing failure to grasp the real portent of the Milky Way came to a close with the bafflingly obscure question: "If it [the Milky Way] be situated at the same altitude as the stars which are seen through it, why may not stars be scattered in the milky way itself, as well in the rest of the ether?"[79] Perhaps he meant to say, why, of course?

While Bacon was certainly fertile in outlining scientific research projects and in formulating questions to be investigated, he failed to pursue any of his research programs, large or small. Entirely different was the case of Descartes, the boldest—and perhaps the most rationalistic—systematizer of modern times. He not only performed anatomical research, but also formulated principles on the basis of which one could readily explain, so he claimed, why the world and its parts looked and moved exactly as they did.[80] The world as it came from the hands of the Creator was, according to Descartes, an indefinitely large lump of viscous liquid composed of many small particles. That primordial substance was immediately subdivided by God into sections having a diameter equal to the average distance of stars. Finally, God caused each small particle to turn around its own center and around the center of its section. This was the essence of Descartes' famous vortex theory, on the basis of which he might have reasonably argued that the fixed stars kept their mutual distances. Nevertheless, the very foundations of the theory could scarcely cope with a huge ring of densely packed stars. Thus, Descartes' silence on the Milky Way should appear curious, to say the least. His only reference to the subject occurs in *Le Monde ou Traité de la lumière,* written around 1630, but published posthumously in 1664. There he noted that many stars appeared very small because of their distance, that many were so remote as to be simply invisible, and that many could only be seen through their combined effect, such as the nebulous patches and the Milky Way, both of which he described as aggregates of many distant stars.[81] To say nothing more of the luminous band of the Milky Way was a baffling performance on the part of the author of *Les météores,*[82] a book broadly patterned on Aristotle's *Meteorologica.* Understandably enough, Descartes made no room for the Milky Way among meteorological phenomena.

The master's studied neglect of the Milky Way turned into a studied ambivalence when Henry de Roy of Utrecht, one of Descartes' early admirers, took up the topic. He described the galaxy as an elongated array of stars which, because of their smallness, could be seen as separate entities only with a "Batavian [Dutch] telescope." The same telescope, he added, also showed that "nebulous stars and spots" visible in the Southern Hemisphere were also hosts of stars. He did not seem to be thinking specifically of the stars of the Milky Way when he added in the next breath: "How most these stars are scattered and distributed across the heavens, can be seen from a look at the celestial sphere, or from its diagram added here."[83] The diagram showed a large circle, with the concentric orbs of the planets inside it. On its circumference there were sixteen contiguous circles, too big to evoke the apparent smallness and enormous number of the stars in the Milky Way.

Pascal, too, appears perplexing with his brief reference to the Milky Way. It occurs in a short essay,[84] which he intended as preface to a treatise on the void. Whether he would have taken up the question of absolutely empty spaces in the cosmos may never be known with certainty. He had, of course, been keenly aware of some questions raised by infinite empty spaces, but he left no statement about the arrangement of stars there. He mentioned the "new" appearance of the Milky Way through the telescope only to call attention to startling developments in science, and to warn against the dangers of traditionalism in thinking. His comment reflected some unfamiliarity with the old consensus on the Milky Way as a congeries of stars, and it certainly failed to intimate the full portent for scientific speculation of a phenomenon which he described as the blending of the light "of an infinity of small stars."

Pascal's case amply illustrates the fact that disregard for the almost obvious can plague even a genius. With Jeremiah Horrocks the same pattern appears even more striking. Unlike Pascal, the stars were for Horrocks an exclusive study. He might have developed into the foremost astronomer of the century had he not died in 1641, at the untimely age of twenty-four. A self-taught genius, Horrocks startled the world of the learned by predicting the transit of Venus in 1639. His mental powers were equally well-displayed by his recognition of the crucial importance of Kepler's contributions. Yet, in his defense of Kepler's astronomy, he neglected several good opportunities to consider the topic of the Milky Way. Even in reference to Galileo's *Starry Messenger* he mentioned only some nebulous stars, but not the Milky Way.[85] One should not, therefore, judge harshly Johann Amos Comenius, the pioneering genius in modern methods of education. In his pedestrian treatment of the Milky Way, he added another proof to the host of evidence that it was not in his competence to write a physics textbook "reformed according to divine light."[86]

The case of Gassendi is curious in that his sole reference to the Milky Way printed during his lifetime, forms an extremely minor detail in his vast list of the erroneous ideas in Aristotle's works. There he emphasized the absurdity of Aristotle's explanation of the Milky Way by recalling that Averroes, "pressed by the truth on one side, and by his loyalty to the infallible Aristotle, on the other," kept wondering about the true meaning of Aristotle's statement.[87] A year later, in 1625, occurred the first of his many chances that could logically have invited a comment on the Milky Way—which Gassendi failed to exploit. In his letter of July 17, 1625, to the "incomparable man," Galileo, he heaped encomium on the "interpreter of celestial things" and on his *Starry Messenger,* but in its contents Gassendi did not recall the Milky Way.[88] Gassendi's voluminous diary of astronomical observations,[89] in which every year from 1618 to 1655 contains at least one entry on planets, solar spots, comets, eclipses, the variations of the shape of Saturn, the occultations and conjunctions of planets and stars, there is not a single one on the Milky Way. He found no place for it even in his famous synthesis of astronomical science, the *Institutio astronomica* (1647), although on more than one occasion he considered the topic of the fixed stars.[90]

What Gassendi, a professor of mathematics at the Collège Royale in Paris from 1645 to his death in 1655, really thought of the Milky Way, is known only from his lecture notes covering logic, ethics, natural philosophy, physics, and astronomy, published posthumously under the title, *Syntagma philosophicum.* There, in the section on physics, which included astronomy,[91] he provided insights which, surprisingly enough, made no impact either on succeeding generations of astronomers or on historians of astronomy. Following a detailed account of the history of speculations on the Milky Way in classical antiquity (Gassendi ignored the medievals), he presented Galileo's discovery as a vindication of Democritus.[92] This remark of his also implied Democritus' idea of an infinite number of worlds (stars): "Democritus would indeed say," wrote Gassendi, "(or anyone else, who like him would contemplate innumerable worlds communicating with one another) that this infinity of things is like a forest of trees. They are so arranged, that those to the left and right, over a very long stretch, are extremely dense, forming a girdle, but they are very sparse to the front and the back, and outside the girdle. Thus, someone placed inside that girdle would see in the forward direction a meadow with few trees, but would find them very numerous to the right and left."[93]

The analogy of a dense, ring-like forest was a momentous step forward in man's understanding of the Milky Way. Even more so was Gassendi's tentative exploration of the structure of the cosmos with the help of that analogy: "Therefore, if in the immensity of things, you would imagine a girdle of worlds, or rather stars, lying not only to the right and left, but also above and below; and that we with our sun (taking it as one of them) would also be included in that girdle; then you would see thereby innumerable stars encircling you above and below, and on both sides, that is in an orb, and we would have a circle appearing as Milky." Gassendi hastened to add that the same appearance would obtain from any other star within that ring (girdle), and that our sun and some other stars would appear especially noticeable when viewed with a telescope, but other stars, particularly the more remote ones, would appear smallish: "You may understand now," Gassendi noted, "that why outside that tract or girdle, or rather in the forward and backward directions [radially towards and away from its center], we see stars much less in number, as if scattered through more open spaces of the immensity." He was, of course, wrong on the crucial detail of his otherwise fine reasoning. The optical appearance of that ring of stars would not be a great circle but two roughly triangular whitish patches at opposite points of the horizon (Illustration V). But Gassendi turned out to be correct in his concluding and highly original remark that nebulous stars or patches seen in very clear nights were so many Milky Ways: "Therefore it may be that there should occur similar stellar girdles and tracts; when now and then, here and there, some of the stars exhibit themselves in groups (as trees cluster in the middle of forests or in fields), these shine as nebulous stars especially on clear, moonless nights."[94]

One could only wish that Gassendi had not left the subject with the remark: "But why should we pursue dreamy speculations?" Actually, he felt that questions about the reason for the ring-shaped arrangement of stars in the Milky Way were beyond the competence of human intellect. The Milky Way represented a pattern whose explanation rested in the mind and sovereign will of the Creator.[95] Such intellectual resignation was not altogether consonant either with some of Gassendi's well-posited principles of scientific research, or with the perspective from which he criticized astrologers for their silence about the Milky Way: "Is it not strange that although the Milky Way is so conspicuous in the sky, no special influence is attributed to it? Was not special influence ascribed to Aselli, two small stars in Cancer, but also to Praesepe in it, and to other nebulous stars, which are nothing but very small aggregations of most minute stars? Does not the Milky Way, a similar aggregation, span the whole sky and overlap even the zodiac?"[96] By much the same argument, one could establish the primary significance of the Milky Way for cosmology as well. In this connection Gassendi was almost as shortsighted as Galileo, whose famed diagram in the *Starry Messenger* he probably had in mind as he referred to Aselli and to the nebulous star, Praesepe, between them.

Gassendi's immediate target among astrologers was Jean Baptist Morin, a younger colleague of his in the chair of mathematics. Several of Morin's views were criticized in 1648, in a long letter by Gassendi,[97] to which Morin replied in a pamphlet two years later.[98] Neither touched upon the question of the Milky Way. Ten years earlier, in his textbook of astronomy, Morin referred in passing to the *Starry Messenger,* but he mentioned only the immense number of stars and the moons of Jupiter.[99] In view of this fact there is a cryptic touch to Morin's remark, "we have already discussed the nature of the Milky Way elsewhere," which occurred in his massive *Astrologia gallica*,[100] the last major effort by a reputable man of science to vindicate astrology with seemingly rational arguments. For the purposes of astrology, a spherically closed world seemed indispensable, and Morin argued zealously in support of a solid sphere of fixed stars.[101] Imbedded in that sphere of pure crystalline ether were the stars, composed in part of ordinary matter.[102] That the stars were mixed bodies could be deduced, so Morin argued, by naked-eye observation of their different colors. Clearly, he had no high regard for the telescope, which broke the sphere of stars wide open and forced on astrologers a measure of precision which their art could not withstand. As one would expect,

Morin devoted only a few words to the Milky Way. Its immutability proved to him that, contrary to Brahe's theory, the sphere of stars could not be liquid.[103] A pathetic stance, but logical for a captive mind who, in the 1650's, still rejected Galileo's views on the roughness of the moon's surface[104] and on the solar spots.[105]

The finiteness of the world, by which Morin set such store, was also of great interest to Father Marin Mersenne, although for very different reasons. In 1624 he tried to vindicate it against the "impious" and the "libertines" of his time.[106] Some of his arguments were based on the actual dimensions and measures of the world machine, which he considered "contingent," that is, underivable from other scientific considerations. He might, therefore, have utilized very well the "contingent" position and size of the Milky Way. That he did not, seems all the more curious, since he certainly must have been familiar with Kepler's efforts along these lines. The Milky Way did not retain Mersenne's interest in his "Cosmographia," a smallish essay first published in 1626.[107] It is in one of his first works, perhaps the most monumental of them all, his commentary on Genesis,[108] that one finds his views concerning the Milky Way. Yet, what he said there did not reflect his usually encyclopedic information. To Mersenne, the great multitude of stars in the Milky Way merely provided an important argument against those who still pictured the sphere of stars as solid. Kepler's arguments served, however, two of four proofs submitted by Ismael Boulliau (1605–1694) on behalf of the finiteness of the world.[109] In fact, he seemed to suggest that Kepler had already claimed that "the Milky Way which is located in the orb of the fixed stars, and which appears finite, as we see it from the inside, is most likely finite also from the outside." He added nothing to this remark laden with much cosmological portent.[110]

The discussion of the Milky Way by Giambattista Riccioli typified the sad shortsightedness of a most industrious mind to support the cause of geocentrism four decades after the *Starry Messenger.* In Riccioli's *Almagestum novum*[111] one finds a wordy account of minor details about the Milky Way. He acknowledged that Democritus was proven right by the telescope, but he also claimed that the author of the old *Almagest* held that the Milky Way was a great frequency of stars. Not surprisingly, he also belabored the point that Clavius would have recognized the truth of Democritus' opinion, had he possessed a telescope while preparing for press the "earlier" editions of his commentary on Sacrobosco.[112] A defense like this was unlikely to reflect credit on either Riccioli's objectivity or on Clavius' mental prowess. Riccioli, who liked to offset substantive proofs by the number of contrary opinions, devoted ample space to the latter-day defenders of Aristotle's theory of the Milky Way, although the number of these was negligible. He made far less effort to provide a representative sampling of the advocates of the opinion which the telescope vindicated.

Riccioli listed six defenders of Aristotle's theory. They were Cremonini, Antonio Rocco,[113] Polo, d'Amici, Cottunio, and Liceto. He overlooked Scipio Chiaramonti, a strange omission, and two much less conspicuous authors. The first of these, Claude Berigard, professor of Aristotelian philosophy in Pisa and Padua, admitted that the telescope proved the presence of countless stars in the Milky Way and in the two nubeculae (the Magellanic Clouds), but nevertheless sided with Aristotle.[114] He argued that there were two Milky Ways, and the lower one, in the zone of fire, did not show a parallax because the zone of fire was by definition invisible. The same pattern of reasoning may also be seen in the commentary on the *Meteorologica* by Francisco Mateo Fernandez-Bejarano.[115] Against Aristotle's definition of the Milky Way as a comet, he noted that if comets were harbingers of physical upheavals, then warfare and epidemics must always occur along a zone of the earth lying directly below the Milky Way. This remark, which could long before have been exploited against astrologers and their traditional silence concerning the Milky Way, was followed by better and more cogent arguments against Aristotle's explanation. For all that, the author rushed to the conclusion that Aristotle was right on the subject. No such inconsistency marred the diagram in his text which illustrated Aristotle's version of Democritus' theory of the Milky Way (Illustration Ia).

The *Opus de universo*[116] of Chiaramonti saw publication seven years before Riccioli published his own, and represented a major effort to salvage some credibility for the pre-Galilean and pre-Copernican world picture. Chiaramonti's effort was not new; in the 1610's he entered the scene as a friend of Galileo, but soon he changed sides. He grossly abused his expertise in astronomy when, in 1621, he defended the sublunary location of all comets in his

Anti-Tycho.[117] This work is remembered now because of the relentless criticism to which Galileo subjected it in the *Dialogue.* As a result, Chiaramonti became one of Galileo's antagonists, battling his discoveries and theories. It should not, therefore, cause surprise that, in his *Opus de universo,* Chiaramonti rejected the idea that the Milky Way was the fusion of the light of many small stars, placed side by side. This was not, of course, the exact opinion of Galileo whose discovery in respect to the Milky Way Chiaramonti recounted, not by a direct quotation from the *Starry Messenger,* but by a passage from Blancano's *Sphaera.*[118] Clearly, if in the Milky Way small stars were literally contiguous, one could ask with Chiaramonti why some big stars could nevertheless be seen through that contiguous wall of stars. Such obviously slanted reasoning recorded well with Chiaramonti's hapless advocacy of the view that the Milky Way was a denser part of the firmament.

Strangely enough, Riccioli started the list of those who held the true or Democritean theory of the Milky Way with Manilius and Ptolemy. They were followed by Albertus Magnus, Stöffler, Kepler, Fromondus, Aversa, Cabeo, and Mastrius. To these Riccioli could have added several dozens if he had wished, but he did not so much as intimate that there were many more. Cabeo and Mastrius were authors of very recent commentaries on the *Meteorologica*—but so was Girolamo Trimarchi. With these three works, the *Meteorologica* appeared for the last time as a text which shaped the explanation of the phenomena of the sublunary world. The most massive of them was that by Niccolo Cabeo, Jesuit professor at the Collegio Romano.[119] He illustrated the optical situation of the Milky Way with two cardboards, one black, the other white. The former appeared white from a distance and the latter black when many dots of the opposite hue were painted on them. This was interesting, but rather belabored the obvious. It was also evident why Cabeo failed to mention Galileo by name in this and in two even more glaring contexts. In fact, his only reference to the author of the *Starry Messenger* appeared when he pointed out the arbitrary connection between some mythical personalities and some constellations in the zodiac. They had, so Cabeo argued, as little to do with one another as the princely family with Jupiter's four moons which Galileo named "Medici stars."

Bartholomeo Mastrius de Meldula, a Minorite, was co-author, with his confrère, Bonaventure Belluti, of a compendium of philosophy representing the tenets of Duns Scotus; the volume also contained discussions on various works of Aristotle.[120] The two authors were rather well informed concerning recent literature on the Milky Way, and rejected Polo, Rocco, Cremonini, and d'Amici. Actually, Riccioli might well have taken his list of Aristotle's protagonists from this work. His list of Aristotle's critics did not contain the name of Girolamo Trimarchi, whose discussion of the topic is interesting because, although he rejected the exhalation theory, he claimed that it did not originate with Aristotle.[121] In proof he referred to a very old Greek text in the library of the Minimite Convent in Messina. He did not see that text but, according to his superior (who enjoined him to write accordingly), it was already "corrupt" so far as the Milky Way was concerned. Trimarchi had, therefore, rejected the Coimbrenses who attributed decisive importance to the Greek texts of the *Meteorologica* to argue that Aristotle authored the exhalation theory. The old Greek codex proved for Trimarchi that the "corruption" had set in long before the Latin translations appeared on the scene, an argument that failed to present Trimarchi's acumen in a favorable light.

When it was a question of the Milky Way, there seemed to be an end to the acumen displayed by Athanasius Kircher in archeology, hieroglyphics, subterranean lore, and optics, to mention only a few of the fields where he made a name for himself. In his *Iter extaticum coeleste,* a discussion of the planets and the stars in a dialogue during an imaginary journey from the earth to the sphere of the fixed stars, twice the enormous number of stars in the Milky Way was recalled in an almost ecstatic style.[122] But this was all that Theodidactus (Kircher) "heard" on the point from Cosmiel, the angel who carried him from planet to planet.

Compared to the *Iter extaticum,* Bishop Wilkins' *The Discovery of a World in the Moone*[123] could appear a sober, explanatory marshalling of scientific evidence and considerations in support of the subtitle, "A Discourse tending to prove that 'tis probable there may be another habitable world in that planet." Wilkins made much of the sighting of mountains on the moon by the "eye-witness," Galileo, and he went to great lengths to show that the moon also had oceans and atmosphere similar to ours. It was in connection with atmospheric and meteoric phenomena that Wilkins considered the topic of the Milky Way. He illustrated with a diagram the failure of the Aristotelian theory because of the parallax (Illustration IIIb), and, a few pages afterward, he took issue with Tycho's opinion. Its fault, according to Wilkins, was that a condensation of the ether should of necessity gravitate downwards: "for if there had beene so great a condensation as to make them shine so bright, and last so long, they [the condensed parts] would then sensibly have moved downwards towards some center of gravity, because whatsoever is condenst must necessarily grow heavier, whereas these rather seemed to ascend higher, as they lasted longer."[124] Wilkins said not a single word about Galileo's "eye-witnessing" the Milky Way. This was all the more curious because, two years later, in the third impression of the work, Wilkins replaced that passage with a short quotation from Fromondus' *Meteorologicorum libri sex* which he gave both in the original Latin and in his own English translation: "The milky way is nothing else but the pale and confused light of many lesser starres, whereby some parts of the heaven are made to appear white."[125]

There was no such slighting of Galileo in Pierre Borel's sometimes entertaining, sometimes bewildering argument that all planets were inhabited.[126] Thus, according to him, the "paradise-bird" (*oiseau de Dieu*) was proof of this because nobody had yet found its nest or eggs on earth.[127] Borel set much greater store by Galileo's sighting the moons of Jupiter and the moon's rough surface. It was in this connection that he added: "That great *Galileus,* who seemed onely to be in the World for to resolve the doubts in Astrologie, hath discovered with his admirable invention of Prospective-glasses, which immortalize his name, by the discovery of what is contained in the Stars; he is the first who hath directed his *Telescopes* or Prospective-glasses towards Heaven, and by help of them, that the milky line were small stars, which by reason of their proximity and great number do confound their light."[128] In his polemic work on the true discoverer of the telescope, whom he named as Zacharias Joannides of Middleburg, Borel ignored Galileo in connection with the Milky Way. He did so in an interesting chapter on the discoveries already made and still to be made with telescopes. His remark that the problem of the nature of the galaxy made "miserable" the philosophers in ancient times,[129] showed only that by 1655 the true record was no longer common knowledge.

It was rather exceptional what one could find on the Milky Way in the charmingly smallish astronomy textbook written by Egidius Strauch, historian, theologian, classicist, and teacher of astronomy in Wittenberg. In 1659 he not only quoted directly the passage on the Milky Way in the *Starry Messenger,* but brought his remarks to a close with the words: "Those who wish to know more about the Milky Way might consult Gassendi."[130] This displayed a remarkable alertness to the latest on the subject, for Gassendi's study of the question had seen print only two years earlier. By and large, however, the Milky Way was referred to, forty or fifty years after the *Starry Messenger,* in short, trite phrases, if at all. In this respect budding geniuses were no better than the representatives of the scientific rear guard to which J. B. Duhamel certainly belonged. Since he published a book in 1660 with the title, *Astronomia physica,*[131] he should logically have delved into the topic of the nature of the Milky Way, for it had been traditional to assign this and similar questions to the "physical" part of astronomy. But Duhamel brought up the question of the Milky Way merely to solve the problem of the provenance of the nova of 1572. He cast his vote for the Tychonian theory of the Milky Way on the basis that an entity which is fertile—that is, can give rise to new entities such as stars—should be preferred to barren congeries of stars.

When a year earlier Huygens published his *Systema Saturnium,* he gave immediate proof not only of his unusual telescope-making ability, but also of his masterly talent for observations. This, however, went only so far as Saturn, its rings, and its satellites were concerned. His most interesting report about a peculiar-looking area in Orion, resembling the head of a horse, was as brief as possible. He spoke of it as an opening into a deeper, more lucid region of the cosmic spaces, because he could not resolve it into stars. Therefore, for all its nebulous appearance, the "Horsehead" nebula, as it was later called, was not classed by him with other nebulae or the Milky Way. These, "when viewed with a telescope," Huygens remarked, "are found entirely free of nebulosity. They show nothing but an agglomeration and a heap of many stars."[132] When Huygens wrote this, he was thirty, and at the beginning of a career equally creative in theoretical mechanics and optics, and in observational astronomy. About forty years later, his career ended with a posthumous work, which, as will be seen in the next chapter, exemplified the strange myopia with respect to the Milky Way.

Something of that myopia was manifested in 1653 in a little work on comets by Seth Ward, first Savilian professor of astronomy at Oxford. He mentioned the Milky Way only to dismiss the theory that comets originated there.[133] Four years later, Christopher Wren, usually remembered as one of the great architects of all time, gave his inaugural address as professor of astronomy at Gresham College. He was only twenty-five and he surveyed, with the enthusiasm of youth, the tremendous advances of astronomy, especially since Kepler announced the elliptic orbit of planets and Galileo introduced the telescope. In a poetical vein Wren evoked Seneca's prophecies and the likely envy of Seneca's contemporaries, had they learned from him that times would come when men could "discover two thousand times as many stars as we can [see with the naked eye]; and find the galaxy to be myriads of them; and every nebulous star appearing as if it were the firmament of some other world, at an incomprehensible distance, buried in the vast abyss of intermundious vacuum."[134] Yet, while the moderns could boast with Wren that they could now "stretch out their eyes, as snails do, and extend them fifty feet in length," not even the greatest of them were to extend their mental vision for another hundred years in order to catch a glimpse of the true picture of the Milky Way.

Wren's striking utterance on nebulae as other Milky Ways provides an example of simultaneous discoveries. It came when Gassendi's similar statement, made years earlier, was printed for the first time. Both remarks suffered the same fate: they elicited no echo in the world of science. It was not from a scientific ambience or through scientific interest that, in 1665, Ludolph Georg Lünde's dissertation[135] caused a noticeable break in the silence surrounding the subject of the Milky Way. His work might just as well have been sponsored by a professor of classics rather than Sigismund Hosemann, then professor of mathematics at the University of Helmstadt. More than three-fourths of the thirty-nine pages of the dissertation (Illustration VI) were devoted to opinions on the Milky Way voiced in classical antiquity. Typically enough, the dissertation came to a close with a lengthy quotation from Pontano, not an author to be keenly remembered by mathematicians. The truly valuable part of the work consisted of its introductory chapters, where Lünde listed some eighty authors, from Albertus Magnus to Gassendi, who wrote about the Milky Way. The list omitted some important names, such as Diacceto and Boccadiferro, and was replete with references to wholly insignificant authors, whose views were not stated by Lünde. Still, his list was by far the most informative that had been offered until that date, or even for the next three hundred years, so far as the history of the question prior to 1665 was concerned. Compared to Lünde's dissertation, the one by Jonas Hertzberger[136] appears rather insignificant. Although his purpose was to write a commentary on the chapter on the Milky Way in Aristotle's *Meteorologica,* he merely embellished a paraphrase of it with staple references to the ancients and to some moderns, such as Meurer, Scaliger, Kepler, and Galileo. The best detail he offered was a diagram illustrating Aristotle's refutation of the theory which the latter attributed to Democritus (Illustration Ib).

Although both Lünde and Hertzberger dealt but briefly with comets, quite probably their choice of topic for a dissertation was prompted by the comets of 1663, 1664, and 1665, which produced ample literature, largely written by minor authors.[137] The only notable astronomer, Johannes Hevelius, who published on the comets of those years,[138] in fact did the right thing when he ignored the Milky Way in that context. The situation is somewhat less understandable when one considers his silence on the Milky Way in his systematic and historical treatment of comets.[139] A profound problem emerges, however, when one examines his gigantic and magnificently printed star catalogues. He spent decades in carefully registering the position of thousands of stars, many of them in the belt of the Milky Way. None of these could be observed without seeing, in their immediate neighborhood, many, much smaller stars. The 350 folio pages of Hevelius' *Catalogus stellarum fixarum*[140] and the 56 magnificent plates of constellations in his *Firmamentum Sobiescianum sive Uranographia*[141] remind one of the man who saw all the trees but not the forest. At any rate, Hevelius did not mention the Milky Way even in his *Prodromus astronomiae,*[142] where he had more than one obvious opportunity to do so. His enthusiasm for the telescope (he built one with a focal length of 200 feet on the seashore near Danzig) contrasts sadly with his indifference toward the Milky Way. The sole reference to it in his huge

Machina caelestis was to praise the telescope and to pity the ancients: "After all, how poorly did they philosophize about most ethereal [celestial] phenomena before the discovery and construction of the telescope! They did not even have correct knowledge about the Milky Way."[143] He reserved for the readers of his monograph on the moon the scarcely novel or enlightening information that the Milky Way, through the telescope, appears as an immense number of "very small" stars.[144]

Details about the "correct knowledge" of the Milky Way could readily be found in the now largely forgotten "cosmographical" books published in the 1650's and 1660's. They were not pretentious, since their authors tried to serve the practical needs of students of navigation. In chronological order, the first of these is Nicholas Kaufmann (Mercator) who died in Paris, in 1687, as Danish ambassador after a long and varied career during which he even attained membership in the Royal Society. In his *Cosmographia* of 1651, he traced the Aristotelian error on the Milky Way to "rude ignorance" about the parallax.[145] In the *L'Introducteur à la cosmographie* by Gaston J. B. de Ranty (Renty), the truth concerning the Milky Way was credited to "lenses [telescope] made in Italy," as an outgrowth of the question whether or not all stars shine by their own light.[146] Andreas Cellarius, schoolmaster at Hoorn, Holland, reviewed with the pedantry of his profession the old errors, especially the myths, about the Milky Way, in a work which claimed to exhibit the "general and new cosmography."[147] Still longer was the section on the Milky Way in a similar work by Joseph Moxon (1627–1700), although only because he quoted in full the almost eighty-year-old passages of Thomas Hood on the topic, adding, of course, the results "more lately found out"[148] by means of the telescope.

While all this was more scientific an attitude toward the Milky Way than Hevelius' often unimaginative industry, neither he nor the cosmographers reflected the true scientific spirit of the 1670's. The real scientific advances were not to be found, either, in the pages of Jacques Rohault's *Traité de physique,* published in 1671.[149] Long after Newton's *Principia* appeared in 1687, Rohault's work dominated classrooms, and kept alive Cartesian physics and cosmology. It shed no light whatever on the Milky Way, although Rohault had adequate opportunity to consider it. One such occurred during his discussion of the number of fixed stars, where he pointedly referred to the telescope.[150] In a chapter on comets, he took issue with the opinion of "some philosophers before Aristotle," who explained comets as the random concurrence of the motion of many small stars. Rohault readily admitted, with a reference to the telescopic evidence, that there were many more small stars than the theory required. Its fault lay, he felt, in the fact that the path of stars did not display the kind of relation that comets apparently did.[151] Tellingly enough, Rohault's discussion of Aristotle's theory of comets lacked the detail that, according to Aristotle, comets and the Milky Way had the same physical origin.[152] Again, in rebutting some claims about the influence of stars, Rohault failed to recall that the Milky Way presented an insuperable problem to astrologers.[153] It plagued the Cartesians no less.

This is best seen in the manner in which Claude Gadrois, author of a book on the influence of stars according to Descartes' principles,[154] evaded the topic of the Milky Way. He went to great lengths to argue that Descartes did not describe the starry vortices as having almost impenetrable physical boundaries. Rather, they had, in Gadrois' view, many small pores through which the refined material ejected from the bodies of other stars could reach the central part of any other vortex. Moreover, he illustrated the influence of the stars of the Small Bear on the various planets with a diagram[155] which might well have displeased Descartes himself. Gadrois singled out the many small stars in the Pleiades and the Hyades as causes of rainy weather.[156] Logic demanded that something now be said about the many small stars in the Milky Way, but it remained undiscussed here and also in another work of Gadrois, a treatise on the system of the world.[157]

Rohault and Gadrois were not astronomers, but neither were those who, in the 1670's offered some speculations on the Milky Way that came tantalizingly close to the true explanation. Otto von Guericke was an engineer by training, and served as mayor of his town, Magdeburg, for over four decades. His work, *Experimenta nova (ut vocantur) Magdeburgica,*[158] is usually remembered for its chapters on the vacuum, but it is also a storehouse of information in many other respects. Almost the entire second half of the work deals with astronomical and cosmological questions which culminate in that of the true dimensions of the universe. Guericke painted a graphic picture of the Milky Way where, he believed, the nebulosity was caused by the atmosphere of planets which distance caused to appear very close to their suns. But what makes Guericke's remarks about the Milky Way valuable does not lie in such details. He showed great interest in the question of whether the universe was finite or infinite.[159] He did not accept strict infinity as a demonstrated truth, though he warned against setting a limit to the distribution of stars in a practically infinite space. The analogy by which he illustrated his point is worth quoting in full: "Just because stars cannot be observed at infinity, it does not follow that they do not exist there; one cannot assert by the same token that there is an end to the forest because no more trees can be seen [distinctly]; therefore it seems to be more likely that even though one should see beyond the field of stars observed through the best telescopes, no distance would be reached where all stars would evanesce, and no more such celestial torch would beckon its light."[160]

Rarely in the history of science was the true import of an excellent analogy so badly overlooked. It contained, as if by inspection, the clue to recognizing Olbers' Paradox. Clearly, what Guericke meant by the analogy was simply that, because trees cannot be individually distinguished in the depths of a forest, one must not argue that there are no trees beyond. The analogy might equally well be paraphrased in a form stating that the fusion of distant trees was clear proof of many trees beyond the apparent limit of vision set by the uninterrupted wall of trees. Application to the immense depth of the starry fields would have been most natural. After all, Guericke in the next breath considered the merging into one field of the light of individual stars in the galaxy, but he failed to apply the analogy of the forest either to the Milky Way or to the structure of the universe. Concerning the latter point he overlooked another major issue when he blandly asserted that the dense grouping of stars in a special section of space only proves the richness of the creative planning of God who "chose to depart from strict uniformity in arranging the stars."[161] This is not to suggest that Guericke had evaded a problem by referring it to the Almighty. He simply did not see the gravity of the problem which the Milky Way's singularity posed in a supposedly homogeneous and infinite universe of stars.

The frustration is equally keen, although for a different reason, when one reads the young Halley's famous *Catalogue* of the stars of the southern sky. It was the product of his one-man scientific expedition to the island of Saint Helena. He went there in November, 1676, at the age of twenty, driven by an overpowering enthusiasm for astronomy. It had seized him six years earlier when he tasted, to quote his own words, "the delights of astronomical studies to a degree which anyone inexperienced therein could hardly believe."[162] For a better understanding of the Milky Way nothing could have been more propitious than the combination of youthful zeal, unquestionable mental prowess, and a new vantage point from which to observe the skies. At Saint Helena the weather was more than disappointing, but Halley succeeded in establishing the position of some 350 stars. He was more than a routine observer and tabulator of data; his penchant for speculation was strikingly revealed when he noted that some stars looked fainter than indicated in the catalogues of Kepler, Tycho, and Ptolemy. To explain this, Halley boldly suggested that the change was caused by a transformation in the bodies of the stars themselves. The most fascinating remark in his *Catalogue* had a direct bearing on the Milky Way. With his telescope, he could distinguish many stars in the Magellanic Clouds, and he drew a parallel between the cause of their whiteness and that of the Milky Way.[163] Yet, the shape of those Clouds failed to awaken his curiosity about the shape of the Milky Way.

Upon his return to England in May, 1678, Halley began his steady rise in that scientific circle which was already feeling the animosity between Isaac Newton and Robert Hooke, Gresham Professor of Geometry and Cutlerian Lecturer at the Royal Society. Hooke's frame rests not on what he saw of the very distant, but what he unveiled with his microscope about the very small. Logically, his *Micrographia* should have come to an end with its fifty-seventh chapter on "Vinegar worms," but Hooke's favorite interest, the cosmic realms, would not be denied. On the meager pretext that the microscope revealed much about the density of various substances, he went on to discuss the relevance of the variation in the density of air (about which his microscope revealed nothing) on telescopic observations. His two main examples were the moon's surface and the swarm of stars in the Pleiades. He also mentioned Huygens' observations of stars in the Horsehead nebula, which Hooke described as "that small milky Cloud."[164] Of the "milky Way," he failed to offer a single remark.

This was all the stranger, for Hooke plainly revealed his deep interest in the speculative aspects of the starry realm when, in his *Method of Improving Natural Philosophy,* he proposed for investigation such weighty questions as "How many fixt Stars? In what Order plac'd? . . . Whether constant in Light and Heat and Influence? . . . Whether included in Orbs, or swimming in the Aether?"[165] The same list also contained questions into which he probed with great determination. Witness his *Lectures of Light,* where he courageously faced the problem of "infinitely infinite" light rays coming to every point in an infinite universe of stars.[166] Truly, the problem could not have been more effective for guiding his eyes toward the real clue to the appearance of the Milky Way; namely, the special arrangement of stars in space. In actual fact, his only statement on the Milky Way appeared in the very context which might also have obliged him to recognize the optical paradox of an infinite universe.[167] This statement can now be read only with a deep sense of frustration: "It was by some of the Ancients conjectured that the *Galaxia* or Milky Way was nothing else but a great number of Stars, so small as that they could not be distinguished: Telescopes have discovered the Truth of that Conjecture, and manifested it to be so, and we have henceforth no more reason to doubt that it is so, than we have to doubt whether there are any Stars at all in the Heavens."[168]

This was as incontrovertible as the remark of Jean-Dominique Cassini that the whiteness of a huge luminous band, which appeared in two branches from Taurus to Aries in 1683, resembled that of the Milky Way.[169] He left the matter there. Taciturnity on the subject of the Milky Way remained, indeed, a little-noticed characteristic of the astronomical dynasty which he founded at the Paris Observatory. It was the typical attitude of the professionals —Gassendi and Guericke providing the sole exceptions. Those who waxed somewhat verbose on the topic did not come from the ranks of creative minds. In the 1576-page conversational account of the starry realm by that prolific German baroque author, Erasmus Francisci-Finx, the Milky Way occupied one page, with brief mention of Democritus, Aristotle, Averroes, and some Dutch admirals.[170] No new horizons emerged in the small, stereotyped dissertation on the Milky Way which Olaf Wexstedt presented in 1686 "in illustri ad Varnum academia."[171] That place in the Baltic was as little known as Wexstedt's mentor, Georg Funck of Königsberg. Dissertation, author, mentor and school were, however, a perfect match for that mental myopia which Galileo bequeathed in respect to the Milky Way.

References

[1] *Sidereus nuncius magna, longeque admirabilia spectacula pandens . . . quae a Galileo Galileo . . . perspicilli nuper a se reperti beneficio sunt observata in lunae facie, fixis innumeris, lacteo circulo, stellis nebulosis, apprime vero in quatuor planetis circa Iovis stellam . . . circumvolutis . . .* (Venice: apud Thomam Baglionum, 1610). This first edition of 28 small octavo leaves has two unnumbered leaves [16b] and [16c]. The section on the Milky Way is on f. 16cr, and the one on "nebulous" stars is on f. 16cv. The facsimile reprint of the manuscript appears in Vol. III, Part I, of *Le opere di Galileo Galilei. Edizione Nazionale,* by A. Favaro (Florence: G. Barbera, 1892). It will be referred to as *Ed. Naz.* III. The reference to the Milky Way in the title is omitted in the otherwise excellent translation, *Starry Messenger,* by Stillman Drake, in his *Discoveries and Opinions of Galileo* (Garden City, N.Y. Doubleday, 1957), pp. 21–58.

[2] *Starry Messenger,* p. 49.

[3] *Ibid.,* p. 47.

[4] See translation by Stillman Drake (Berkeley: University of California Press, 1962), pp. 319–20, 326 and 382–83. It should also be noted that, to Galileo, inertia along straight lines held valid only for terrestrial motion. The inertial motion of heavenly bodies was to him circular.

[5] Galileo's neglect of the Milky Way is amply mirrored in the great Galileo bibliographies. No specific studies on any aspect of his findings about the Milky Way are listed there. The outpouring of studies on Galileo in connection with the fourth centenary of his birth in 1964 continued the already hallowed silence.

[6] The original Latin is reprinted in *Ed. Naz.* III, pp. 100–26. Quotations are from the profusely annotated translation by E. Rosen, *Kepler's Conversation with Galileo's Sidereal Messenger* (New York: Johnson Reprint Corporation, 1965).

[7] *Ibid.,* p. 36.

[8] *Ibid.,* p. 35.

[9] *Ibid.*

[10] In 1604, in his *Ad Vitellionem paralipomena quibus astronomiae pars optica traditur,* Cap. I, prop. ix, in *Werke,* vol. 2, p. 22.

[11] On the question in general, see my *The Paradox of Olbers' Paradox* (New York: Herder and Herder, 1969), chapters 5 and 7. The specific question of Olbers' independence of Chéseaux is discussed in two recent articles of mine, "New Light on Olbers' Dependence on Chéseaux," *Journal for the History of Astronomy* 1 (1970): 53–55, and "Olbers als Kosmologe," in *Nachrichten der Olbers Gesellschaft* Nr. 79. Oct. 1970: 5–13.

[12] Rosen, *Kepler's Conversation,* p. 34.

[13] In Kepler's *Werke,* vol. 4, p. 343.

[14] *Ibid.,* vol. 7, p. 44.

[15] *Ibid.,* p. 45.

[16] Published in 1610. In *Ed. Naz.* III, pp. 133–45.

[17] *Ibid.,* p. 135.

[18] In Kepler's *Werke,* vol. 10, p. 343.

[19] *Narratio de observatis a se quattuor Iovis satellitibus erronibus* (1611), in *Ed. Naz.* III, pp. 181–88.

[20] *Quatuor problematum . . .* (1610), in *Ed. Naz.* III, pp. 147–78; on the Milky Way, see especially pp. 162–64.

[21] *Epistola apologetica . . .* in *Ed. Naz.* III, pp. 193–200.

[22] *Ed. Naz.* III, pp. 188–90.

[23] *Ed. Naz.* III, pp. 291–98.

[24] The letter is dated March 8, 1611; see *Ed. Naz.* III, pp. 68–69.

[25] The title of the Appendix reads: "Accedit Appendix Observationum, quae per Radium artificiosum habitae sunt circa Saturnum, Iovem, et lucidiores stellas affixas," (Typis Gotthardi Vogelini, 1612), pp. 36–50.

[26] *Ibid.,* p. 44.

[27] *Ibid.,* p. 45.

[28] *Dianoia astronomica, optica, physica . . .* (1611), *Ed. Naz.* III, pp. 203–250.

[29] *De phoenomenis in orbe lunae . . .* (1612), *Ed. Naz.* III, pp. 311–99; see especially p. 374.

[30] *Apologia dictorum Aristotelis de via lactea. De facie in orbe lunae* (Venice: apud Thomam Balionum, 1613).

[31] *Ibid.,* p. 40. The words *gaiaxia, cometae, via lactea, stella* are not listed in the detailed index attached to Cremonini's *Apologia dictorum Aristotelis de*

quinta caeli substantia adversus Xenarcum, Ioannem Grammaticum et alios (Venice; apud Rubertum Meiettum, 1616).

[32] *Mundus Iovialis anno MDCIX detectus ope perspicilli belgici, hoc est quatuor Iovialium planetarum* . . . (Nürnberg: sumptibus et typis Johannis Lauri civis et bibliopolae, 1614).

[33] *Ibid., f.*)( )(3v.

[34] *Institutionum astronomicarum libri duo* . . . (Groningen: ex officina Ioannis Sassij, 1616), p. 55.

[35] *Institutiones astronomicae et geographicae, fondamentale ende grondelijcke onderwysinghe van de sterrekonst* . . . (Amsterdam: Willem Jansz [Blaeu], 1621), p. 5.

[36] *Institutiones physicae juventutis Lugdunensis studiis potissimum dicatae* (Leiden: ex officina Iacobi Patii, 1614), see Lib. VII, "De Meteoris", chap. VII [VIII], "De via lactea et spectris in sublimi apparentibus." The book, which went through several editions, failed to carry Galileo's name in this connection even in the "latest and most corrected" edition of 1644 (Amsterdam: apud Ludovicum Elzevirium), p. 222.

[37] *Tweevovdigh / Onderwiis / van de Hemelsche en Aerdsche Globen* . . . originally published c. 1620. See edition of 1634 (Amsterdam: Willem Blaeu), p. 27. The work is better known in its Latin translation by M. Hortensius, *Institutio astronomica* . . . (Amsterdam: apud Gulielmum Blaeu, 1634). The dozen or so lines on the Milky Way (see p. 26) are a close rendering of the Dutch text.

[38] *De mundo quaestiones et controversiae nobiliores* . . . *Accessit brevis Uranologiae summa* . . . (Copenhagen: ex calcographia Sertoriana, 1617); see chap. iii of *Uranologia,* "De coelo in specie . . . itemq; de via lactea in coelo," especially ff. 14r–15r. Bartholin also mentioned the resolution of "nebulous stars" into a congeries of stars, and admitted that he had been a defendant of Tycho's view on the galaxy.

[39] See *The Controversy on the Comets of 1618* [writings by Galileo, Grassi, Guiducci, and Kepler], translated by Stillman Drake and C. D. O'Malley (Philadelphia: University of Pennsylvania Press, 1960).

[40] *Ibid.,* p. 201.

[41] *Hemelsche trompet Morgentwecker ofte Comet met een Langebaert etschenen Anno 1618* (Groningen: Hans Sas, 1618), f. Bv. There is no reference to the Milky Way in Eberhard Welpern's *Observationes auss dem Stand und Lauff des im nechstabgeloffenen 1618 Jahrs im Monat Novembri erschienen grossen Cometens* (Strasbourg: Christoph von der Heyden, 1619) and in Willebrord Snellius' *Descriptio cometae qui anno 1618 mense Novembri primum effulsit* . . . (Leiden: ex officina Elzeveriana, 1619). I could not locate the booklet by H. Rasch(ius), *Disputatio de cometis et de*

via lactea (Regiomonti, 1619), listed in the Catalogue of the books of the Pulkovo Observatory, published in 1845 by F. G. W. Struve (see *Librorum in bibliotheca Speculae Pulcovensis contentorum Catalogus systematicus* [Petropoli, typis academiae scientiarum, 1845], col. 197).

[42] In his *Kurtze und grundliche Beschreibung eines newen ungewohnlichen Sterns oder Cometen . . . 1618 Jahr erschienen* (Strasbourg: bey J. Carolo [1618]), a booklet of 70 pages. See especially "Das Ander Capitel: Von der Cometen Ursprung, Natur, Art, und Eygenschaften," pp. 6–17. There is no reference to the Milky Way in Habrecht's *Planiglobium coeleste ac terrestre* (2d ed.; Nürnberg: typis Christophori Gerhardi, 1666), originally published in 1628.

[43] *De cometa anni 1618 dissertationes Thomae Fieni . . . et Liberti Fromondi* . . . (Antwerp: apud G. a Tongris, 1619), pp. 79–140.

[44] *Ibid.*, p. 134.

[45] Amsterdam: ex officina Guiljelmi I. Caesii. It was reprinted in 1640, also in Amsterdam, by Joh. et Corn. Blaeu; references are to the first edition.

[46] The title of the appendix is "De ascititiis coeli phaenomenis, nempe stellis novis et cometis"; see especially its chap. 5, "De causa materiali novorum phaenomenum coelestium," pp. 7–11.

[47] *Ibid.*, p. 10.

[48] *Ibid.*, pp. 10 and 9.

[49] *Recreation mathematique, composee de plusieurs problemes plaisants et facetieux* . . . (Paris au Pont-à-Mousson: par Jean Appier Hanzelet, 1624), p. 71 (Problème 73).

[50] *Sphaera mundi seu cosmographia demonstrativa* . . . (Bologna: typis Sebastiani Bonomij, 1620).

[51] *Ibid.*, p. 310. Curiously enough, there was no mention of the Milky Way in the book of the German Jesuit, Adam Tanner, on the origin, substance, and annihilation of the heavens, *Dissertatio peripatetico-theologica de coelis in qua de coelorum ortu, interitu, substantia . . . disseritur* (Ingolstadt: ex typographeo Gregorii Haenlin, 1621), although a long chapter (pp. 115–68) was devoted to problems raised by new phenomena, that is, comets, novae, variable stars, and the great discoveries described in the *Starry Messenger*, with the exception of the Milky Way. Tanner still defended the sublunary location and origin of comets (pp. 149–50).

[52] *De novis astris et cometis libri sex* (Venice: apud Io. Guerilium, 1622).

[53] *Ibid.*, pp. 97–100.

[54] *Ibid.*, p. 99.

[55] Antwerp: ex officina Plantiniana, 1627. See Lib. II, cap. 5, art. ii, "De galaxia," pp. 82–86.

[56] *Ibid.*, p. 84. For his definition of the galaxy, see p. 86.

[57] *Philosophia metaphysicam, physicamque complectens . . . Tomus secundus in quo de entibus corporeis ac spiritualibus distincte disseritur* (Rome: apud Iacobum Mascardum, 1627); see Q. XXXIV, sec. x, "De albedine quae per gyrum coeli conspicitur et via lactea nuncupatur," pp. 160–65.

[58] *In Aristotelis libros De Caelo et mundo dilucida textus explicatio et disputationes* (Naples: apud Secundinum Roncaliolam, 1626), see Tractatus IV, qu. viii, "De via lactea, seu Galaxia, quid sit et an sit aliqua pars orbis," pp. 213–21.

[59] *Ibid.*, p. 216. In all likelihood, d'Amici referred to the Venetian Georgius Raguseus, who is identified as theologian, physician, and professor of philosophy at Padua on the title page of his *Epistularum mathematicarum seu de divinatione libri duo* (Paris: sumptibus Nicolai Buon, 1623). I have not been able to locate a copy of his commentary on the *Meteorologica.*

[60] *Federici Bonaventurae Urbinatis Opuscula . . . De lactea via Arist, sententiae explicatio et defensio . . .* (Urbino, ex typographia Marci Antonii Mazzantini, 1627), pp. 183–234.

[61] *Ibid.*, pp. 210–11.

[62] *Ibid.*, p. 232.

[63] *Ibid.*, p. 226.

[64] *Lectiones Joannis Cottuni Veriensis . . . in primum Aristotelis librum de meteoris . . .*, edited by Innocentius Cremonius (Bologna: typis Nicolai Tebeldini, 1631), pp. 316–29, "Lectio Trigesima IV. Caput Octavum. De Galaxia sive de Via Lactea."

[65] *Ibid.*, p. 328.

[66] *Ibid.*, p. 326.

[67] *Urania cum geminis filiabus: hoc est astronomia et astrologia speciali* (Frankfurt: J. Bringerus, 1615).

[68] *Ibid.*, pp. 62–70. His procedure contrasted sharply with his reference to the Milky Way as a "congeries of many small stars that cannot be distinguished because of the distance of the celestial sphere" in his *Cosmographiae seu Sphaerae mundi descriptionis Hoc est Astronomiae et Geographiae Rudimenta* (Lemgoviae: impensis Justi Grothaus Bibliopolae, 1603), f. C2r.

[69] Frankfurt: impensis Godefridi Tampachii, typis Erasmi Kempfferi, 1622. The work centered on the question of whether Galileo's manner of philosophizing was contrary to the Bible or not.

[70] *Ibid.*, p. 38.

[71] Edited by Tobias Adam and published for the first time in Frankfurt: apud Egenolphum Emmelium, impensis Godefridi Tampachij, 1620.

[72] *Ibid.*, p. 209.

[73] *Roberti Fludd alias de Fluctibus Philosophia sacra et vere Christiana, seu meteorologia cosmica* (Frankfurt: in officina Bryana, 1626).

[74] *Examen philosophiae Roberti Fludd . . .* (1628), in *Opera omnia* (Lyons: sumptibus Laurentii Anisson et Joan. Bapt. Devenet, 1658), vol. 3, pp. 211–68.

[75] In his lectures on physics and astronomy published posthumously as *Syntagma philosophicum,* in *Opera,* vol. 1, p. 729.

[76] In English translation in *The Works of Francis Bacon,* edited by J. Spedding, R. L. Ellis and D. D. Heath (Boston: Taggard and Thompson, 1864), vol. 10, pp. 403–60.

[77] *Ibid.*, p. 418.

[78] *Ibid.*, p. 457.

[79] *Ibid.*, pp. 457–58.

[80] This claim immediately precedes, in par. 45 of Part III of his *Les Principes de la philosophie,* his account of the primordial differentiation in the universe. See *Oeuvres de Descartes,* edited by Charles Adam and Paul Tannery, vol. IX–2 (new imprint; Paris: J. Vrin, n.d.), p. 124.

[81] In Descartes, *Oeuvres,* vol. 11, p. 107.

[82] Originally published with his *Discours de la méthode* in 1637; in *Oeuvres,* vol. 6, pp. 229–366.

[83] *Philosophia naturalis in qua tota rerum universitas per clara et facilia principia explanatur* (Amsterdam: apud Ludovicum et Danielem Elzevirios, 1661), p. 157. This book was a much enlarged version of his *Fundamenta physices* (1646) which rightly suggested to Descartes some plagiarism on du Roy's part.

[84] "Préface pour le Traité du vide," in *Oeuvres complètes,* edited by Jacques Chevalier (Paris: Gallimard, 1954), pp. 529–35; for the passage on the Milky Way, see p. 534.

[85] *Astronomia Kepleriana defensa et promota,* in *Opera posthuma* (London: typis Gulielmi Godeid, 1673), p. 63. There Horrocks erroneously claimed that Captain Thomas James (c. 1593–c. 1636) observed during the very clear night of January 30–31, 1632, an unusually large number of stars in the Milky Way. In *The Strange and Dangerous Voyage of Captaine Thomas James . . .* (London: J. Liggatt for J. Partridge, 1633), there is reference in this connection only to the Pleiades and the Crab nebula (p. 69).

[86] *Physicae ad lumen divinum reformatae synopsis* (Amsterdam: apud Ioh. et Iod. Ianssonios, 1643), p. 92.

[87] *Exercitationes paradoxicae adversus Aristoteleos* (1624), in *Opera omnia,* vol. 3, p. 141.

[88] In *Opera omnia,* vol. 6, pp. 4–6.

[89] *Commentarij de rebus coelestibus,* in *Opera omnia,* vol. 4, pp. 75–498.

[90] *Ibid.*, pp. 3–74.

[91] Physicae Sectio II, Liber I, cap. ii, "Quid in Caelo Cyaneus color, Lacteusque Circulus sit?" in *Opera omnia,* pp. 504–08.

[92] According to Gassendi, Aristotle was alone in his version of Democritus' opinion about the Milky Way as caused by the shadow of the earth, *ibid.,* p. 507.

[93] *Ibid.,* p. 508.

[94] *Ibid.*

[95] *Ibid.,* p. 805.

[96] Physicae Sectio II, Lib. VI, cap. iv., "De vanitate astrologorum circa specialia Placita ad mutationes aeris attinentia," *ibid.,* p. 729.

[97] *Petri Gassendi Apologia in Jo. Bap. Morini librum. cui titulus "Alae telluris fractae"* . . . (Lyons: apud G. Barbier, 1649).

[98] *Response à une longue lettre de Monsieur Gassend . . . touchant plusieurs choses belles et curieuses de Physique, Astronomie et Astrologie . . .* (Paris: [J. Lebrun] 1650). Equally silent was Morin concerning the Milky Way in his *Defensio suae dissertationis de atomis et vacuo adversus Petri Gassendi philosophiam . . .* (Paris: apud Petrum Menard, 1657), although he mentioned the *Starry Messenger* and Jupiter's moons as he defended astrology (p. 106).

[99] *Astronomia iam a fundamentis integre et exacte restituta . . .* (Paris: apud Ioannem Libert, 1640), p. 210.

[100] The Hague: A. Vlacq, 1661, p. 193.

[101] *Ibid.,* Lib. III, Sec. IV, cap. i, p. 94.

[102] *Ibid.,* Lib. IX, Sec. III, cap. v, pp. 188–89.

[103] *Ibid.,* Lib. IX, Sec. III, cap. ix, pp. 192–93.

[104] *Ibid.,* Lib. IX, Sec. II, cap. vii, p. 173.

[105] *Ibid.,* Lib. IX, Sec. II, cap. x, pp. 179–80.

[106] *L'impiété des déistes, athées et libertins de ce temps . . .* (Paris: chez Pierre Bilaine, 1624), reprinted in one volume under the title *Questions rares et curieuses, théologiques, naturelles, morales, politiques et de controverse resolues par raisons tirées de la philosophie et de la théologie* (Paris: chez Pierre Billaine, 1630); see especially sections V and VIII.

[107] Originally published in Mersenne's *Synopsis mathematica* (Paris: ex officina Rob. Stephani, 1626), pp. 209–40; reprinted in his *Universae geometriae mixtaeque mathematicae synopsis . . .* (Paris: apud Antonium Bertier, 1644), pp. 257–72.

[108] *Quaestiones celeberrimae in Genesim cum accurata textus explicatione* (Paris: sumptibus Sebastiani Cramoisy, 1623), cols. 839–42.

[109] In his *Astronomia philolaica* (Paris: Simeon Piget, 1645), pp. 12–13.

[110] *Ibid.,* p. 12. On the other hand, Boulliau raised two objections, which seemed to be rather original, against the Aristotelian universe. They could be considered as the equivalents of "Olbers' Paradox" in the universe of Aristotle, and were largely overlooked before Boulliau. He argued, first, that a crystalline sphere had to reflect the light of the sun and of the stars, so that there could be no darkness at night. Second, there had to be a continuous, extremely high temperature in the universe if the sun's heat was indeed due to friction caused by its motion. *(Ibid.,* p. 9).

[111] Bologna: ex typographia haeredis Victorij Benatij, 1651. See Lib. VI, cap. xxiii, "De galaxia seu lacteo circulo," pp. 474–76.

[112] *Ibid.,* p. 476.

[113] In a commentary on the *Meteorologica* which I was unable to locate.

[114] In his "circuli" or discussions of Aristotelian philosophy first published in 1643 under the general title, *Circulus Pisanus Claudii Berigardii . . . de veteri et peripatetica philosophia Aristotelis* (Utini: ex typographia Nicolai Schirattii, 1643). In the second and corrected edition of the work, dated 1661, the section on the *Meteorologica* is printed with the special title page: *Circulus Pisanus Claudii Berigardii . . . de veteri et peripatetica philosophia in Aristotelis libros Meteorologicos. Pars Quinta* (Padua: typis Pauli Frambotti, 1661), pp. 537–583. The Milky Way is discussed in the last ten lines of that section, on p. 583.

[115] *Super quatuor libros Meteororum Aristotelis philosophorum principis quaestiones* (Lyons: sumptibus Petri Prost, 1643), pp. 69–74.

[116] Cologne: apud Iodocum Kalcoven, 1644. On the Milky Way, see pp. 204–12.

[117] In clear evidence of his obstinacy, Chiaramonti still argued the same point in 1636 in his *De sede sublunari cometarum opuscula tria in supplementum Anti-Tychonis cedentia* (Amsterdam: apud Iohannem Ianssonium, 1636).

[118] *Opus de universo,* p. 211.

[119] *In quatuor meteorologicorum Aristotelis commentaria . . .* (Rome: typis haeredum Francisci Corbeletti, 1646). According to the rather lengthy title of the work, it also gave an almost complete exposition of experimental philosophy. On the Milky Way, see vol. 1, pp. 215–46, and especially p. 246.

[120] *Philosophiae ad mentem Scoti cursus integer. Tomus tertius. Continens disputationes ad mentem Scoti in Aristotelis Stagiritae libros De Anima, De Generatione et corruptione, De Coelo et Metheoris.* Editio novissima et mendis expurgata (Venice: apud Nicolaum Pezzana, 1727), pp. 495–98. I was unable to consult the original edition of 1640, nor Belluti's *Disputationes in libros de caelo et mundo et meteoris* (Venice, 1640) ascribed to him by Ideler, *op. cit.,* vol. 1. p. xxxi (see p. 27, note 18), if it is indeed different from the previous work.

[121] *Disputationes in libros Aristotelis Meteororum* (Genova: excudebant P. I. Calenzanus & I. M Farronus soc., 1637), pp. 77–87. According to the subtitle of the work, it was useful not only for students but also for professors.

[122] First published in 1656 and re-edited by C. Schott with the approval of the author (Würzburg: sumptibus Joh. Andr. et Wolffg. Jun. Endterorum haeredibus, 1660); see pp. 25 and 335.

[123] London: printed by E. Griffin, 1638.

[124] *Ibid.,* p. 179.

[125] *A Discourse concerning a New World and Another Planet in 2 Bookes. The First Book, The Discovery of a New World* (London: printed for John Maynard, 1640), p. 178. The argument and wording remained unchanged in the further revised but posthumous fourth edition of the work published, twelve years after Wilkins' death, in London in 1684, p. 136.

[126] *Discours nouveau prouvant la pluralité des mondes que les astres sont des terres habitées et la terre est une estoile, qu'elle est hors du centre du monde dans le troisieme ciel et se tourne devant le soleil qui est fixe et autres choses très-curieuses* (Geneva, 1657).

[127] *Ibid.,* f. E2r (chap. xxiv).

[128] Passage is quoted from the English translation, *A new Treatise proving A Multiplicity of Worlds . . .* (London: printed by John Streater, 1658), pp. 78–79. Probably the translator was D. Sashott, whose name was printed at the end of the dedicatory epistle addressed to Frederick Claudius.

[129] *De vero telescopii inventore cum brevi omnium conspiciliorum historia. . . .* (The Hague: ex typographia Adriani Vlacq, 1655), p. 50.

[130] *Astrognosia, synoptice et methodice in usum Gymnasiorum academiarum adornata . . .* (Wittenberg: impensis Jobi Wilhelmi Fincelii, 1659), p.

114. The section on the Milky Way (pp. 111–14) remained unchanged in the fifth edition (Wittenberg: impensis Joh. Lud. Quenstedii, 1694), pp. 136–41. By comparison, the treatment of the Milky Way was rather pedestrian in the *Institutiones physicae* (Lübeck: typis Gothofridi Jegeri, 1647), by Johann Sperling, another professor at Wittenberg; see pp. 516–17.

[131] *Astronomia physica seu de luce, natura et motibus corporum coelestium libri duo* (Paris: apud Petrum Lamy, 1660); on the Milky Way, see p. 76. It should be to Duhamel's credit that comets and Milky Way were excluded from his *De meteoris et fossilibus libri duo* (Paris: apud Petrum Lamy, 1660).

[132] *Oeuvres complètes de Christiaan Huygens,* vol. 15 (The Hague: Martinus Nijhoff, 1925), p. 239.

[133] *De cometis ubi de cometarum natura disseritur . . . Praelectio Oxonii habita a Setho Wardo* (Oxford: excudebat L. Lichfield, 1653), p. 24. There is no reference to the Milky Way in Ward's critical discussion of Boulliau's work, *Inquisitio in Ismaelis Bullialdi astronomiae philolaicae fundamenta* (Oxford: excudebat Leon Lichfield, 1653).

[134] Quotations are from the English draft of the inaugural discourse, delivered in Latin in somewhat different form but with the passages on the Milky Way and other galaxies left intact. See James Elmes, *Memoirs of the Life and Works of Sir Christopher Wren . . . with an Appendix of Authentic Documents* (London: Priestley and Weale, 1823), pp. 37 and 50 of the Appendix.

[135] *Discursus de via lactea quem . . . praeside Sigismundo Hosemanno . . . examini commilitonum sistit author Ludolphus Georgius Lunde* (Helmstadt: typis Jacobi Mulleri, 1665).

[136] *Theorian Viae Lacteae ad Arist. cap. viii. Lib. Meteor. I. sub praesidio M. Andreae Glauchi Lips. . . . instituit publice Jonas Hertzbergerus* (Leipzig: typis Coleranis, 1663). 16 pages in 8°.

[137] Examples are: Johann Matthias Schneubern, *Umständliche Beschreibung des grossen Cometen . . .* (Strasbourg: bey Johann Pastorius, 1665), an opuscule of 52 pages; Abdias Trew, *Gründiche . . . Beschreibung des jüngst zu End des Jahrs 1664 . . . erschienen Cometen* (Nürnberg: Michael und Johann Friderich Endkern, 1665) of 64 pages; and P. Petit, *Dissertation sur la nature des comètes au Roy avec un discours sur les prognostiques des eclipses et autres matières curieuses* (Paris: Th. Jolly, 1665), where the Milky Way is mentioned only in connection with Tycho's theory on the origin of comets (p. 151). No significant reference to the Milky Way occurs in the two former works.

[138] *Descriptio cometae anno aerae Christi MDCLXV exorti* (Danzig: imprimebat Simon Reiniger, 1666), and *Prodromus cometicus quo historia*

cometae anno 1664 exorti . . . exhibetur (Danzig: imprimebat Simon Reiniger, 1665).

[139] *Cometographia* . . . (Danzig: imprimebat Simon Reiniger, 1668).

[140] Danzig: typis Johannis Zachariae Stollii, 1687.

[141] By the same printer in Danzig, 1690.

[142] By the same printer in Danzig, 1690.

[143] Danzig: imprimebat Simon Reiniger, 1673, p. 381.

[144] *Selenographia* . . . (Danzig, 1647), p. 29.

[145] Danzig: A. Hünefeldt, 1651, p. 11.

[146] Revised and augmented by Louis Coulon (Paris: chez Gervais Clouzier, 1657), vol. 1, p. 50.

[147] *Harmonia macrocosmica, seu Atlas universalis et novus totius universi creati cosmographiam generalem et novam exhibens* . . . (Amsterdam: apud Johannem Janssonium, 1661), pp. 184–85.

[148] *A Tutor to Astronomie and Geographie* . . . (2d ed.; London: printed by S. Roycroft, 1670), pp. 23–25. The first edition was printed in 1659, the fifth in 1699.

[149] *Traité de physique* (Paris: chez la Veuve de Charles Savreux Libraire Juré, 1671), two volumes bound in one.

[150] *Ibid.,* vol. 2, pp. 6–7.

[151] *Ibid.,* vol. 2, p. 106.

[152] *Ibid.,* vol. 2, p. 107.

[153] *Ibid.,* vol. 2, pp. 111–18. The Latin translation, *Tractatus physicus,* by Th. Bonet(um) (Geneva: sumptibus Ioannis Hermanni Widerhold, 1774), was amplified by an 84-page index with, of course, no entries such as *via lactea,* or *galaxia.*

[154] *Discours sur les influences des astres selon les principes de M. Descartes* (Paris: chez Jean Baptiste Coignard), 1671.

[155] *Ibid.,* p. 65. In the diagram, the stars of the Small Bear were shown as small circles on the circumference of a large circle representing the sun's vortex, and their influence on Mars and the Earth was indicated by straight dotted lines. It escaped Gadrois that, on the same scale, no more than forty stars could be allocated around the sun's vortex, and that according to Descartes the passage of bodies (comets as well as light corpuscles) was not strictly straight.

[156] *Ibid.,* p. 82.

[157] *Le système du monde selon les trois hypothèses . . .* (Paris: chez Guillaume Desprez, 1675).

[158] First published in Amsterdam in 1672; references are to the reprint edition (Aalen: Otto Zeller Verlagsbuchhandlung, 1962). The Milky Way is discussed in Book II, chap. 2, "De magnitudine et multitudine stellarum," pp. 226–29.

[159] *Ibid.*, and see also Book I, chaps. 29 and 34.

[160] *Ibid.*, p. 228.

[161] *Ibid.*

[162] *Catalogue des estoilles australes ou supplement du Catalogue de Tycho* (Paris: chez Jean Baptiste Coignard, 1679). The 36 pages of the Preface, in parallel Latin and French columns, are not numbered. For quotation, see p. [4].

[163] *Ibid.*, p. [32].

[164] *Micrographia* (London: printed by Jo. Martyn and Ja. Allestry, 1665), p. 242.

[165] In *The Posthumous Works of Robert Hooke,* edited by Richard Waller (London: Sam. Smith and Benj. Walford, 1705), p. 29.

[166] *Ibid.*, pp. 71–148; see especially pp. 76–77 and 121.

[167] See, on this, my *The Paradox of Olbers' Paradox,* pp. 56–58.

[168] *The Posthumous Works,* p. 98.

[169] "Nouveau Phenomene rare et singulier d'une Lumière Celeste, qui a paru au commencement du Printemps de cette année 1683," in *Journal des Sçavans* 11 (1683): 131–44. For reference on the Milky Way, see p. 131.

[170] *Das eröffnete Lust-Haus der Ober- und Nieder-Welt . . .* (Nürnberg: Wolffgang Moritz Endter und Johann Andreae Endters Sel. Erben, 1676), pp. 338–39.

[171] *De galaxia seu circulo lacteo, amplissimi philosophorum ordinis consensu in illustri ad Varnum academia,* Disputabunt, Praeses, M. Georgius Funccius, Regiomonte-Prussus, & respondens Olaus Wexstedt/Wexionia Svecus. (Rostock: imprimebat Jacobus Richelius, 1686). 13 pages in 8°.

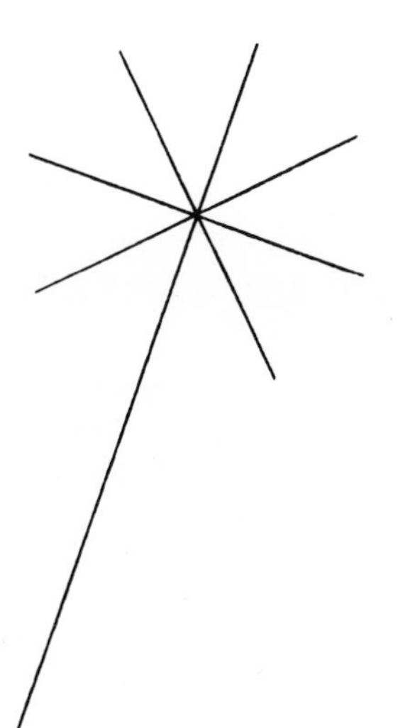

CHAPTER FIVE

Newtonian Distraction

In 1686, when Wexstedt's trivial discourse on the Milky Way saw print under the foggy skies of the Baltic Sea, there appeared in the City of Light a book, *Entretiens sur la pluralité des mondes,*[1] in which a brilliant new light was shone on the topic. In the Preface the anonymous author claimed that he did not wish "to instruct but only to entertain." In this he certainly succeeded. Before long, the book had become the most eagerly read science popularization, and had obliged its author, Bernard de Fontenelle, to abandon his studied anonymity. A novelist of twenty-nine when the *Entretiens* first appeared, Fontenelle, largely because of the enormous success of his book, won membership in the Académie Française in 1691, and, six years later, the powerful position of Perpetual Secretary of the Académie des Sciences.

The avowed objective of the *Entretiens* was to strengthen belief in denizens of other planets. The scientific service which the book performed consisted of influencing public opinion in support of Copernicanism, although Fontenelle's science, steeped in Cartesian vortices, was somewhat amateurish. But five evenings of pleasant dialogue between Fontenelle and a marchioness did what dry arguments could not do.

At any rate, during the fifth evening Fontenelle owed nothing to Wilkins' *Discovery of a World in the Moone,* his chief source for speculation about other planets and their alleged denizens, when he carried the topic beyond the planets to the realm of stars, and described the Milky Way as a region where the skies were studded day and night with the disks of myriad small suns:

> You see yonder, white part of the Heavens that is called the *Milkey Way*. Can you form to yourself any idea of what it is? An infinite number of little Stars, invisible to the eye, because of their smallness, and set so near each other, that they appear as one continued whiteness. I wish you could with a telescope visit this ants nest of Stars: this cluster of Worlds, if I may be allowed the expression. They in some sort resemble the *Maldivia* Isles, which are twelve thousand little islands or banks of sand, only separted [sic] by a canal of the Sea, that one might almost leap over. The little Vortexes of the *Milky Way* are sown so thick, that it appears as if from one World to another they might speak or even shake hands together. At least, I believe that the birds of one World may easily pass into another, and that they may train them up like pigeons to carry letters, as they do here in the *Levant* from one town to another. These little Worlds apparently vary from the general rule, that a *Sun* in his Vortex effaces the light of all other Suns. If you were in one of the little Vortexes of the *Milky Way,* you would find your *Sun* scarce nearer to you, nor would he sensibly have any more force on your eyes, than a hundred thousand other *Suns* of neighbouring little Vortexes. You would therefore see your Heavens shine with an infinite number of fires, which are very near to each other, and little distant from you. When you lose sight of your particular *Sun,* there will remain sufficient for you, and your night will not be darker than your day, at least the difference may not be sensible; and for speaking more justly, you never have any night. The people of those Worlds would be greatly astonished, accustomed as they are to perpetual light, if they were informed that there are unhappy people, who have very dark nights, who fall into profound darkness, and who, when they enjoy light, even then see only one Sun. They would look on us as beings, disgraced from Nature, and our condition would strike them with horror.[2]

In the context, this graphic description of the Milky Way served to document Fontenelle's claim that the number of suns in the universe was infinite. He gallantly parried his companion's anxiety about the specter of infinite worlds with the remark that all of them were outshone by charming eyes and beautiful lips. His flattery effectively cleared the ground for scientific proof of infinitely numerous worlds. In the Milky Way alone, he noted, a telescope could unveil an uncounted number of stars surrounding our solar system. Therein lay the catch to a universe of vortices. According to Descartes and the Cartesians, each sun was the center of a vortex. The actual shape of stellar vortices had never been clearly described by Cartesian physicists, a fact which became all too evident in the objection of the marchioness whose mind was not, apparently, numbed by flatteries: "But do you know, that in multiplying these Worlds so liberally, you have given rise to a very great difficulty? The Vortexes, whose Suns we see, touch the Vortex where we are. The Vortexes are round, is not this true? And how can so many Vortexes touch one only?"[3] Fontenelle's reply tried to resolve this question with the help of the many, very small star-vortices forming the Milky Way. He urged his companion to think of the edge of our vortex as not being perfectly circular, but actually having many small sides.

As far as the plane of the Milky Way was concerned, such a solution seemed to possess some plausibility. In fact, it merely sowed the seeds of greater problems. The principal one of these seemed to be the crowding of so many small star-vortices into one plane. Why was it that, in a three-dimensional, homogeneous, infinite world, the "touching" of big stars on our vortex in any plane other than that of the Milky Way was not "smoothed" by a great number of small stars? This was the real issue concerning the Milky Way and the actual appearance of the visible world, but Fontenelle failed to carry the logic of a Cartesian to the perhaps "bitter" end, although he enjoyed referring to the sweep of logic on which science depended. He told his companion that "we reason in mathematics [science] as we reason in love."[4] Since she obviously knew the decisive role of a first step in the latter, she could readily understand the ultimate portent of a first assumption. Lovers and mathematicians were much alike. "The more you give to these two sorts of people," Fontenelle remarked, "the more they always take."[5] This was certainly true, but the Milky Way still failed to receive an intellectual love that was ready to go all the way.[6]

Not that 1686 lacked intellects capable of extraordinary pursuits. After all, Fontenelle's book, classic evidence of the popularity of Cartesian vortices, is best remembered by historians of science because of the devastating blow which Cartesian physics suffered the next year. This blow was administered through a book which received more acclaim than any other scientific publication in history—a well deserved distinction. Newton's *Principia,* still the greatest single advance in science, opened up, as if by magic, astonishing insights into the working of the physical universe. The fall of an apple was now revealed as a case basically identical to the "falling" of the moon toward the earth, and what is more, the explanation could be put within reach for the interested layman. For scientists, the *Principia* held many additional and far more intricate—if not distracting—revelations, such as the explanation of various aspects of the motion of the lunar nodes.

Most of these revelations appeared in the Third Book of the *Principia,* which Newton called "The System of the World," a rather ill-chosen title. The Third Book consisted of two parts, one technical, the other popular. Both, in their respective styles, were replete with accounts of dazzlingly new discoveries pertaining to a world which hardly extended beyond the solar system. The starry realm was only once discussed there at length, in the popular section, in a chapter entitled, "On the Distance of the Stars." Its contents were burdened with some exasperating oversights. Newton's starting point was the well known fact that the fixed stars could not be closer than the distance determined by the "smallness of their annual parallax."[8] Since astronomers of his time could easily measure angular separations of the order of one minute of arc, the minimum distance "of the fixed stars is," to quote Newton's words, "above 360 times greater than the distance of Saturn from the Sun."[9] With a separation of twenty seconds of arc, the minimum distance had to be set at "above 2000 times" the sun-Saturn distance. By comparing the luminosities of Saturn and of the sun, Newton could readily compute the distance at which the sun would appear as a first-magnitude star. The figure was $10{,}000\sqrt{42}$ times the sun-Saturn distance, but Newton hastened to add a proviso. Some, he recalled, "may, perhaps, imagine that a great part of the light of the fixed stars is intercepted and lost in its passage through so vast spaces, and upon that account pretend to place the fixed stars at nearer distances."[10]

Newton did not name anyone in particular, nor did he seem to agree. On such a basis, he wrote, "the remoter stars could be scarcely seen," a point which he took pains to illustrate quantitatively: "Suppose, for example, that ¾ of the light is lost in its passage from the nearest fixed stars to us; then ¾ will be lost twice in its passage through a double space, thrice through a triple, and so forth. And, therefore, the fixed stars that are at a double distance will be 16 times more obscure, viz., 4 times more obscure on account of the diminished apparent diameter; and, again, 4 times more on account of the lost light. And, by the same argument, the fixed stars at a triple distance will be 9·4·4, or 144 times more obscure; and those at quadruple distance will be 16·4·4·4 or 1204 times more obscure." From this it followed that "so great a diminution of light is no ways consistent with the phenomena and with the hypothesis which places the fixed stars at different distances."[11]

Strangely enough, he did not take up the possibility of a much weaker absorption of light in the ether, which could have explained "the phenomena," though only because thermodynamics was still not even in its infancy. Moreover, he left undeveloped an important point; namely, the question of the distribution of the stars in the universe. His quantitative example was equivalent to a distribution of stars in concentric shells, the width of each shell being equal to the mean distance separating the stars from one another. It is very likely that he perceived this, because, as will shortly be seen, two scientists closely associated with him later developed the same idea in considerable detail.

Such a distribution of stars was also equivalent to an infinite, homogeneously distributed universe of stars, a notion which was to become synonymous, not without very good reason, with Newton's idea of the universe. Newton himself was to spell out this concept more definitely. Such a distribution of stars did not admit an absolute center for the universe. One could imagine, around any star, an infinite number of concentric shells containing the stars which were visually of the first, second, third . . . magnitudes. But the same distribution was fraught with the optical and gravitational paradoxes of an infinite, homogeneous universe. Both were brought to Newton's attention, as will be discussed, but he ignored the former and talked away the latter.

Most revealing, he did not realize that a distinctly non-homogeneous distribution of stars was strongly suggested by the most grandiose phenomenon of the night sky, the star-strewn band of the Milky Way. Almost certainly he observed its multitude of stars with his reflecting telescope, one of his finest discoveries. Yet, Newton spoke but briefly of the Milky Way. He did so when he discussed "the uncertain agitation" of the tails of the comets. It might have arisen, according to him, "from parts of the Milky Way which might have been confounded with and mistaken for parts of the tails of the comets as they passed by."[12] Of course, Newton did not associate comets and the Milky Way as did some old Aristotelians. Still, he could not have added a more disappointing paragraph to the record of the dicta on the Milky Way.

But who could see this strange blemish in the flood of dazzling light radiating from the *Principia?* One of the first to exploit that light was Richard Bentley, who in his Boyle Lectures,[13] preached between March and December, 1692, made much of the *Principia* to discredit the atheists of his day. The account of the planetary system in the *Principia* clearly reflected a grandiose design, and necessitated continual intervention by the Creator so that the stability of the world machine would be maintained.

Before publishing the text of his Sermons, Bentley sought Newton's approval, which was generously granted.[14] The only point where a major disagreement arose between Newton and Bentley concerned the latter's claim that the universe could not be infinite. In such a universe gravitational attraction would not operate, so Bentley argued, because attraction from every side would be of exactly the same magnitude.[15] In the ensuing dispute about the "adding and subtracting of infinities" Newton earned only a superficial victory by lecturing Bentley on some basic rules in dealing with infinities.[16] The theory of gravitational potential, with which Bentley could have clinched his argument in rigorous form, was still generations away. However, he was unquestionably shortsighted in failing to perceive as false the assumption concerning the homogeneous distribution of stars on which the idea of an infinite, homogeneous universe largely rested. One needed only refer to the Milky Way to give the lie to that assumption, and raise serious doubts about the infinity of the universe.

In his zealous search for design in the universe, Bentley again failed to perceive in the Milky Way one of the most astonishing proofs of that design. His sole reference to it was apologetic. He cautioned his readers lest they should think that he had succeeded in finding the humanly purposeful aspect of every phenomenon in the universe: "We dare not undertake to shew what advantage is brought to us by those innumerable stars in the galaxy and other parts of the firmament, not discernible by naked eyes."[17] Newton said even less about the galaxy in the *Principia.* The members of his "principate" (Newton used his influence rather vigorously to secure the best scientific appointments for his friends and adulators) followed suit so far as the Milky Way was concerned. And so did those who became his disgruntled foes.

Intellectually the most towering figure of that "principate" was Samuel Clarke, chaplain at Westminster, twice Boyle lecturer, and Newton's spokesman, around 1714, in the famed controversy with Leibniz about the "perfection" of the world machine. But long before that, Clarke labored to distinguish himself as a champion of Newtonianism. In 1697, he published a Latin translation of Jacques Rohault's *Traité de physique.*[18] It was not Rohault's reputation that he wished to enhance; the translation was embellished with many notes in which Clarke constantly reminded the reader of the many major improvements which Newton's system offered over that of Rohault, whose text was widely used at Cambridge. On one point there could be no improvement. To Rohault, Newton, and Clarke too, the Milky Way, for all practical purposes, was non-existent. And so it remained to Huygens, the most scientific of all Cartesians, to the very end of his life in 1697. His last work, published posthumously a year later, was the *Cosmotheoros,* an unbridled speculation on the inhabitants of other planets.[19] Huygens thought that one could plausibly infer such details as the feet, the hands, and stature of other planetarians, their navigation and the various keys in which their music was composed.[20] He was not as certain and prolific about the characteristics of the denizens of planets circling other stars. In speaking of the stars, he stressed their immense, if not infinite, number and did his best to discredit Kepler's argument about the finiteness of the world.[21] Kepler could easily have been proven wrong in 1698, but Huygens strangely left unsaid that part of Kepler's argument which concerned the Milky Way.

Huygens' silence on the Milky Way was ironic in that his *Cosmotheoros* started with the complaint that the "ingenious French author of the Dialogues about the *Plurality of Worlds*" had not carried any farther the topic of the inhabitants of the universe.[22] Hardly more was said about Fontenelle by the astronomer members of Newton's "principate." In view of the immense popularity of the

Entretiens in England, they had almost certainly read its section on the Milky Way, but it did not prompt them to further reflection. Thus, in David Gregory's famous compendium of astronomy, which upon its publication in 1702[23] became the first such work to endorse in detail the new astronomy of the *Principia,* an account of the Milky Way appeared which was distinctly pedestrian. It started with a reference to Manilius, of all "authorities," to document that some of the ancients had already surmised that the Milky Way was "a Collection of innumerable little Fix'd Stars."[24] This was followed by a detailed description of the position of the Milky Way among the constellations, as for Gregory it was but a phenomenon that "comes under this Head of Constellations." The conclusion, a mention of Halley's findings about the Magellanic Clouds, was equally lacking in novelty.

This lack of "in-depth" consideration of the Milky Way was all the stranger in Gregory's case, since in the immediately preceding pages he pondered the merits of a shell model[25] of the distribution of stars in the universe. The essence of that model was a presumed strict correspondence between the number of stars that could be geometrically arranged in successive shells of unit radius (the average distance of stars from one another) and the number of stars as counted by their observed magnitudes. For the first and second shells, the hypothetical numbers (13 and 4 × 13 = 52) agreed closely with the actual figures. Beyond that, as Gregory remarked, "the Matter does not go on so well."[26] The reason was very simple and is better given in his own words: "Upon the first cast of our Eyes upon the Heavens, some Tracts of the Firmament appear fill'd with innumerable Fix'd Stars, whereas others are found to be almost empty and void of any."[27] But the "innumerable Fix'd Stars" were, as Gregory pointed out three pages later, in the Milky Way. If they were "little" it was not because telescopes had revealed any difference between the apparent diameters of "big" and "little" stars. They were "little" because of their optical magnitude, which, Gregory recalled, one group of astronomers viewed as an indication of distance. He seemed to wish that such an opinion were true as he noted: "And this Opinion is very much favoured by the Number of the Fix'd Stars of the first and second Magnitude."[28] He did not suspect that he held in his hands all the parts needed to construct a true picture of the Milky Way.

The true picture was not, of course, the ring model which Jean-Dominique Cassini suggested with studied diffidence in 1705, in his account of his observations of the satellites of Saturn and its ring. In his own words, he would not "dare to think that this ring [of Saturn] could have been formed as a chain of small satellites which might produce on Saturn an appearance analogous to that made by the Milky Way by an infinity of small stars of which it is constituted." To this he merely added that whereas the Milky Way showed no parallax, the ring should display a very large parallax when observed from Saturn. Other difficulties inherent in the ring-model of the Milky Way seem to have eluded Cassini completely. He gave no credit to Kepler and Gassendi as he made his timid suggestion, which although carried to every corner of the learned world in the pages of the *Mémoires* of the Académie des Sciences,[29] produced no noticeable echo.

Chief among those corners was Cambridge, where William Whiston lectured on astronomy as Newton's handpicked successor to the Lucasian chair. In his *Praelectiones astronomicae,* published in 1707, Whiston spoke explicitly on the problem of whether or not the "smallness" of stars in the Milky Way was due to their distances. He felt that one "must refer the Decision of the Matter to further Observations."[30] Nevertheless, he went on to discuss topics highly germane to the problem. One of these was a consequence of the ever greater number of stars revealed by telescopes. Were these stars distributed across a space with a radius even "Twenty-fold" the distance of the sun? Whiston was not ready to go that far. On the other hand, he showed no hesitation in dismissing the opinion that "the Fixed Stars seem to be dispos'd in the Heavenly Spaces in no certain Order, but as it were by Chance only."[31] However, his proof against a chance distribution of stars rested solely on an analogy with the orderly paths of planets and comets. The ordering of stars which did not appear "when they are seen from this Earth," might be very evident "when they are beheld from some other Place."[32] He even conjured up the possibility of sets of proportionally arranged star-systems still "wholly unknown to us." Rarely in the history of science was the "unknown" so within reach and the "proper Place" so close. The Milky Way, which Whiston previously described as "the Path" where the telescope showed a "vast Multitude of little Stars," was indeed an elusive road. By 1707, the blame for this lay at the door of astronomers and not with the age-old "Ignorance or Weakness of frail Man," where Whiston let matters rest.[33]

Whiston could not have been briefer concerning the Milky Way in his *Astronomical Principles of Religion Natural and Reveal'd,*[34] although what he said was not without interest. In speaking about some "lately discover'd Spaces of Light, or Starry Mists," he took the view that "they are a Company of very small Fixed Stars, as invisible to us with our ordinary Telescopes, as the known Telescopick Stars in the Milky Way are to our natural Eyes, which give such an irregular Appearance of indistinct Light also."[35] The work itself showed the prevailing tendency of Whiston's mind which, in some of his other publications, produced somewhat extravagant ideas on the harmony between Sacred History and science. The trend was still restrained in his *Astronomical Lectures,* which were used to good advantage by William Derham in his *Astro-theology,*[36] or, as the subtitle explains it, *A Demonstration of the Being and Attributes of God from a Survey of the Heavens.* It was clearly Whiston whom Derham followed when he took up the question of whether the stars were arranged in the universe according to some design.[37] Their appearance indicated the contrary, but the reality had to be otherwise in a universe which, in every other respect, reflected some distinct pattern. The disorderly arrangement of stars was, according to Derham, an optical illusion. Were we able to look at them at close range, their true co-ordination would immediately become obvious. This rather curious reasoning he capped by the analogy of an army drilling at a distance. Its soldiers "would appear to us in a confused manner, until we come near and had a regular prospect of them, which we should then find stand well in rank and file."[38] This unfortunate reversal of perspective must have blocked insight when Derham applied the analogy to the starry realm: "So doubtless, if we could have an advantagious prospect of the Fixt Stars, we should find them very commodiously and well set in the Firmament in regard of one another."[39]

The prospect from the earth could not, in fact, have been more advantageous, as witnessed by the magnificent band of the Milky Way. Derham spoke of it twice in the *Astro-theology,* and in both cases he might as well have passed over the subject. First, in the Preface, he argued that since the Milky Way was "the fertile place of *New Stars,"* the whiteness there could not be "caused by the bare Light of the great number of Fixt Stars in that place, as hath commonly been thought, but partly by their Light, and partly (if not chiefly) by the Reflections of their Planets; which stop and reflect, intermix and blend the Light of their respective Stars or Suns, and so cause that Whiteness the *Galaxy* presents us with; which hath rather the colour of the Reflected Light of our Moon, than the Primary Light of our Sun."[40] This was sheer pedantry at best, even by the scientific standards of the early 1700's. Nor did science profit from his short, formal discussion of the Milky Way. It said, in essence, that in contemplating the incredible richness of stars in the Milky Way, one "cannot be but struck with Amazement at such a multitude of God's glorious Works."[41] The scientific part of his dictum was that the great multitude of stars caused "that remarkable whiteness there."[42]

Two decades later he was still preoccupied with the hue of that whiteness—but hardly in a constructive sense. After observing, with his eight-foot reflecting telescope, some nebulae listed in Hevelius' *Prodromus astronomiae,* the venerable canon of Windsor came to the conclusion that nebulae formed a class basically distinct from the Milky Way.[43] The latter could always be resolved into stars; the former could not. The *Nebulosae* were not, he advised the Royal Society, *"Lucid Bodies,* that send their Light to us, as the Sun and Moon." Nor were they "the *combined Light* of *Clusters* of Stars, like that of the *Milky-Way."* They were, his verdict went, *"vast Areae,* or *Regions of Light,* infallibly *beyond the Fix'd Stars, and devoid of them."*[44] In support of this conclusion, he referred to the consensus of "the Learned in all Ages (both Philosophers and I may add Divines too)" with respect to the existence of a Region beyond the fixed stars. He called it the *"Third Region* and *Third Heaven."* Whether the *Nebulosae* were "particular Spaces of Light," or "Openings into an immense Region of Light, beyond the Fix'd Stars," he left it "to the great Sagacity and Penetration of this Illustrious Society."[45] As will shortly be seen, sagacity did not necessarily mean alertness.

Frustrating as Derham's utterances were, he kept alive the discourse about the possible arrangement of stars, even though as a topic apart from the Milky Way. There was no such concern in the astronomy manual of John Keill,[46] a disciple of David Gregory and Newton, and the most ardent defender of the latter in the controversies with Leibniz. What Keill said of the Milky Way reflected in its brevity the mesmerizing inadequacy which astronomers of the time as a rule offered on the topic:

> The *Galaxy* or *Milky way* is also to be reckoned among the Constellations: This is a broad Circle of a whitish Hue, like Milk; in some places it is double, but for the most Part it consists of a single Path, and goes round the whole Heavens. The Great *Galileus* with his Telescope, discovered that the Portion of the Heavens which this Circle passes thro', was every where filled with an infinite multitude of exceeding small *Stars,* which tho' they cannot, by reason of their smallness, be seen distinctly by the naked Eye, yet with their Light they all combine to illustrate that Region of the Heavens where they are, and diffuse thro' it a shining whiteness.[47]

A hundred years after the *Starry Messenger,* Keill's dictum was the classic picture of a perfect standstill, initiated by the "Great Galileus" himself.

Edmond Halley, scientifically the most prominent figure of Newton's "principate," was also the one who might most naturally have become the discoverer of the Milky Way's true shape. His scientific career, as shown in the previous chapter, began with a striking comparison of the Magellanic Clouds with the Milky Way. His interest in the nebulae was still strong thirty-six years later, when he communicated to the Royal Society a paper on "An Account of several Nebulae or lucid Spots like Clouds, discovered among the Fixt Stars by help of the Telescope."[48] Some of them had a central star in them; others did not. About these latter, Halley was quick to remark that they were "the Light coming from an extraordinary great pace in the Ether; through which a lucid Medium is diffused, that shines with its own proper Lustre."[49] He thought this might help to resolve the objection often raised against the Mosaic account of creation, in which light came before stars existed. Scientifically more interesting was the conclusion of his short paper wherein he pointed out the immense size, comparable to the whole solar system, which these nebulae must occupy. "In all these so vast Spaces it should seem that there is a perpetual uninterrupted Day, which may furnish Matter of Speculation, as well to the curious Naturalist as to the Astronomer."[50]

The ring of the expression "uninterrupted day" clearly echoed Fontenelle, and was worthy of further speculation. Truly speculative, in fact, were the two papers which Halley read, in 1720, before the Royal Society on the arrangement of stars in the universe and on their total light[51]—a topic most germane to the question of nebulae and the Milky Way. In those two papers, Halley considered the total optical effect of rows of stars placed behind one another in endless concentric shells. He clearly recognized that the logical consequence of such a model was a sky that blazed at every point with the brilliance of the sun. His solution to this problem was to enclose the paradox of concentric shells of stars in the shell of a still unexplained phrase.[52] No less puzzling was his failure to consider, in relation to his model, that pale whitish band, the Milky Way. Instead of circular shells he should have envisaged two parallel planes and only a finite number of rows of stars.

Puzzlingly, the concepts of Halley's paper failed to penetrate the minds of those present, among them Newton, who was in the chair. The optical paradox of an infinite universe as initially described by Halley occasioned no comment, demonstrating the prevailing faith that nothing could be wrong with an infinite, homogeneous universe. This now served as the clearest insight against which no objections would be considered. A firm belief in that insight could only turn with instinctive suspicion from a vast cosmic singularity like the Milky Way. But the most immediate reason for the neglect of the Milky Way in Newton's "principate" lay in the intoxicating potentialities of the *Principia* for further research in celestial dynamics.

Such research presupposed a foundation—specifically, a list of the positions of stars, observed with greater accuracy than ever. The Newtonian Renaissance of British astronomy ended, in part, with a careful cataloguing of details with no reference to their totality or to any major groupings of them. The work was carried out largely by the first Astronomer Royal, John Flamsteed, a fearless critic of Sir Isaac. Although he preceded Halley in that office, Flamsteed is mentioned here because the second and much enlarged edition of his monumental *Historia coelestis britannica*[53] was published posthumously in 1725, and the twenty-six magnificent plates of his *Atlas Coelestis*[54] not until 1753. Although Flamsteed reserved a book-length Preface in his *Historia coelestis* for the history of astronomy, with him the Milky Way seemed to be without a memorable past and without a present. An assiduous and dedicated worker, the first Astronomer Royal's achievements were not made in the field of speculations and insights. Nor was he greatly interested in conceptual innovations, as may be seen from his scornful remark on attempts aimed at introducing the new, vernacular nomenclature in star catalogues. He insisted on giving the names "secundum Ptolemaeum,"[55] an aside that also applied, ironically, to his neglect to investigate the nature of that band where the stars were most numerous. He was a celestial forester who knew each tree, but never pondered over the forest itself.

The Milky Way did not benefit from the work of James Bradley,[56] Savilian Professor of astronomy at Oxford from 1721, and, from 1742, Halley's successor in the office of Astronomer Royal. Bradley was the last notable figure of British astronomy in Newton's time and also a perfect example of the "distraction" triggered by Newton's *Principia.* Bradley's discovery of the aberration of light caused by the motion of the earth was made in 1729, during his continual attempts to detect stellar parallaxes, the experimental touchstone of heliocentric theory. Equally indicative of the prevailing "distraction" of astronomers was Bradley's discovery of nutation and his work on Jupiter's moons to enable the determination of longitude from Greenwich, Lisbon, and New York. Bradley also performed valuable work by improving the theory and use of optical instruments. Similar in purpose was the standard 18th-century English treatise on optics, Robert Smith's *Compleat System of Opticks,* written, in 1738, principally for astronomers.[57] In its two volumes nearly every question of astronomy found a forum, but the Milky Way was treated most cursorily: "The Milky Way, which had puzzled the ancient philosophers for many ages, was found at last to be nothing else but a prodigious number of very minute stars, so close to one another that the naked eye can only perceive a whitish mixture of their faint lights. This was Galileo's discovery, who found also that those faint stars, which Astronomers call Nebulosae, appeared through his telescope to be small clusters of very minute stars."[58] It was from Smith's two volumes, years later, that William Herschel taught himself the intricacies of astronomical optics. Judging by the telescopes he built, those volumes must have been extremely instructive. He could learn almost nothing there of the Milky Way.

It was not books by English authors that, from the 1730's on, contained the lion's share of advances in planetary dynamics, the principal preoccupation and "distraction" of astronomers for the rest of the century. The names of Leonhard Euler, Alexis Claude Clairaut, and Jean Le Ronde d'Alembert, to mention only the principal ones of the mid-18th century, clearly demonstrate the leadership of the Continent. Theirs were extraordinary efforts to come to grips with the three-body problem, with the perturbations of Jupiter and Saturn, and with some apparent anomalies in the motion of the moon, to say nothing of many other minor, though almost equally perplexing details of the subject. Stars were at best discussed incidentally. A classic case is Euler's discussion, in 1750, of the various degrees of the light of the sun and of the stars. The work showed his mastery of the mathematics of the question, but was utterly devoid of any cosmological consideration, although few topics could have been more germane. The stars were foster children even in the three volumes of his *Lettres à une princesse d'Allemagne,*[60] in which the aging Euler waxed garrulous on a wide assortment of topics, ranging from sense perception to universal attraction. The Milky Way failed to stir his interest to the end.

D'Alembert found time for the Milky Way only when the *Encyclopédie,* of which he was co-editor, required an article on the subject.[61] He wrote approximately 300 words on it, which were published in 1757, and covered the etymology of the term, the position of the galaxy among the stars, and the statement that its whiteness was the optical effect of many small stars. This was hardly in keeping with the "revolutionary" character of a work that was to make history. "It is pretended," d'Alembert added dubiously, "that the whiteness of nebulae is also produced in the same manner." His final remark on the topic will be discussed later.

Second-ranking figures of mid-18th-century French astronomy repeated no more than the same platitudes about the Milky Way. Jacques Cassini, successor to his father, Dominique, at the Observatory of Paris, simply mentioned the Milky Way in his famous textbook with the remark that nebulae were also resolvable into stars.[62] Nicolas de Lacaille, perhaps the most industrious French astronomer, earned fame with his ephemerides, star catalogues, and analysis of the refraction of starlight in the atmosphere. A man like Lacaille, known for his pursuit of accuracy, could only be dismayed by the fact that one knew nothing, as he put it in his *Leçons élémentaires d'astronomie géométrique et physique,* about the position in depth of stars in space. Nevertheless, this was hardly a good reason to say nothing of the Milky Way.[63] Jean Jacques Dortous de Mairan is best remembered as Fontenelle's successor at the Académie des Sciences. As such, he delivered the eulogy on Halley, in which, understandably, he made no mention of Halley's brief references to the Milky Way, nor of Halley's efforts to solve the optical paradox of an infinite universe.[64] All this reflected de Mairan's own disinterest in the Milky Way. In his great work on the *aurora borealis,*[66] de Mairan spoke only briefly of the Milky Way, and almost stopped, as Halley had, at comparing the whiteness of nebulae with that of the Milky Way: "The Milky Way, viewed with the telescope, appeared to me in a number of places nothing else but the texture of similar luminous spaces interspersed with stars, just like Orion."[66] He had already made the suggestion that those "luminous spaces" could very well be replicas on a major scale of the "atmosphere" of the sun, easily observable during eclipses.[67]

Distinctly more could be expected on the Milky Way from Pierre Louis Moreau de Maupertuis. He earned fame with his proof of the oblateness of the earth, which he accomplished by measuring the length of one degree of the meridian in Lapland in 1736–1738. He was also the author of a *Discours sur les différentes figures des astres,*[68] which did go beyond the planetary world although dealing largely with the "stars" around the sun—that is, the planets. Curiously enough, Maupertuis, after devoting the first five chapters to the planetary system, passed directly to the nebulae.[69] What he said was not without interest. He speculated on their formation from a luminous substance, and on two grounds: one was the universal effectiveness of gravitational attraction, the other a discovery which is best rendered in his own words: "Famous astronomers having applied themselves to observing those celestial appearances which are called nebulae and which were formerly explained as the fusion of the light of many small stars very close to one another, and having made use of stronger lenses, have discovered that at least several of these appearances not only were not caused by these heaps of stars as it was imagined, but that they did not even contain any; they did not seem to be but some great, luminous, oval air, or a light brighter than the rest of the heavens."[70] Maupertuis named Huygens and Halley, thus adding by implication a nuance not really contained in their dicta. As to the shapes of the nebulae, he clearly wished to see in some of them replicas of the flattened disk of the solar system on a much larger scale. He suggested that they were probably at various distances from us, and attributed their different "appearances," at least in part, to the fact that their "flattened ovals" might have equatorial planes lying in a great variety of directions with respect to us. His insistence that the variety of nebulae mirrored the variety everywhere evident in nature was a distinct anticipation of a favorite theme of Herschel. But, for Herschel, this and similar considerations about the nebulae always had a direct bearing on the Milky Way. They were, in fact, rooted in his analysis of it. Maupertuis, however, spoke of the nebulae without saying a word about the Milky Way. As his *Essai de cosmologie* shows, he was a cosmologist who thought that the job could be completed with only "last-minute" references to the stars.[71]

This was scarcely philosophical in the sense of being critically intelligent, as the term "philosophe" was now used in France. The most renowned figure of the "philosophe" movement, Voltaire, also looked for scientific laurels, at least at the beginning of his career. The spread of Newtonianism in France owed much to Voltaire's facile pen, but in his exposition of the Newtonian system, or in his related writings on science and on its philosophy, the universe seemed to be confined to the planetary system. In Voltaire's *Elemens de la philosophie de Newton,* the planets were followed by the comets, the topic of the penultimate chapter.[72] Although the last chapter concerned universal attraction, the reader was not carried beyond the comets into the realm of stars, but was instead instructed about the validity of that attraction between the smallest parts of matter.[73] Voltaire's two discussions of the sky in his *Dictionnaire philosophique* offered nothing on the stars, but were full of venom on the credulity of ancients and of Christians about a solid firmament.[74] In the same work his discussion of the infinity of the world also showed the same instinctive bent of his mind away from science and toward metaphysics. Only a brief paragraph of that discussion pertained to the real universe of stars and planets, in the form of questions that displayed no depth.[75] His *Lettres philosophiques* contained only one reference to the stars in relation to the precession of equinoxes,[76] the explanation of which was a signal triumph for Newton's terrestrial and planetary dynamics. Voltaire's real star was none other than Sir Isaac Newton, a figure powerful enough to distract even Voltaire's mind from the realm of stars.

In England, where criticism moved on paths less abrasive than Voltaire's cutting wit, the topic of the Milky Way was mentioned only twice in that mirror of the times, the thousands of pages of *The Spectator,* but in neither case for its own sake. In discussing the benefits, power, and purposefulness of eyesight, Steele, the co-editor, described how the "sight can travel even beyond the planets, to the stars," and added: "But the Enquiries of the *Sight* will not be stopp'd here, but make their Progress through the immense Expanse to the *Milky Way,* and there divide the blended Fires of the *Galaxy* into infinite and different Worlds, made up of distinct Suns, and their peculiar Equipages of Planets; till unable to pursue this Track any farther, it deputes the Imagination to go on to new Discoveries, till it fill the unbounded Space with endless Worlds."[77] The famed astronomers of Addison's and Steele's England said in essence nothing more and at times far less.

The topic of the Milky Way would have admirably served F. Algarotti, author of an explanation of "Newtonianism" for ladies,[78] since the work emphasized optical phenomena. But Algarotti was only a skillful imitator of Fontenelle, who had already pre-empted the topic of the Milky Way. Algarotti offered a story verging on the anecdotal. It was about some Englishmen "who searched for the Northwest Passage, and were forced to spend the winter on an island not much more to the north than London. Their shelter became a block of ice, and so did their vessel, and even themselves, or almost. Even their strongest wine had to be chopped up with an axe."[79] Optics and the Milky Way entered at that momentous point. Algarotti had already said a good deal about the change in the refractivity of air caused by temperature and humidity. Now in that area where every drop of humidity froze itself out of the air, refractivity produced curious phenomena: "the moon appeared to them as a very long and compressed oval, the sun twice as wide as long, often the air was so pure that they spotted two thirds more stars than one usually could see under milder skies; and they recognized without a recourse to spyglasses that the Milky Way was an immense 'ant hill.' In that land there was no need of a Democritus to divine this amidst the dreams of ancient philosophers, or of a Galileo to verify it with the telescope."[80] In all this, the element of suspense outweighed the scraps of information which did not even recount, to the probable chagrin of Algarotti's ladies, the deliverance of those Englishmen turned into ice.

Mental inertia about the Milky Way was also exemplified in the article on it in that monumental enterprise known as the *Universal Lexicon.*[81] It was to contain, as its subtitle stated, all "the sciences and arts which up to now have been invented and improved by human mind and ingenuity." Ingenuity *(Witz)* meant, however, also humor and when, in 1739, the twenty-first volume carried the article, "Milchstrasse," there was something humorous about it. The 200 or so words devoted to the Milky Way recalled its position in the sky, the old myths and speculations about it, and closed with the remark: "What occurred to Democritus and Ptolemy was demonstrated by Galileo in the *Starry Messenger* as soon as man began to look at the sky with a telescope, namely, that the Milky Way was to be counted among the stars because it consists of an innumerable quantity of small stars."[82] Actually, the article was original only in its reporting of some mythological details about the Milky Way. Its scientific content was borrowed from the first volume of the *Vollständiges Mathematisches Lexicon,* published five years earlier.[83] There, too, Ptolemy's *Almagest* was referred to by chapter and verse, in clear proof of how errors could linger undetected.

A mitigating circumstance, in this case, is the fact that, in the 1730's, one could learn very little about the Milky Way, let alone its history, from contemporary writings. An informative dissertation on the Milky Way, dated 1713, which will be discussed later, already gathered dust on the few shelves it had reached. The folio volumes of Gassendi's *Syntagma philosophicum,* in which the history of the question was given in sufficient detail, had already become a fossil. Neither the discovery of Galileo, nor the penetrating remarks of Gassendi and Guericke, were enough to generate historical consciousness of the Milky Way. It was not to have a history—or rather historiography. Lexicographers and essayists in the 1730's had no astronomer of standing to consult on the subject. The young genius, Tobias Mayer, the future glory of Göttingen, was still in his teens. Later, he was to be preoccupied with the "Newtonian" aspect of astronomy. The vistas were no brighter farther north, in Copenhagen. There Peder Horrebow had already made a notable though not particularly enduring name for himself with his *Copernicus triumphans* (1727), or the claim that he had measured the parallax of stars. In his textbook on astronomy, the *Clavis astronomiae* (1725), there was no discussion of the starry realm, and consequently no clues to the Milky Way.[84]

In Stockholm, where the misty skies allowed an even rarer view of the Milky Way, matters seemed to become clearer at first glance or, more typically, on the basis of some secondhand information. The "first glance" can be readily reproduced:

> The common axis of the sphere or starry heaven seems to be the galaxy, where we perceive the greatest number of stars. Along the galaxy all the vortices are in a rectilinear arrangement and series, and cohere as to their poles; in like manner, they are there more intimately associated, and have spires of greater curvature. The other solar or stellar vortices afterwards proceed from the axis, and are bent in different directions; but nevertheless all have reference to that axis. This is a consequence of the preceding observations. For the greater the number of active centres in the same space, the closer and more interior is the association of the spirals. Their greatest number is in the milky way, and there also their reciprocal conjunction is the strongest. For this reason the ancients believed the palace and hall of their deities to be there. When the gods ascended into heaven, they went by this path and across this arch, and each passed to his own halls and his own kingdom. For here lies the chain and magnetic course of the whole of our sidereal heaven.[85]

The least one can say about this passage is that it should give second thoughts to anyone seeking a clear understanding of it. It was not the only enigmatic paragraph that came from the pen of Emanuel Swedenborg, assessor at the Board of Mines since 1724. What Swedenborg, the mystic, wrote is of no direct concern here, although it should be borne in mind that mystical proclivities denote a type of thinking rather different from the type useful in science. The two can, of course, co-exist in one person with no harm to his science, as the case of Pascal shows all too well. Swedenborg's studies in civil and mining engineering have a clear, factual air. The same cannot be said of the whole of his *Principia rerum naturalium,* of which the foregoing passage forms a part. The *Principia* was the first of the three volumes of his *Opera philosophica et mineralia,* published in Dresden in 1734. The second and third volumes contained valuable information on the mining and processing of various metals, chiefly iron and copper, and were received with considerable interest. The first volume was a different matter. Its First Part came from Descartes' *Principia,* but changed Descartes' clear though erroneous statements on the successive differentiation of matter into a painfully obscure set of propositions. The Second Part of Swedenborg's *Principia* owed much to Musschenbroek's experiments on magnets, with Swedenborg's own observations thrown in for good measure.

As the passage quoted indicates, Swedenborg interpreted the galaxy as the axis of a "magnetic sphere." The meaning he gave to the idea of "magnetic spheres" forms, therefore, the logical clue to the true aspect of the picture he painted of the Milky Way. In chapter I, Second Part of the *Principia,* devoted to the topic of magnets, Swedenborg discussed at length the "lines of forces" emanating from the poles of magnets. These lines had been observed for some time by means of iron filings. He called those "lines" the paths of the magnetic effluvium, or, in a more Cartesian vein, the set of chains of "vorticles." The totality of these chains, which had their origin in the two ends of a bar magnet, was a "sphere." "By a sphere," he wrote, "we understand the entire combination and series of vorticles connected with the figure extending round the magnet."[86] In the same breath he identified the axis of that sphere with that of the bar magnet: "The axis of the sphere, or the common axis of the vorticles, lies parallel with the common axis of the element itself, so as to be exactly accommodated to it."[87] But all this represented only the natural or normal state: "Inasmuch as the motion of all the vorticles has its termination, and ends in a common arrangement of particles or axis, it will consequently, if free, always be directed and fall into this arrangement by common effort."[88] External factors or constraints could readily change not only what Swedenborg persistently called the spherical form of the set of chains of vorticles, but also the straightness of the common axis: "The axis of the whole sphere, or the common axis of the vorticles, may be curved in a similar manner; and together with it the sphere itself, from one pole to another, may also undergo any change; or what amounts to the same, the sphere itself may from one pole to the other undergo a change, and in accordance with that change the polar axis will be bent."[89]

In this statement one may readily catch a glance of the apriorism of Swedenborgian or rather Cartesian physics, replete with declarations about the physical reality with no basis in observational evidence. The inversion of the proper sequence can be seen in Swedenborg's dictum about the mutual relationship of a sphere and axis. "The axis derives its origin from the sphere."[90] Therefore, he also had to state that "if the sphere form an axis according to its figure, then if the figure of the sphere be changed, the axis also is changed or curved and contrariwise."[91] Now, while it was, for example, easy to produce a change in the lines of iron filings around a bar magnet by means of another magnet, a change in the axis of the element could not be observed.

Still less was there any trace of a disk-shaped set of "vorticles" around any magnet. Nor can such an interpretation be given to Swedenborg's remark "that the axis of the vorticle is equal to its entire breadth; it is not formed by merely a brief motion from its centre, but extends throughout the whole of the breadth occupied by the vortical motion as in fig. 1 [Illustration VII] where a vorticle is represented by a circular form."[92] The figure showed a larger set of concentric magnetic lines around a small, brick-shaped magnet, and a narrow, tightly wound spiral which intersected the small magnet and the circles as a common diameter. Neither this nor the other figures offered by Swedenborg suggested a disk-shaped set of "vorticles" or lines of force—a basic requisite, if he were to be considered the man who anticipated the ideas of Wright, Kant, Lambert, and Herschel.

Swedenborg was in a class of his own, as shown by his curious manipulation of the concept of an axis. It revealed a type of thinking ready to fashion the universe from preconceived notions and imaginations. In that first chapter where he spoke of the "magnetic sphere with its vorticles" as "type and small image of the starry heaven,"[93] the axis of the sphere was almost invariably spoken of as a straight line or a bent modification of one, either in the form of a regular arc or of a modified, narrow spiral. From this, one could derive only a completely unsatisfactory explanation of the Milky

Way—and this is what Swedenborg did in the Third Part of the *Principia* in the passage previously quoted. It permits only one interpretation; namely, that he took the Milky Way for the straight, column-like axis of a cosmic bar magnet. There is not even a hint that he at least pictured it as a half ring, an equally unsatisfactory hypothesis. Had he spoken of a ring, he still could not have claimed originality for what had already been proposed repeatedly. The scientifically worthless nature of Swedenborg's statement on the Milky Way can also be inferred from his readiness to recall fondly an old myth in which the Milky Way was a road. No road, straight, curved, or spiraling, ever bore any resemblance to a disk, or could serve as a path to understanding the Milky Way.

Swedenborg's contemporaries, who did not see him through a haze of legends and national hero-worship, praised the *Principia* for its information on iron and copper. The recension of the work in the *Nova Acta Eruditorum*[94] commended his speculations on God, the infinite, and the "physical influx" by which the soul operated on the body. No mention was made of the Third Part of the *Principia,* the least worthy of consideration. Typically enough, Kant, who was familiar with both Swedenborgs, the scientist and the visionary, lampooned the latter in a booklet.[95] He never saw the necessity of vindicating the originality of his own speculations on the Milky Way in comparison with Swedenborg's dictum.

The first man of science of any serious standing to consider Swedenborg as the originator of the disk theory of the Milky Way was, rather naturally, his countryman, Magnus Nyrén, first astronomer at Pulkovo. In an essay on Swedenborg's cosmology,[96] he quoted the statement about a magnetic axis corresponding to the whole circle of vortical action, and amplified it in the sense that, according to Swedenborg, the axis was a cylinder encompassing the whole sphere of the lines of force of a magnet. He then concluded that Swedenborg pictured the Milky Way "as a cross section of the axis cutting perpendicularly through the middle of the visible stellar realm and this is what we should call an equatorial cross section of the celestial sphere."[97] Such was a strange distortion of the meaning of a cylinder by flattening it into a thin disk, a feat which not even the most unpredictable Swedenborg tried to perform. Characteristically, Nyrén neither unconditionally stated the correctness of his interpretation of Swedenborg's words on the cylinder, nor did he wish to abandon it: "But, since as it seems to me, no other meaning can be given [to Swedenborg's statement] than that the Milky Way is the equatorial cross section (zodiac) of our whole visible sky, one cannot but accord priority to Swedenborg on the interpretation of the galactic system of stars."[98] Truly a strange sleight of hand, resting solely on the belief that Swedenborg had to be right.

The same characterizes the evaluation of Swedenborg's cosmology by Svante Arrhenius,[99] Sweden's first Nobel laureate. The occasion was the return from England of Swedenborg's remains, aboard the Swedish battleship *Fylgja* in 1908. But what Arrhenius said as a cosmologist in highly publicized books, failed to equal in value and rigor what he accomplished as a physical chemist. Thus he found it strange that Swedenborg did not compare the Milky Way to a ring-shaped magnet to explain at least the visual image of the Milky Way.[100] No touch of criticism was evident in the approving tone with which Arrhenius recalled Nyrén's interpretation of the axis as an equatorial plane.[101] Equally incredible was Arrhenius' phrase, "the suns are arranged around the milky way," in which he summed up the alleged anticipation by Swedenborg of the ideas of Wright, Kant, and Lambert. His phrasing,[102] an unwittingly good rendering of Swedenborg's worthless explanation of the Milky Way, was adopted by Knut Lundmark, a prominent figure in the animated discussion about the status of nebulae in the 1920's. According to him, Swedenborg thought that the stars "are arranged around the Milky Way."[103] Arrhenius and Lundmark were not the last of Swedenborg's respectable champions who betrayed themselves by careless expressions. More recently, Sygne Toksvig spoke of Swedenborg's *Principia* as a work which "belongs largely to the pure realm of mathematics."[104] The fact is that it contains not even elementary algebra. But its Third Part offers sufficient foretaste of the pseudo-scientific mysticism which ensnared Swedenborg during the second half of his life. In that respect alone, as will be clear in the next chapter, was Swedenborg a forerunner of Thomas Wright of Durham.

The era covered in this chapter provides three more details that are equally symbolic of Wright's "discovery" of the Milky Way in 1750. During the previous six and a half decades, only two small booklets were written on the Milky Way, and they fell almost immediately into oblivion. Such a fate was well deserved in the case of the eight-page dissertation by Johannes Fendt, which he defended at the University of Copenhagen in 1707, under the aegis of Georg Christian Bang.[105] In the first part of his treatise Fendt disproved Aristotle, by that date an unremarkable accomplishment. In the second part, he tried to prove that the Milky Way was beyond the realm of the zone of "meteors"—that is the zone of fire—and that it constituted part of "the stellar realm created on the fourth day."[106] The list of those whom he quoted in support was exceedingly meager. He even failed to mention Galileo in his eagerness to show, in a truly Aristotelian manner, that "the Milky Way was a corporeal substance endowed with the essential parts of substance and form, and with the properties of quantity, quality, motion, and place."[107] He showed himself no more modern by using the word *astrologica* instead of *astronomica* in the title of his dissertation, and by closing it with Brahe's dictum on the Milky Way.

On the other hand, had the *Dissertatio physico-astronomica de galaxia* by Johann Paul Schneider[108] been remembered, it would have effectively kept on record the best statements on the galaxy between Galileo and Thomas Wright. Schneider read Stöffler, Riccioli, and Gassendi, who familiarized him with the opinions of classical authors which he discussed, somewhat verbosely, in more than half of his essay of thirty pages. His principal targets were some latter-day defenders of Aristotle, such as Chiaramonti and Fortunio Liceto. He paraphrased the opinion of the overwhelming majority of moderns about the galaxy in a phrase which lacked any hint about the measure of depth in which the stars could be located there.[109] The constituent stars of the galaxy were, for Schneider, exceedingly small, and he followed Gassendi's lead in his reluctance to discuss the cause of that peculiarity of the starry realm.[110] But Schneider discussed the question of the possible presence of nebulous matter amidst the stars of the galaxy, expressing sympathy with the dicta of Brahe, Longomontanus, and others. The most significant part of his essay was its concluding page, a quotation in full of Fontenelle's portrayal of the Milky Way, concerning which professional astronomers should have spoken more fully.

Rare as Schneider's dissertation is today, copies of it had probably reached other German universities. Its contents might have even come to the attention of Christian Wolff, the most influential, controversial, and encyclopedic figure in German academic life during the 1720's and 1730's. In his famed "Deutsche Physik,"[111] which dealt with the whole universe and particularly with the purposefulness of all its parts and arrangements, there was no reference to the Milky Way. Wolff's erudition was more apparent in his descriptive account of the physical universe,[112] where he briefly pictured the Milky Way in Galileo's own words. Wolff's main interest was not, however, such as to be kindled by reading Schneider's historical material. More difficult to explain is the absence of any reference to Schneider's dissertation in the first history of astronomy of any consequence, published in Wittenberg in 1741. In that work of more than 600 pages by J. Weidler,[113] the *via lactea* was mentioned only three times (Democritus, Aristotle, and Galileo),[114] but the topic itself was not found worthy of listing in a lengthy subject index. For another century the Milky Way remained a topic without a monograph, and its history so remained for yet another.

The detail that should most appropriately close this chapter is connected with the year 1746, when Keill's *Introductio ad veram astronomiam* was translated into French by Pierre Charles Lemonnier, an industrious but unoriginal astronomer, who added the following words to Keill's description of the Milky Way as a congeries of stars: "This common [vulgaris] opinion has been repeated in an infinite number of places; but for all that it has not been adopted by all astronomers; the reason being that by using long telescopes of 15 to 25 feet, one does not discover there more stars than in other regions of the sky. One merely notes in the Milky Way a whiteness that could originate from a material of which the nebulae are composed."[115]

Lemonnier was not alone in this reservation, but there was something curious in d'Alembert's reference to him in an article from which lack of space should have eliminated all but the most essential points. Lack of information about the history of the question had, however, its own rules of selection. The galaxy contained nebulous matter, although not in the sense assumed by Lemonnier; his opinion harked back to the past—as did most of his views. In Delambre's voluminous *Histoire de l'astronomie au dix-huitième siècle,* [116] which contains no reference to the Milky Way, Lemonnier is characterized as one always slightly behind the times. This was especially true of the discovery of the Milky Way. Possibly no other major discovery came as late as Wright's book in 1750, or a year later than the flash of insight that crossed Lambert's mind. When the latter spoke of it fifteen years later, he voiced his puzzlement that Sir Isaac Newton had not thought of it in the midst of his lengthy speculations on the mutual attraction of all celestial bodies. Universal gravitation evidently was a topic attractive to the point of distraction.

References

[1] [Bernard de Fontenelle], *Entretiens sur la pluralité des mondes* (Paris: chez la Veuve C. Blageart, 1686). It was simultaneously published in Lyons (chez T. Amaulry). The dual publication was authorized by the royal permit issued on January 9, 1686. For details on the book's popularity, see the Introduction to the critical edition by R. Shackleton (Oxford: Clarendon Press, 1955).

[2] This quotation and those following are from the excellent English translation, *Conversations on the Plurality of Worlds,* by a Gentleman of the Inner Temple (London: 1767), pp. 272–74. (For the passage in the original, see pp. 317–22).

[3] *Ibid.,* p. 274.

[4] *Ibid.,* p. 269.

[5] *Ibid.*

[6] If there was any loving interest given to any topic in Everhard G. Happel's three-volume cosmography, it was not the Milky Way. He granted it only seven lines although herrings, for instance, were discussed throughout a dozen pages. See vol. 1, p. 3 and pp. 266–79, in his *Mundus mirabilis tripartitus oder Wunderbare Welt in einer kurtzen Cosmographia fürgestellt* (Ulm: M. Wagner, 1687–89). Happel merely stated that the Milky Way consisted of many small stars as proven by the telescope. To this, only the opinions entertained in antiquity were added by Vincenzo Coronelli, professor of geography at Padua, in his *Epitome cosmografica o compendiosa introduttione all' Astronomia, Geografia et Idrografia . . .* (Colonia: ad istanza di Andrea Poletti in Venetia, 1693), pp. 156–57. A year later, Johann Sperlette, rector of the Gymnasium Fridericano-Gallicum in Berlin, offered a few perfunctory lines on the Milky Way in his *Physica nova, sive philosophia naturae* (Berlin: impensis Johann. Michael Rüdigeri, 1694), p. 172, a work in which he claimed to explain all phenomena of nature as mechanical effects. The far more illustrious Hooke kept to his old, commonplace utterances on the Milky Way, when he remarked in a lecture given in February 1792 on telescopes and microscopes that Galileo detected with his telescope "the *Galaxia* to be an infinite *Congeries* of small Stars; as also the cloudy Stars, to be of a like Composition." See *Philosophical Experiments and Observations of the late Eminent Dr. Robert Hooke, S.R.S. and Geom. Prof. Gresh. and Other*

Eminent Virtuoso's in his Time, publish'd by W. Derham, F.R.S. (London: printed by W. and J. Innys, 1726), p. 258.

[7] See Andrew Motte's translation of the *Principia,* revised and annotated by Florian Cajori, *Sir Isaac Newton's Mathematical Principles of Natural Philosophy and his System of the World* (Berkeley: University of California Press, 1962), vol. 2, pp. 596–97.

[8] *Ibid.,* p. 596.

[9] *Ibid.*

[10] *Ibid.*

[11] *Ibid.,* pp. 596–97.

[12] *Ibid.,* p. 525.

[13] First published as *Eight Sermons Preached at the Honourable Robert Boyle Lectures in the First Year MDCXCII* in London in 1693. It was reprinted under the title, *Matter and Motion Cannot Think: or a Confutation of Atheism,* in *The Works of Richard Bentley,* edited by Alexander Dyce (London: Francis Macpherson, 1838), vol. 3, pp. 1–200.

[14] See especially the first paragraph of the first of his four letters to Bentley, in *The Works of Richard Bentley,* vol. 3, p. 203.

[15] Of Bentley's four letters only one is extant. For its text, see *The Correspondence of Isaac Newton, Volume III, 1688–1694,* edited by H. W. Turnbull (Cambridge: The University Press, 1961), pp. 246–53.

[16] See Newton's second letter, in *The Works of Richard Bentley,* pp. 208–09.

[17] In his Eighth Sermon, *The Works,* p. 174.

[18] This translation went through a fourth edition by 1718. Antoin le Grand, who published Clarke's annotated translation with his own notes *(Jacobi Rohaulti Tractatus physicus* [Amsterdam: apud Joannem Wolters, 1708]), also left the Milky Way unmentioned. The same was true in the third edition of the translation into English of Clarke's translation, *Rohault's System of Natural Philosophy illustrated with Dr. Samuel Clarke's Notes,* done into English by John Clarke (London: printed for James, John and Paul Knapton, 1735). The Cartesian silence in respect to the Milky Way was not pointed out in the most acid criticism of Descartes' world view, G. Daniel's *Voyage du monde de Descartes* (Paris, 1691).

[19] An English translation immediately followed, under the title, *The Celestial Worlds Discover'd* (1698), reprinted in 1968 (London: Frank Cass & Co., Ltd.)

[20] *Ibid.,* pp. 55, 71, 73 and 89.

[21] *Ibid.*, p. 147.

[22] *Ibid.*, p. 3.

[23] *Astronomiae physicae et geometricae elementa* (Oxford: e Theatro Sheldoniano, 1702), pp. 161-62; quotations are from the second, revised edition of the English translation, *The Elements of Physical and Geometrical Astronomy,* by Edmund Stone (London: D. Midwinter, 1726).

[24] *Ibid.*, p. 293.

[25] *Ibid.*, pp. 289–90.

[26] *Ibid.*, p. 290.

[27] *Ibid.*

[28] *Ibid.*, p. 289.

[29] *Histoire de l'Académie Royale des Sciences. Année MDCCV. Mémoires de mathématiques et de physique* (Paris: chez Jean Boudot, 1706), pp. 14–24, "Reflexions sur les Observations des Satellites de Saturne et de son Anneau," par M. Cassini. For quotation, see p. 18.

[30] See the English translation, *Astronomical Lectures* (London, 1715), p. 38.

[31] *Ibid.*, p. 41.

[32] *Ibid.*, p. 42.

[33] *Ibid.*

[34] London: printed for J. Senex, 1717.

[35] *Ibid.*, p. 80. Some of them were discovered by Halley, whose paper on the subject is discussed later in this chapter.

[36] The first edition (1715) was immediately followed by the second, London: printed for W. Innys, 1715.

[37] See *ibid.*, Book 3, chap. 3, "Of the nice Proportions of the Distances of the Heavenly Bodies," pp. 56–61.

[38] *Ibid.*, p. 57.

[39] *Ibid.*

[40] *Ibid.*, p. XLVI.

[41] *Ibid.*, Book 2, chap. 1, "A General View of the Numbers of the Heavenly Bodies," pp. 30–32; for quotation, see p. 32.

[42] *Ibid.*

[43] "Observations of the Appearances among the Fix'd stars, called Nebulous Stars," *Philosophical Transactions* 38 (1733–34): 70–74.

[44] *Ibid.*, p. 73.

[45] *Ibid.*, pp. 73–74.

[46] *Introductio ad veram astronomiam* (Oxford: e Theatro Sheldoniano, 1718), p. 53.

[47] Quotation is from the English translation, *An Introduction to the True Astronomy* (London: B. Lintot, 1721), p. 50.

[48] *Philosophical Transactions* 29 (1714–16): 390–92.

[49] *Ibid.*, p. 390.

[50] *Ibid.*, p. 392.

[51] "Of the Infinity of the Sphere of Fix'd Stars," and "Of the Number, Order, and Light of the Fix'd Stars," *Philosophical Transactions* 31 (1720): 22–24 and 24–26.

[52] For details on this, see my *The Paradox of Olbers' Paradox* (New York; Herder and Herder, 1969), p. 79.

[53] London: typis H. Meere, 1725. Since relations were rather tense between Flamsteed and Halley, the latter's name was omitted at the head of the list of stars of the Southern Hemisphere.

[54] The path of the Milky Way across the stars is not indicated on any of the plates.

[55] *Historia coelestis*, p. 49. There is no reference to the Milky Way in Francis Bailey's *An Account of the Rev. John Flamsteed, the First Astronomer-Royal, compiled from his own manuscripts, and other authentic documents never before published to which is added his British Catalogue of Stars corrected and enlarged* (London: Printed by Order of the Lords Commissioners of the Admiralty, 1835).

[56] See his *Miscellaneous Works and Correspondence,* edited by S. P. Rogaud (Oxford: University Press, 1832).

[57] Cambridge: University Press, 1838.

[58] *Ibid.*, vol. 2, p. 447.

[59] "Reflexions sur les divers degrés de lumière du soleil et des autres corps célestes," pp. 130–52, in *Leonhardi Euleri Commentationes opticae,* vol. 1, edited by David Speiser (Thurn: Orell, Füssli, 1962).

[60] First published in three volumes in 1768–72.

[61] *Encyclopédie ou Dictionnaire raisonné des sciences, des arts, et des métiers,* vol. VII (Paris: chez Brisson etc., 1757), p. 429, col. 1.

[62] *Elemens d'astronomie* (Paris: Imprimerie Royale, 1740).

[63] The second and revised edition of the work (Paris: chez H. L. Guerin & L. F. Delatour, 1755) was not in this respect an improvement on the first, dated 1746.

[64] His eulogy of Halley is reprinted in the *Correspondence and Papers of Edmond Halley,* arranged and edited by Eugene F. MacPike (Oxford: Clarendon Press, 1932), pp. 15–27.

[65] *Traité physique et historique de l'aurore boréale* (Paris: de l'Imprimerie Royale, 1733).

[66] *Ibid.,* p. 249.

[67] *Ibid.,* p. 245. The second, revised edition of the work (1754) contained no changes in this connection.

[68] First published in 1742. In *Oeuvres de Mr. de Maupertuis* (new ed.; Lyons: Jean-Marie Bruyset, 1756), vol. 1, pp. 79–170.

[69] *Ibid.,* "Des taches lumineuses découvertes dans le ciel," pp. 142–48.

[70] *Ibid.,* p. 143.

[71] First published in 1750. In *Oeuvres,* vol. 1, pp. 1–78. Only the last four pages deal with stars and nebulae.

[72] First published in 1738. See chapter 13 of Part III, in *Oeuvres complètes de Voltaire* (Paris: Garnier Frères, 1877–85), vol. 22, pp. 572–77.

[73] *Ibid.,* pp. 577–82.

[74] In *Oeuvres,* vol. 18, pp. 182–90.

[75] In *Oeuvres,* vol. 19, pp. 455–59.

[76] "Lettre dix-septième: Sur l'infini et sur la chronologie," in *Oeuvres,* vol. 22, pp. 143–48.

[77] See the issue of Monday, September 1, 1712; in *The Spectator,* edited with an introduction and notes by F. Bond (Oxford: Clarendon Press, 1965), vol. 4, p. 170. The other, very insignificant reference to the Milky Way is in the July 9, 1714, issue, *ibid.,* p. 529.

[78] *Le Newtonianisme pour les dames,* translated from the Italian (1737) by Du Perron de Castera (Amsterdam: aux depens de la Compagnie, 1741).

[79] *Ibid.,* p. 225.

[80] *Ibid.,* pp. 225–26.

[81] Known also as Zedler's Lexicon after the publisher, Johann Heinrich Zedler of Dresden.

[82] *Ibid.,* cols. 163–64.

[83] Leipzig: bey Johann Friedrich Gleditschens sel. Sohn, 1734, cols. 846–47.

[84] These two works are in volumes 3 and 1, respectively, of *Petri Horrebowii operum mathematico-physicorum* . . . (Copenhagen: sumtibus Jacobi Preussii, 1740–41).

[85] *The Principia,* translated by James R. Rendell and Isaiah Tansley (London: The Swedenborg Society, 1912), Part III, chap. 1, par. 8, pp. 159–60. The Latin original was published in Dresden and Leipzig, sumptibus Friderici Hekelii, 1734.

[86] *Ibid.,* Part II, chap. 1, par. 23, p. 258.

[87] *Ibid.,* par. 24, p. 258.

[88] *Ibid.,* par. 23.

[89] *Ibid.,* par. 24.

[90] *Ibid.*

[91] *Ibid.*

[92] *Ibid.,* Par. 22, pp. 257–58. In the Latin original, the figure in question is Tab. III. Principp. Fig:37., between pp. 134 and 135. In the translation, the numbering of figures starts anew with each Part of the work.

[93] *Ibid.,* par. 22, p. 256.

[94] In the issue of December 1735, pp. 556–59.

[95] *Träume eines Geistersehers erläutert durch Träume der Metaphysik* published anonymously in 1766, but Kant's authorship was an open secret.

[96] "Ueber die von Emanuel Swedenborg aufgestellte Kosmogonie," in *Vierteljahrschrift der Astronomischen Gesellschaft* (Leipzig) 14 (1879): 80–91.

[97] *Ibid.,* p. 83.

[98] *Ibid.,* p. 86.

[99] "Emanuel Swedenborg as a Cosmologist," in *Emanuel Swedenborg as a Scientist,* edited by Alfred H. Stroh (Stockholm: Alfonbladets Tryckeri, 1908), pp. 59–77.

[100] *Ibid.,* p. 65.

[101] *Ibid.*

[102] *Ibid.,* p. 66.

[103] *Studies of Anagalactic Nebulae. First Paper* (Uppsala: Almquist & Wiksells Boktryckeri A. B., 1927), p. 7.

[104] *Emanuel Swedenborg: Scientist and Mystic* (London: Faber & Faber, 1948), p. 71.

[105] *Dissertatio astrologico-physica de galaxia* (Copenhagen: ex Typographeo Regiae Majest. & Universit., 1707).

[106] *Ibid.*, p. 5.

[107] *Ibid.*

[108] Altdorf: literis Magni Danielis Meyeri, 1713. The dissertation was defended with Professor Johann Heinrich Muller presiding. Since the presiding professor's name was printed more prominently on the title page of dissertations than the author's, the latter could be overlooked by bibliographers ready to accept secondhand or hearsay information. This happended repeatedly to Schneider's dissertation and to those of Lünde, Hertzberger, and Wexstedt, discussed in the previous chapter. At times, the same dissertation was quoted under two different names.

[109] *Ibid.*, pp. 23–24.

[110] *Ibid.*, p. 26.

[111] *Vernünftige Gedancken von den Absichten der natürlichen Dinge,* reprinted five times between 1724 and 1752.

[112] *Vernünftige Gedancken von den Wuerckungen der Natur* (3d ed.; Halle im Magdeburg: Rengerische Buchhandlung, 1734), p. 159.

[113] *Historia astronomiae sive de ortu et progressu astronomiae liber singularis* (Wittenberg: sumtibus Gottlieb Heinrici Schwartzii, 1741).

[114] *Ibid.*, pp. 102, 110, and 424.

[115] *Institutions astronomiques, ou Leçons élémentaires d'astronomie* (Paris: chez H. L. Guerin & J. Guerin, 1746), p. 60 note.

[116] Paris: Bachelier, 1827, p. 179.

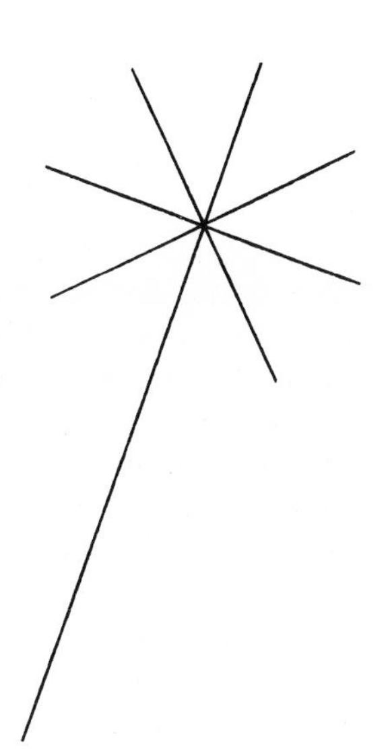

CHAPTER SIX

Wright's Wrong

In the world of science, Thomas Wright (1711–1786) has until recently remained a figure known mostly through second-hand accounts. There was no noticeable excitement when his *An Original Theory and New Hypothesis of the Universe* saw print in London in 1750 in an edition of a few hundred copies.[1] Still, the book marks a turning point in the history of man's speculations on the Milky Way, although, contrary to the generally accepted belief, it did not contain the "true," that is, the disk or grindstone theory of our galaxy. The ultimate steps leading to the correct explanation of the Milky Way proved to be as elusive a process for the historiography of science during the last two hundred years as the Milky Way had been for the men of science for over two millenia.

Here again, mitigating circumstances are present. The rarity of Wright's *Original Theory,* its convoluted style, and its cumbersome reasoning make it a difficult objective even when these factors are not coupled to ingrained reluctance to consult the original. On the other hand, a strange misconception of Wright's idea of the Milky Way might not have imposed itself, if two of his other works had not remained in manuscript until very recently. The recent publication of Wright's *Second or Singular Thoughts upon the Theory of the Universe,* written in the early 1770's,[2] and of his "Elements of Existence, or a Theory of the Universe", written in 1734,[3] now make it clear that the *Original Theory* represented only the middle stage of a conceptual development that exercised his mind intermittently for almost four decades. Its initial and concluding points reflect a most peculiar and unsatisfactory notion of the Milky Way. Awareness of this fact might have prevented, as time went on, the facile projection of an unexceptional picture of the Milky Way into the *Original Theory*. The content of those long-hidden works also shows the extent to which Wright's cosmological speculation was dominated by non-scientific considerations. The real purpose of Wright's search was an ordering of stars, and systems of stars, that would serve as a mirror image of a moral and spiritual order centered in God. In his "Elements of Existence," Wright set the Sacred Throne of God in the center, immediately surrounded by the sphere of Heaven. Enveloping this was the layer of stars, including the sun. The third shell extended indefinitely outward in the form of total darkness embodying Hell. In that framework, the Milky Way received only a short and rather unsatisfactory account:

> At a certain distance from ye Sun equal to a vissual ray of ye smallest visible star is a faint circle of light terminating the utmost extent of ye visible creation, in a finite view from ye Earth; and all within this sphere is more or less inlighten^{d} by ye rays of our Sun according to their distance from him.[4]

Far more valuable was the drawing (Wright was a fine draftsman) that illustrated, on a long and narrow sheet (18 feet by 1 foot), "several thousand worlds and systems, and a great number of emblematical figures." Wright included this proud reference in his *Original Theory*,[5] wherein speculations about the Milky Way are the culmination of the scientific aspect of his message. There is no reason to doubt that, as he put it, the luminous circle of the Milky Way had often engrossed his thoughts, and that it had for a while occupied all his "idle hours."[6] He partially misrepresented his own book when he wrote that its "chief design" was "an attempt towards solving the Phenomena of the *Via Lactea*."[7] The last two of the nine Letters composing the work indicate more than his effort to construct through that solution "a regular and rational Theory of the known Universe, before unattempted by any."[8] Wright's understanding of "regular and rational" is somewhat unusual, notwithstanding his insistence, in Letter V, that the question of the site of the *Sedes beatorum* should be solved on the basis of revelation.[9] Wright made it clear at the outset that he introduced "the Glory of the Divine Being" as "the principal Object in View." His meaning was not the customary advance from nature to nature's God. Instead, his book was to be "instrumental towards the advancing the Adoration of the Divine Being in his infinite Creation of higher Works."[10] The significance of the last words will be clear shortly.

It is indeed on distinctly theological grounds that the argument of the *Original Theory* begins. To prove that the universe should be infinite because the power of the Creator can admit no bounds, Wright offers lengthy quotations from Bruno, Newton, and Derham, to name only his principal authorities. Letter I is almost exclusively a chain of excerpts from "authorities," although in Letter II Wright admits that such proof has only the value of overwhelming probability and not of strict demonstration. Besides "infinity," Wright chooses "order in endless variety" as the other pillar on which to rest his world concept. Letter III illustrates this with a verbose account of the Copernican ordering of planets, all different from one another. The infinite variety of stars forms, in Letter IV, another example of "order in endless variety." Characteristically, Wright takes care to note at the end of Letter IV that the variety of stars, together with the planets of each, provides "manifest Mansions of Rewards and Punishments, suited no doubt most equitably to all Degrees of Virtue, and to every Vice."[11]

Wright is, however, fully aware that three-dimensional infinity can only be shapeless. For him, the universe of stars is an "infinite shapeless Universe" or an "endless Orb of Immensity."[12] Consequently, ordering or form ceases with the supreme constituent parts of the universe, the galaxies.[13] Of these, only one is visible, our own stellar world, the strictly finite "sidereal Creation,"[14] which is first discussed in Letter V: "This amazing Phenomenon which has been the occasion of so many *Fables,* idle Romances, and ridiculous Opinions amongst the Antients, still continues to be unaccounted for, and even in an Age vain enough to boast Astronomy in its utmost Perfection."[15] The Letter is largely taken up by the description of the position of the Milky Way in the sky, by an enumeration of nebulae and new stars, by conjectures on the distances of stars, and, last but not least, by opinions on the Milky Way formed in classical antiquity. Rather curiously, Aristotle is not mentioned at all, a clear indication of Wright's scant familiarity with the history of the question. His source of past opinions on the Milky Way was Edward Sherbourne's annotated English verse translation (1675) of Manilius' *Astronomica*. This explains Wright's silence on medieval and Renaissance writers. Among the moderns, he referred solely to Galileo and Kepler, and only in the sense that their observations confirmed Democritus' opinion.

The question of the structure or shape of the Milky Way is not touched upon explicitly even in Letter VI, where Wright goes to great lengths to show that the stars must be in circular motion around some center, a point which is not indispensable for correct understanding of the appearance of the Milky Way. Wright's concern for that circularity is, however, important because it prompted him to hint for the first time at his own preference as to which of the various possible shapes is the most likely: "To a Spectator placed in an indefinite Space, all very remote Objects appear to be equally distant from the Eye; and if we judge of the *Via Lactea* from Phaenomena only, we must of course conclude it a vast Ring of Stars, scattered promiscuously round the celestial Regions in the Direction of a perfect Circle."[16]

That the Galaxy is a *ring of stars* is one of the two possibilities entertained by Wright in Letter VII, the climax of the work from the scientific point of view. The other possibility is a spherical shell. In both cases, Wright argues, the actual appearance of the Milky Way is a natural result for an observer placed in the ring or in the spherical shell, provided his sight is directed at a right angle to the line joining him to the center. (In the case of the ring, the direction of the view should also be parallel to the plane of the ring.) Needless to say, such a picture is distinctly at variance with almost every account of Wright's idea in the astronomical and cosmological literature. Consequently, a detailed analysis of the contents of Letter VII should cast some light on an astonishing oversight in the modern history and historiography of science. In the introductory part of Letter VII, Wright concerns himself with the choice of a proper vantage point from which the Milky Way must appear as it actually does. This all-important idea is, however, nearly buried under an avalanche of words—as are almost all major statements of the book. One cannot help feeling that, even in 1750, Wright's manner of developing his ideas would not have been classified as straightforward. He often becomes lost in analogies, although these are frequently very much to the point. Thus, he argues the importance of the position of the observer by a reference to the respective merits of viewing the solar system from the earth or from the sun. This leads to the crucial section not only of the Letter but of the whole book, and is introduced with the warning that the Milky Way should be considered as being "formed of an infinite Number of small Stars."[17]

The section in question consists of three parts. In the first, Wright analyzes the optical effect on an observer situated equidistant from two parallel, endless planes between which the stars are scattered in random fashion. The observer's distance from each plane is specified as the distance of the faintest star from the sun. Clearly, when looking parallel to the planes, the observer would see incomparably many more stars than at right angles to that direction. In the former case, the optical effect is the blending of stars into one continuous zone of light, while in the latter instance one would see relatively few stars scattered at uneven distances from one another. In Wright's own words:

> Let us imagine a vast infinite Gulph, or Medium, every Way extended like a Plane, and inclosed between two Surfaces, nearly even on both Sides, but of such a Depth or Thickness as to occupy a Space equal to the double Radius, or Diameter of the visible Creation, that is to take in one of the smallest Stars each Way, from the middle Station, perpendicular to the Plane's Direction, and, as near as possible, according to our Idea of their true Distance.
>
> But to bring this Image a little lower, and as near as possible level to every Capacity, I mean such as cannot conceive this kind of continued Zodiack, let us suppose the whole Frame of Nature in the Form of an artificial Horizon of a Globe, I don't mean to affirm that it really is so in Fact, but only state the Question thus, to help your Imagination to conceive more aptly what I would explain. Plate XXIII will then represent a just Section of it.[18]

Then, with a diagram, which he mistakenly engraved as Plate XXI instead of Plate XXIII[19] (Illustration VIII), Wright helped his reader to visualize the situation:

> Now in this Space let us imagine all the Stars scattered promiscuously, but at such an adjusted Distance from one another, as to fill up the whole Medium with a kind of regular Irregularity of Objects. And next let us consider what the Consequence would be

> to an Eye situated near the Center Point, or any where about the middle Plane, as the Point A. Is it not, think you, very evident, that the Stars would there appear promiscuously dispersed on each Side, and more and more inclining to Disorder, as the Observer would advance his Station towards either Surface, and nearer to B or C, but in the Direction of the general Plane towards H or D, by the continual Approximation of the visual Rays, crowding together as at H, betwixt the Limits D and G, they must infallibly terminate in the utmost Confusion. If your Opticks fails you before you arrive at these external Regions, only imagine how infinitely greater the Number of Stars would be in those remote Parts, arising thus from their continual crowding behind one another, as all other Objects do towards the Horizon Point of their Perspective, which ends but with Infinity: Thus, all their Rays at last so near uniting, must meeting in the Eye appear, as almost, in Contact, and form a perfect Zone of Light; this I take to be the real Case, and the true Nature of our *Milky Way,* and all the Irregularity we observe in it at the Earth, I judge to be intirely owing to our Sun's Position in this great Firmament, and may easily be solved by his Excentricity, and the Diversity of Motion that may naturally be conceived amongst the Stars themselves, which may here and there, in different Parts of the Heavens, occasion a cloudy Knot of Stars, as perhaps at E.[20]

One might wish that Wright had stopped at this point. But his obsession with the circular character of the created whole drew him toward considerations that derailed, although not of necessity, a most promising start. The stars of the "visible Creation" had to be contained within a circular shape because, Wright reasoned, they could and must move only in such orbits. A linear motion for stars, Wright remarked, was contrary to "all the Laws and Principles we at present know of."[21] Wright, therefore, submitted that the circular shape of the Milky Way could only be twofold if its actual appearance was also to receive a satisfactory explanation. One of the two shapes was an orbital arrangement of stars around the center, possibly a huge opaque body—in close analogy to the motion of planets in one general plane around the sun. The other arrangement was a spherical shell. In respect to the former case, Wright almost immediately went off at a tangent, speculating how the motion of stars in primary, secondary, tertiary, etc., orbits around the "central body" would produce a counterpart of the precession of equinoxes on a higher level. The crucial passages of the second stage of Wright's theory of the Milky Way read as follows:

> It only now remains to shew how a Number of Stars, so disposed in a circular Manner round any given Center, may solve the Phaenomena before us. There are but two Ways possible to be proposed by which it can be done, and one of which I think is highly probable; but which of the two will meet your Approbation, I shall not venture to determine, only here inclosed I intend to send you both. The first is in the Manner I have above described, i.e. all moving the same Way, and not much deviating from the same Plane, as the Planets in their heliocentric Motion do round the solar Body.[22]
>
> The second Method of solving this Phaenomena, is by a spherical Order of the Stars, all moving with different Direction round one common Center, as the Planets and Comets together do round the Sun, but in a kind of Shell, or concave Orb. The former is easily conceived, from what has been already said, and the latter is as easy to be understood, if you have any Idea of the Segment of a Globe, which the adjacent Figures, will, I hope, assist you to.[23]

To illustrate the former case Wright offered Plates XXI and XXII[24] (Illustrations IX and X), where the letters A, B, and C designate the center of the sidereal system, the sun, and the earth, respectively. The second, or shell model, was illustrated by Plates XXVI and XXVII (Illustrations XI and XII), whose explanation began the third and concluding stage of Wright's theory.[25] He argued that many more stars would crowd into one's vision if one looked along a chord rather than toward the center. He took it for granted that the optical effect would then be exactly like the Milky Way. In the spherical shell model, the constituent stars could move in an infinitely large variety of planes, a possibility which did not really please Wright as much as the motion of planets in one general plane. Had he not, in fact, offered his analogy of the two parallel planes to help those who find it hard to visualize "this kind of continued Zodiack"? Here, however, he added a note of explanation which dissipated at one stroke the absence of distinct contours in his model of two parallel planes. It now became clear that what he meant by the reference to motion of "primary Planets . . . in a general Zone or Zodiack," was the case represented by the rings of Saturn, or perhaps by a system of concentric rings, or, in sum, by "one vast Ring."[26] To make all this more palatable for the purpose—the confinement of so many stars, like so many specks of dust, within one vast ring—Wright submitted that the rings of Saturn were composed of "an infinite Number of lesser Planets inferior to those we call his Satellites."[27] This last, most revealing and most ignored section of Wright's theory of the Milky Way, should be read in his own words:

> But we are not confined by this Theory to this Form only, there may be various Systems of Stars, as well as of Planets, and differing probably as much in their Order and Distribution as the Zones of *Jupiter* do from the Rings of *Saturn,* it is not at all necessary, that every collective Body of Stars should move in the same Direction, or after the same Model of Motion, but may as reasonably be supposed as much to vary, as we find our Planets and Comets do.
>
> Hence we may imagine some Creations of Stars may move in the Direction of perfect Spheres, all variously inclined, direct and retrograde; others again, as the primary Planets do, in a general Zone or Zodiack, or more properly in the Manner of *Saturn*'s Rings, nay, perhaps Ring within Ring, to a third or fourth Order, as shewn in *Plate* XXVIII [Illustration XIII]. Nothing being more evident, than that if all the Stars we see moved in one vast Ring, like those of *Saturn,* round any central Body, or Point, the general Phaenomena of our Stars would be solved by it; see *Plate* XXIX. *Fig.* 1. and 2. [Illustration XIV] the one representing a full Plane of these Motions, the other a Profile of them, and a visible Creation at B and C, the central Body A, being supposed as *incognitum,* without the finite View; not only the Phaenomena of the *Milky Way* may be thus accounted for, but also all the cloudy Spots, and irregular Distribution of them; and I cannot help being of Opinion, that could we view *Saturn* thro' a Telescope capable of it, we should find his Rings no other than an infinite Number of lesser Planets, inferior to those we call his Satellites: What inclines me to believe it, is this, this Ring, or Collection of small Bodies, appears to be sometimes very excentric, that is, more distant from *Saturn*'s Body on one Side than on the other, and as visibly leaving a larger Space between the Body and the Ring; which would hardly be the Case, if the Ring, or Rings, were connected, or solid, since we have good Reason to suppose, it would be equally attracted on all Sides by the Body of *Saturn,* and by that means preserve every where an equal Distance from him; but if they are really lit-

> tle Planets, it is clearly demonstrable from our own in like Cases, that there may be frequently more of them on one Side, than on the other, and but very rarely, if ever, an equal Distribution of them all round the *Saturnian* Globe.[28]

In the next or Eighth Letter of the work, Wright tried to impress on his reader the enormous dimensions of the "visible Creation," that is, our Milky Way, and the endless crowding in every direction of an infinity of galaxies beyond. He put the distance of the edge of the galaxy from our sun at the distance of a ninth-magnitude star, which he estimated at 4.32 million million miles. This meant that the thickness of the ring or shell was twice as much. Then he stated that the distance to the center of the sphere or ring, that is, to the location of the *Ens Primum,* should be somewhere between 30 and 100 million million miles.[29] He gave no direct clue to his derivation of this figure, but a look at his Plates reveals that he invariably drew the width of the shells or rings as between $1/6$ to $1/10$ of the radius. This is important to note because, in such a shell or ring model, the actual view of the Milky Way will hardly appear.

The vision of endless rows of galaxies beyond ours only shows the degree to which Wright's mind was closed to the problems arising in a densely packed, infinite universe of galaxies, shown on some of his Plates, such as XXXI and XXXII, (Illustrations XV and XVI), representing spherical systems of stars. He blissfully spoke of a true center in an infinite universe in both a physical and a spiritual sense. Wright's mental world indeed held a strange mixture of rational, poetical, mystical, and thinly disguised pathological elements. There is undoubtedly something weird in the eyes that stare at the reader from the center of each closely-spaced galaxy on Plate XXXII (Illustration XVI) which depicts both the omnipresence of God and the infinite variety of eternal abodes. Such was the theme of the last or Ninth Letter and Wright's intended climax to the whole work. He had no doubt that his explanation of the Milky Way provided a "scientific" foundation for his mystical vision of the universe. Or, as he put it:

> I must ingenuously confess to you, that nothing is wanting to convince me intirely of the Certainty of what I here advance by way of Conjecture to you . . .
>
> But if I have been so happy as to come so near the Mark, as to

> border upon Truth, I believe you will allow me to carry my Conjectures a little further, and point out some farther pleasing Consequences, which I begin to perceive may naturally follow.
>
> Should it be granted, that the Creation may be circular or orbicular, I would next suppose, in the general Center of the whole an intelligent Principle, from whence proceeds that mystick and paternal Power, productive of all Life, Light, and the Infinity of Things.
>
> Here the to-all extending Eye of Providence, within the Sphere of its Activity, and as omnipresently presiding, seated in the Center of Infinity, I would imagine views all the Objects of his Power at once, and every Thing immediately direct, dispensing instantaneously its enlivening Influence, to the remotest Regions every where all round.
>
> . . . it is highly probable, there may be some one Body of siderial or earthy Substance seated there, where the divine Presence, or some corporeal Agent, full of all Virtues and Perfections, more immediately presides over his own Creation. And here this primary Agent of the omnipotent and eternal Being, may sit enthroned, as in the *Primum Mobile* of Nature, acting in Concert with the eternal Will. To this common Center of Gravitation, which may be supposed to attract all Vertues, and repel all Vice, all Beings as to Perfection may tend; and from hence all Bodies first derive their Spring of Action, and are directed in their various Motions.
>
> . . . Here we may not irrationally suppose the Vertues of the meritorious are at last rewarded and received into the full Possession of every Happiness, and to perfect Joy. The final and immortal State ordain'd for such human Beings, as have passed this Vortex of Probation thro' all the Degrees of human Nature with the supream Applause.[30]

Rationality was not his main inspiration when, twenty years later, he pieced together his *Second or Singular Thoughts upon the Theory of the Universe*. There, preference for the spherical not only prevailed over shapeless infinity, but fire and vulcanism were taken as the primary element and mechanism of the universe. The precipitating factor in the cosmology of the *Second Thoughts* was a gigantic event. The Lisbon earthquake of 1755 not only buried tens of thousands of people, but also unsettled many convictions and many minds. In this last respect, one of its victims seems to have been Thomas Wright. By his own admission, the catastrophe of Lisbon led him to suspect "some other latent cause more probable than the ordenary one generally advanced towards the solution of that allarming effect and unacounted oporation of nature."[31] Just as the case of the Milky Way in the *Original Theory,* the solution of the earthquake also had to form part of general cosmological synthesis. Wright concluded that the recurring tremors of earthquakes could best be explained if the earth were pictured as consisting of a very solid magnetic core surrounded by a liquid layer, heavier than the ordinary soil and water floating on it. The interior liquid layer could readily be imagined as beginning to move in some fashion. Earthquakes were then produced when the inner liquid thrust itself into the inner cavities of the earth, forcing them apart.

If the spherical arrangement of various layers worked so well for the earth, it was tempting to extend the speculation to areas above the earth's surface, and this is what Wright did. Since the liquid and solid matter were already used up, he had, in true Aristotelian fashion, only air, ether, and fire at his disposal. In fact, the Aristotelian world and Wright's final cosmological model differed only in detail. Wright even restored the solid firmament, beyond which he pictured a realm of fire. The stars were nothing but permanent openings in that firmament for the outflow of cosmic flames. New and variable stars were disposed of as accidental outbreaks, while the nebulae were described as the combination of the fiery outflow and of "circumambient vapor." As the exhalation was to Aristotle, so the vapor was readily available to Wright. The Milky Way was to be explained, he added immediately, by "the same laws and solution."[32]

Again, his own words should portray this appalling reversal of reason in all its shocking nuances:

> The visible heavens or stary firmament might prove to be no other than a solled orb of this stupendious nature and the fixd stars no more than perpetual lumination or vast erruptions & of refulgent or inflammable mater promiscuously distributed as celestial vulcanos allround the starry regions emiting an etherial & intense fire of various magnituds, but remov'd for some infinitely wise purpose of the Creator, to so indefinite a distance as to be far with out ye reach of human arts to ascertain but by ye mental eye of reason only.[33]

> To conclude: all the new extinct and vanishing stars, together with the cloudy and other variable stars, that are frequently perceived

> in ye celestial regions, will by this hypothesis be naturally and almost incontestibly accounted for, and without any of these lame and forcd conjectors which astronomers have been reduced to in attemping to explane them.
>
> The first will be found to be no other than fres[h] erruption, or ye revival of an old vulcano, the second such inflamatory substances totally extinguish^d, the cloudy such as are more invelop^d by circumambient vapour, and the variable, with all periodical ones, no other than repeated erruptions of ye same star at different intervals & sometimes nearly at ye same.
>
> The Via Lactea and all other luminous spaces will likewise come under the same definitions as being universally subject to ye same laws & solution.[34]

What was here merely stated by inference concerning the nature of the Milky Way was spelled out in vivid, concrete terms in the concluding part of the work:

> In this hypotheses, the Via Lactea is lookd upon as no other than a vast chain of burning mountains forming a flood of fire surrounding the whole starry regions, and no how different from other luminous spaces, but in ye number of stars that compose them, or where there are none, in the vast floods of celestial lava that form it. In many of which for a certain time, some brighter parts may appear, such as ye maculae or fiacula in the Sun. And such have also appeared in the floods of lava issuing from volcanos upon Earth.[35]

To cap the comedy, Wright added:

> The idea of a sollid firmament is no new hypotheses, since it was first, & always concivd to be so by ye ancients, especially by Pythagoras, Ptolemy &c.[36]

Fortunately for Wright's reputation, it remained a secret until 1967, that the man acclaimed for at least a century as the discoverer of the true picture of the Milky Way, ended with picturing it as a "vast chain of burning mountains," an idea showing strange similarities with the Aristotelian explanation. Like Aristotle, Wright failed to consider how the invariable shape and hue of the Milky Way could be explained within the suggested framework. For the history of astronomy and its historians, Wright's last words on the Milky Way were embodied in his *Original Theory.* The final words of its famous Letter VII had, however, a historical significance unforeseen by Wright and unrecognized until now. There Wright suggested, with somewhat rhetorical humility, that the real value of his theory of the Milky Way "will hardly be known in the present Century, as in all Probability it may require some Ages of Observation to discover the Truth of it."[37]

Neither centuries nor observations—merely some critical reflections—were needed to realize that Wright's explanation of the Milky Way was wrong. A year after publication of the *Original Theory,* a review appeared in *The Gentleman's Magazine.*[38] It was signed S. L., and began with the acknowledgment that Wright's represented a most meritorious and novel advance toward the right solution. But, the reviewer pointed out, a satisfactory explanation of the Milky Way had to account for three major features, namely, that a) it was a great circle, b) that it was not bounded by any regular line, c) that it had two branches for a considerable part of its length.

These features, the reviewer noted, could not be derived from either of the two models of the Milky Way which Wright proposed. In the ring model "instead of a zone, we should see two nebulose or luminous spots in opposite parts of the heavens, in the direction of tangents drawn from the eye to the interior circumference of the ring, for the remote parts of the ring must be at a distance so inconceivably immense, and the number of stars in any other point of view so much inferior, that the azure vault could not thereby be ting'd with any perceptible degree of whiteness."[39] A double-ring model and an opaque central body, the reviewer added, would produce an appearance even more at variance with reality. Were the stars confined to a spherical shell, we should see, the reviewer stated, "a luminous zone in the heavens, but then that zone would appear of an equal breadth throughout. One border ought to appear regular, streight, and well defined; on the opposite side, the zone should have no precise boundary at all, but the luminousness or whiteness of the zone (which ought to be the greatest on the border) should gradually fade and decrease 'till it became imperceptible."[40] The rest of the review was occupied with the questions of whether the stars were really in motion, whether an orbital motion of all stars was needed for a correct explanation of the appearance of the Milky Way, and whether the rings of Saturn really did consist of very small particles. All these questions were answered in the negative, proving that even the most perceptive critic can misjudge a thing or two. Still, his remarks on Wright's two models show an ingenuity that was not duplicated or recalled for more than 200 years.

The principal blame for this rests with the historians of astronomy and cosmology. It should have been their duty to read the original and to read it carefully. The blame should, however, be shared by the anonymous German writer who, in 1751, published a review of Wright's book in three consecutive installments in the *Freye Urtheile,* a Hamburg semiweekly.[41] The reviewer did no major injustice to the main ideas of the book, which he summarized chapter by chapter—or rather, Letter by Letter. The scientifically informative value of the review consisted in its quoting of many passages, especially from Letters III, IV, V, and VII, each of which received two and a half pages of a total of about fourteen. This indicated that Wright's metaphysical views and aims, which otherwise received high praises from the reviewer,[42] did not dull his sensitivity to the potentially rich content of Wright's scientific message. The reviewer's account of the contents of Letter VII remained, however, confined to an almost verbatim account of Wright's analysis of the optical situation arising in the model of two parallel planes. To this were added Wright's words on the two possible modes of motion, rectilinear and circular, and his remark that the general circular motion of stars could be realized either in essentially a single plane, or within a spherical shell. Contrary to Wright, who developed both models, the reviewer reported only on the former, and rather incompletely at that. The one-plane model "is in the manner, which I have above described," so the reviewer permitted Wright to say, "namely, that they all move along one road, and that they do not deviate much from the same plane, as the planets in their heliocentric motion around the solar body."[43] Since the reviewer completely ignored the remainder of the letter, which contained Wright's specification of this model as a vast ring, closely analogous to the rings of Saturn, nothing was more natural than for a reader of the review to conclude that Wright's first model, and the Galaxy too, resembled a huge flat disk.

This is precisely what happened when the review came to the attention of Immanuel Kant, perhaps shortly after it appeared. At any rate, when in 1755 Kant wrote the Preface to his *Allgemeine Naturgeschichte,*[44] exact details of what he had read in the *Freye Urtheile* no longer persisted in his memory, and he felt it important to note: "I cannot exactly define the boundaries which lie between Mr. Wright's system and my own; nor can I point out in what details I have merely imitated his sketch or carried it out further."[45] Kant gave generous credit to Wright in seeing, for the first time, the Milky Way as the key to the structure of the universe and in taking the "happy step" toward its satisfactory explanation. Nevertheless, he found it "astonishing that the observers of the heavens have not long since been moved by the character of this perceptibly distinctive zone in the heavens to deduce from it special determinations regarding the position and distribution of the fixed stars."[46] Without any equivocation, he considered Wright the originator of the idea that the Milky Way resembled a huge disk, although Kant did not use this expression. He came closest to its equivalent when he extended this picture of the Milky Way to the nebulae, which, as Kant phrased it by quoting Maupertuis, resembled a circle when their plane was facing us, and an ellipse when seen edgewise.[47]

The emphatic presentation of nebulae as additional Milky Ways was one of Kant's outstanding contributions. The other was his belief that Milky Ways could form larger systems, and these, in turn, further super-systems with no limit to continual higher groupings of units of lower rank.[48] Like Wright, Kant emphasized that the stars in the Milky Way had to move in orbits to prevent them from falling into the center.[49] Kant, of course, had no inkling of the gravitational paradox of an infinite universe composed of galaxies and supergalaxies, nor did he guess that their rotation could not be a simple consequence of the gradual condensation of a primitive gaseous chaos. Kant also displayed his scientific amateurism by supposing that the point of highest condensation of matter was the absolute center of the universe which he declared, at the same time, to be homogeneously infinite.[50] He spoke of the evolution, dissolution, and re-evolution of galaxies, but in a manner which indicated naive enthusiasm rather than scientific maturity. He supposed galaxies to form first in the neighborhood of the center and later in more outward regions. Thus, a cross section of his universe could be represented by a circular wave, whose crests stood for galaxies in fully developed form and whose troughs corresponded to galactic ashes out of which new galaxies would arise.[51]

The paradoxical history of studies on the Milky Way would not have remained true to itself if Kant's dicta on the Milky Way had succeeded in securing a proper understanding of it in the world of science. As a matter of fact, for the next generation the scientific public remained almost as unaware of Kant's statements as Wright's work for two additional centuries remained largely forgotten and misinterpreted in one crucial respect. Kant's publisher went bankrupt and his possessions were impounded;[52] however, one copy reached the *Freye Urtheile*. There a very favorable review of it was quickly published in the July 15, 1755, issue.[53] Kant received special credit for his boldness in venturing beyond the solar system, and even his description of the moral characteristics of the denizens of the various planets was judged to be within the limit of reasonable hypotheses.[54] "His style," the final verdict went, "is lively and spirited, he has the skill to present the most difficult propositions of astronomy in an easy fashion."[55] Nonetheless, the Milky Way, a keystone of the whole work, received only four short and placid lines: "He brings the fixed stars into a systematic arrangement similar to that of the solar system, inasmuch as they are found more crowded along a certain plane. This, he believes, is defined by the Milky Way."[56] The reviewer found astonishing the notion that all stars might revolve around a central point located within the galaxy. He did not find it worth mentioning that Kant himself credited the good work of the *Freye Urtheile* in acquainting him years earlier with the contents of Wright's work. Nor did he guess that Kant could learn only an incomplete version of Wright's idea of the Milky Way in the very same journal.

The next notable reference to Kant's book came from Kant himself, eight years later, in 1763, when his work on the only possible proof of the existence of God saw print.[57] The proof in question rested on the orderly arrangement of the small and large parts of nature. As a major example, Kant mentioned the grouping of planets into solar systems, of stars into super-planetary systems or galaxies, and of galaxies into higher systems, suggesting the existence of an endless chain of ever higher systems.[58] Unfortunately, he had to admit that if such ideas were then widely known, the source was not his ill-fated *Allgemeine Naturgeschichte.* For, as Kant noted in a poignant aside, it "had not come to the knowledge of even the famous Mr. J. H. Lambert, who six years afterwards, in 1761, had proposed in his *Cosmologische Briefe* about the systematic conception of the universe at large, about the Milky Way, about the nebulae, exactly the same theory which can be found in my above mentioned theory of the heaven, in the First Part, as well as in the Preface, and of which something is given in a short section, pp. 154–158, of this work. The agreement of the ideas of that brilliant man with those which I had then proposed, and which match even in smaller details, increases my confidence that the whole conception will find further confirmation in the future."[59]

The footnote in which Kant, in a rare departure from his normally impersonal style, unburdened his feelings, before long came to Lambert's attention. In fact, when a friend of Lambert, late in 1765, left Berlin for Königsberg, Lambert seized the opportunity to discuss the problem with Kant directly through a letter.[60] Lambert felt it necessary to state with assurance that he had independently formed the ideas on the structure of the universe as a systematic ordering of galaxies, to which Kant had alluded in his book on the only possible proof of the existence of God. Lambert also disclosed that his notion of the Milky Way as an "ecliptic of stars," described at length in his *Cosmologische Briefe,*[61] antedated not only Kant's *Allgemeine Naturgeschichte,* but also Wright's *Original Theory:* "What I said on p. 149 of the *Cosmologische Briefe* goes back to 1749. Contrary to my habits then, I went into my room after the evening meal, and looked through the window at the stellar sky and especially at the Milky Way. The insight, which I had then, to see it as an ecliptic of the fixed stars, I wrote down on a quarto page and that was all that I had as a written note before me, in much the same way as I am writing this letter [on a quarto page]. In 1761 somebody from Nürnberg told me that a few years earlier an Englishman had put similar thoughts in print in form of letters to other Englishmen, but the time not being ripe, the translation begun in Nürnberg was not completed. I replied that the *Cosmologische Briefe* will make no impact until an astronomer shall discover something in the sky that could not be explained otherwise; and when the system will be found demonstrated a posteriori, the lovers of Greek literature will come and have no rest until they can prove that the whole system had already been known to Philolaus, Anaximander or some other Greek pundit, and that recently it was only rediscovered and embellished. These are the people who find everything in the ancients provided one tells them what they should look for. At any rate, it makes me wonder very much that Newton did not stumble on it [the theory of the Milky Way] while he was thinking about the mutual attraction of stars."[62]

The passage needs no explanatory remarks to rank as a major document in the baffling history of the discovery of the Milky Way's true form. The page mentioned contains such specific detail as to prove, regardless of other evidence, that Lambert independently discovered the true arrangement of stars underlying the phenomenon of the Milky Way. On that page in the *Cosmologische Briefe,* on which Letter XII begins, Lambert describes how he sat one evening (the year is not mentioned) at his window, preoccupied with the problems of the next day. Soon his thoughts were drawn irresistibly to the depths of the stellar realm. "I was astonished over the host of small stars in that arc, and missed them outside it. Those stars cannot be, I thought, so close together as to almost touch one another. They should lie behind one another, and the rows of stars must be in the Milky Way many times deeper than outside it. Should those rows be equally deep everywhere, then the whole sky would appear as bright as the Milky Way. But outside that zone I see but almost empty spaces. In short, the edifice of the fixed stars is not spherical, but flat and very much so."[63]

The specific detail was Lambert's contrast of the whiteness of the Milky Way with the dark background of the rest of the sky, in reference to the distribution of stars throughout space. Lambert had said much the same a year earlier, in 1760, in his *Photometria,*[64] when he discussed the light and distance of the fixed stars. He referred to the Milky Way as proof that the stars were not uniformly distributed in an infinite space. Were such the case, Lambert argued against Chéseaux, a forerunner of Olbers in discussing the so-called Olbers' Paradox,[65] the skies should show the whiteness of the Milky Way everywhere. "What is the origin of the galaxy?" Lambert asked. "According to my opinion," he replied, "the system of the fixed stars, as it appears to us, is not at all spherical but round and flat, and I therefore submit that the Milky Way is a sort of ecliptic of the fixed stars."[66] To this Lambert added that the galaxy was composed of many irregularly distributed star systems, giving it an uneven contour and accounting for its splitting into two branches. The sun, together with most visible stars in and outside the Milky Way, formed one of those star systems, but lay out of the plane of the Milky Way since the latter did not exactly appear as a great circle.

It is indicative of the strength of mental inertia that not even Lambert's stature as mathematician, physicist, astronomer, and philosopher was enough to cause a general adoption of the new picture of the Milky Way in astronomical literature during the next twenty-odd years. It did not even help that, in 1778, Johann Elert Bode, the emerging leader of German astronomy, gave explicit credit to Kant and Lambert for their explanation of the Milky Way in his 656-page "short explanation of astronomy and of sciences related to it."[67] He spoke of the lenticular distribution of stars in space, and compared the resulting whiteness to the fusion of the trees of a forest into a continuous field when viewed from a distance. In words closely resembling those of Lambert, Bode explained why the Milky Way did not appear as a great circle, and concluded a three-page, refreshing discussion of the Milky Way with the words: "Kant in his *Allgemeine Naturgeschichte* and similarly Lambert in his *Cosmologische Briefe* have, as true philosophers, discussed this subject with speculations worthy of the greatness of God."[68]

Outside Germany, the old descriptive accounts of the Milky Way lingered on. At times the situation was truly tantalizing; for example, the appearance in 1754 of John Hill's *Urania,* an astronomical dictionary.[69] In it there was a magnificent article on "Nebulous Stars," that anticipated in a sense the flavor of William Herschel's style and vision. The article, "Milky Way," was short and perfunctory; it could have been written more than a hundred years earlier. It also revealed the oblivion into which Wright's book had fallen almost immediately after publication. Equally perplexing is the manner in which the uniqueness of the Milky Way was almost rationalized away in Daniel Melanderhjelm's compendium of astronomy, published in 1779. He claimed that seeing many more stars in the Milky Way through a telescope seemed to be a feature common to nearly every area of the night sky.[70] In 1780, Charles Messier described sixty-eight nebulae in his famous catalogue without attempting to compare them with the Milky Way, let alone to offer any conjecture as to the reason for its appearance.[71] In 1782, James Ferguson wrote in the fifth and corrected edition of his *Astronomy* that the Milky Way "was formerly thought to be owing to a vast number of very small stars therein: but the telescope shews it to be quite otherwise; and therefore it's [sic] whiteness must be owing to some other cause."[72] He gave no clue to this mysterious utterance. It is not, however, difficult to guess why Kant's *Allgemeine Naturgeschichte* was reprinted three times[73] after Herschel had become a household name. Clearly, it was pleasing to present the star of German philosophy as a forerunner of the great Herschel. The resuscitated work of the sage of Königsberg now effectively began to sow erroneous ideas in many minds concerning Wright's true achievement.

It must, however, be noted that Kant was not mentioned in the notes added by Constantine Samuel Rafinesque (1773–1840) to his edition of Wright's *Original Theory,* which was printed in Philadelphia in 1837.[74] Only Lambert and Herschel were listed among those whose views on the structure of the cosmos had been anticipated by Wright.[75] And, of course, Rafinesque, who asserted that, independent of the *Original Theory,* it had occurred to him that the Milky Way could only be a lentil-shaped agglomeration of stars.[76] A rather erratic naturalist and botanist, the largely self-tutored Rafinesque also claimed the discovery of many new species of fish on the basis of somewhat flimsy evidence. To him, Wright's work was a fine example of an astronomy free of geometrical and mathematical lore, from which, in Rafinesque's opinion, astronomy had necessarily to be liberated.[77] One should not, therefore, be surprised that Rafinesque, a heavy-handed editor of the text, failed to perceive that Wright did not mention the rings of Saturn to emphasize the revolution of stars in the galaxy, but to illustrate the shape of space to which they were presumably confined.[78]

That Wright's work first received thorough attention at the hands of a Rafinesque, merely adds one more curiosity to the hectic history of the Milky Way. The only merit in the case seems to be the limbo into which Rafinesque's edition of Wright's work has fallen; the edition is today a greater rarity than the *Original Theory* itself. Rafinesque's mishandling of Wright is fully explicable. What else could be expected of an aging eccentric, soon to be engulfed in utter neglect, than to see the chief merit of Wright's work in its vision of "Worlds of bliss, where Beings fly through the dense atmosphere, as birds do in our Aerial one, or we conceive Angels may—where they and WE ALSO may assume every variety of lovely shapes, the agency of GOD himself being the vehicle that carries there our immortal souls, through unfathomable Space and Time, in the lapse of a moment, to be happy forever."[79]

Such expectations scarcely clouded the scientific vision of F. Arago, that most powerful figure of the French scientific establishment during much of the first half of the 19th century. In his lengthy account of Herschel's researches,[80] Arago, in 1842, devoted six smallish pages to the history of the Milky Way before Herschel.[81] They provide a classic example of how misleading a picture may be given by a prominent scientist on the history of his topic. According to Arago, Aristotle considered the Milky Way "a luminous meteor in the middle region,"[82] a statement which clearly suggested that he cared neither for the change in meaning of the word "meteor" nor for consulting reliable sources. For the early history of the Milky Way his sole "authority" was Manilius. He learned from Kant about Wright, but he was unable to decide whether the *Clavis coelestis* of Wright, which he found quoted in the printed list of books of the Royal Society,[83] was identical with the work to which Kant referred. Arago did not even trouble to point out that Kant was only familiar with a review of it. Ironically, several years after 1842, he permitted a copy of Wright's work to go to Pulkovo, where it remained, through misunderstanding, as a gift. But one wonders whether Arago entertained enough respect for scientific history to have corrected his lackadaisical report had the copy been recovered.[84] After all, few people in Arago's day had more power than he to obtain a copy of a rare scientific book, if he so desired.

Decidedly more scholarly was the founder of the Pulkovo Observatory, F. G. W. Struve. In 1847 he published the first detailed monograph on the efforts to fathom the depths of the Milky Way,[85] in which the major points of the problem's history were given with careful documentation. Unfortunately, the Library of Pulkovo possessed, at that time, no copies of Wright's works, and

Struve had to rest content with the meager information given by Kant. This was all the more frustrating, since Struve's sharp eyes might very possibly have detected the false notion developing around Wright's conception of the Milky Way. This possibility appears somewhat less promising in view of what happened when, in 1848, Augustus de Morgan published a summary of Wright's *Original Theory* interspersed with lengthy quotations from it.[86] De Morgan, a towering pioneer in mathematical logic, and a fine mathematician, was a man of immense erudition, ready to detect fallacies and misstatements, as may be seen from his still popular *Budget of Paradoxes.*[87] His summary of Wright's work proved to be both a service and a hindrance to the history of astronomy. Since the *Philosophical Magazine* in which it appeared was widely available in the learned world, a detailed picture of the contents of Wright's work was accessible for the first time. On the other hand, De Morgan, perhaps the most keen reader of his time, failed to see that neither of Wright's two models of the Milky Way satisfactorily explained its appearance. De Morgan's shortsightedness is the more puzzling because he quoted in full the passages where Wright emphatically identified his first model as a ring. But De Morgan's mental vision must have been preconditioned both by the generally accepted view that the Milky Way was a disk and by his reading of Kant. In his opinion there was no difference between Kant and Wright, so far as the Milky Way was concerned. Kant's sole contribution lay, De Morgan stated, in the identification of nebulae as other Milky Ways and in picturing them as grouped into higher systems. De Morgan was, of course, wholly mistaken when he wrote that Wright "does not seem to be acquainted with the partial resolution [of the Milky Way into stars] by Galileo."[88] As a matter of fact Wright wrote that "DEMOCRITUS long ago believed them [the stars of the Milky Way] to be an infinite Number of small Stars; and such of late Years they have been discovered to be, first by *Gallaleo,* next by *Keplar,* and now confirmed by all modern Astronomers, who have ever had an Opportunity of seeing them through a good Telescope."[89]

Preconceived ideas must also have obscured the vision of Robert Grant, author of the *History of Physical Astronomy,*[90] a work deservedly famous for a careful documentation and a thorough grasp of the technical details of celestial dynamics. But Grant failed even to state that Wright had proposed two models for the Milky Way, and that, of the two, it was the ring model which he favored. Obviously, Grant felt no need to emphasize the basic inadequacy of such an explanation. Wright's ideas were equally misrepresented in another masterpiece of British astronomical erudition, the *Old and New Astronomy,* by Richard A. Proctor,[91] who wrote that "De Morgan has given a good account of Thomas Wright."[92] No wonder that Proctor followed De Morgan to go on to claim that "Wright does not seem to have known Galileo's previous discovery, or at all events he does not refer to it."[93] Failure to check the original source often leads to ludicrous phraseology, as may be seen in Proctor's assertion that Kant's "ideas respecting the universe of stars were admittedly suggested by Wright's 'Theory of the Universe,' which he had read in a Hamburg journal of the year 1751."[94]

Wright remained an elusive figure even for Agnes M. Clerke, who received an honorary membership in the Royal Astronomical Society for her works on the history of astronomy. In her massive *Popular History of Astronomy during the Nineteenth Century,*[95] she wrote: "with him [Wright] originated what has been called the 'Grindstone Theory' of the Universe, which regarded the Milky Way as the projection on the sphere of a stratum or disc of stars."[96] In the footnote she referred, revealingly, not only to the *Original Theory* but also to De Morgan's summary of it. Equally striking was her quoting of three lines from the *Original Theory* in *The System of the Stars.* The quotation dealt with the irregular contours of the Milky Way, in part explaining them by the diverse motion of the stars within. She was only three or four pages away from the truth of the matter. Immediately preceding the quote was her restatement of the old myth that "the 'disc-theory' of the Milky Way was first propounded by Thomas Wright of Durham in 1750: "He supposed all the stars to be distributed in a comparatively shallow layer, producing an annular effect by its enormous lateral spread."[97]

Even more startling appears the myopia of E. Gore, whose *The Visible Universe* carried in full the crucial pages of Letter VII of the *Original Theory.* Rather ironically, he introduced the long excerpts with this remark: "To avoid any misrepresentation or misapprehension of Wright's views on the subject, it will perhaps be better to quote his words in full, especially as his work is so little known."[98] He also inserted into the excerpts a few critical comments, the first of which occurred immediately after the analogy of the two infinite parallel planes. "Here we have the 'disc' theory clearly propounded."[99] Needless to say, this was an unwarranted projection of the disk theory into Wright's words, which already suggested that Gore would overlook Wright's identification of his first model with a huge ring by analogy with Saturn's rings. "From this it is clear," Gore triumphantly concluded, as the excerpts from Wright's work came to a close, "that Wright's views—contained in the first portion of the preceding extract—are exactly in agreement with the 'grindstone' or 'disc' theory of the Milky Way, afterwards propounded by Sir W. Herschel."[100] Gore's words between the dashes are almost impossible to believe. Do they indicate that the second part had caused him no misgivings, or that after carefully consulting Wright's work he ignored the evidence? Or was his otherwise alert mind unduly influenced by Herschel's stature or by De Morgan's paper which he recalled with approval? At any rate, he looked but did not see.

Such a deplorable situation typifies mental inertia and preconditioning, especially in view of the fact that in England copies of Wright's work, though rare, were available for scholars determined to know the truth at first hand. On the Continent, the misrepresentation of Wright's ideas flourished uninhibited. Humboldt, although very scrupulous about careful documentation, had to rely on De Morgan's account in his *Kosmos*.[101] Humboldt, however, evidently failed to read the passages quoted there, or to read them carefully enough. His sweeping statement was: "Thomas Wright of Durham, Kant, Lambert, and at first William Herschel too, were inclined to see the shape of the Milky Way . . . as a consequence of the flattened form and uneven extensions of the island universe."[102] In Leiden, P. Kaiser, the great developer of the Observatory there, recounted at length the various views entertained in classical antiquity about the Milky Way in his handbook of astronomy.[103] As to Wright, he could only say that "while Wright, Kant, and Lambert were right in claiming that the marked unevenness of the distribution of stars was indicative of the shape of the universe, nothing specific could be derived about the extent of the Milky Way because they confined themselves to superficial philosophizing."[104]

No mention was made of Wright in 1877 in the *Geschichte der Astronomie* by Rudolf Wolf, professor of astronomy at Zurich, and then the leading historian of astronomy on the Continent.[105] As late as 1890, he merely registered the rarity of Wright's *Original Theory* in his famous *Handbuch der Astronomie.*[106] Curiously, the sole bibliographical information given there on the contents of Wright's work was a reference to an article published in 1879 by Magnus Nyrén,[107] mentioned in the previous chapter in connection with Swedenborg. Since he was at Pulkovo, Nyrén was one of the fortunate few who had easy access to a copy of Wright's work. He claimed that Wright anticipated practically everything that Kant stated about the Milky Way in 1755. The second of the two short passages which Nyrén quoted from the *Original Theory* was introduced with the words: "On p. 65 Wright states, following his description of his theory which he held to be most correct about our starry sky," and then followed Wright's own interpretation of his first model as a vast ring.[108] Still, it did not dawn on Nyrén, or on Wolf, that a ring is not a disk. The German astronomer and historian of astronomy, J. J. Mädler, who became director of the Observatory at Dorpat, quoted Wright's work as "the theory of the universe."[109] Clearly, he had no inkling of its contents.

In Brussels, J. C. Houzeau, director of the Observatory there, and co-author of the great *Bibliographie générale de l'astronomie,*[110] spoke of Wright in a small essay[111] on the history of the Milky Way with the evasiveness of a scholar unwilling to admit that he has not read the author: "The first author, who tried to include in the same explanation the formation of our planetary system and at the same time the Milky Way, is Wright of Durham, but his imperfect and vague essay had to be forgotten before long."[112] No further light was shed on the subject when R. A. Sampson, professor at the Newcastle College of Science, and future Astronomer Royal of Scotland, tried to refresh British memories about Wright in a paper read before the Society of Antiquaries of Newcastle-upon-Tyne.[113] Although he had access to two copies of the *Original Theory,* his summary of Letter VII was confined to fifteen lines taken from the section in which Wright described the optical conditions arising from the confinement of all stars between two parallel planes.[114] The records of the meeting show that "thanks were voted to Prof. Sampson by acclamation."[115] Nobody guessed that once again an old myopia had received further applause. Plainly, if a Houzeau and a Sampson handled a major point of the history of astronomical research in such cavalier fashion, their colleagues, devoted solely to the task of observations and calculations, could scarcely be blamed for repeating or rephrasing the myth about Wright's theory of the Milky Way.[116]

Since professional astronomers could so completely overlook the obvious, explicitly stated in a book of astronomy, similar blindness on the part of a divinity professor and a professor of philosophy seems perhaps more excusable, although their sustained attention to the topic is not an alleviating circumstance. The professor of divinity was W. Hastie, of the University of Glasgow, and author of *Kant's Cosmogony* which also served until recently as the most convenient source book on Wright.[117] When Hastie wrote, Kant's reputation as a safe guide in the philosophy and methodology of science was as yet unshaken by relativity and quantum theory. Hastie could, therefore, place Kant's cosmology in the best possible light, a procedure which also allowed some generosity to Wright. The true Wright failed to emerge on Hastie's mental horizon, illuminated, to use his own expression, by the sun of Kant's genius. Not that Hastie did not consult Wright's work, and he was certainly familiar with De Morgan's account of it, reprinted as an Appendix to his own work.[118] But he utilized Wright's original text only to replace the often free translation of a handwritten copy of its review in the *Freye Urtheile*. He found this in the library of the University of Edinburgh, but he failed to compare it with the printed original.[119] Luckily for his arguments, the shortcomings of the manuscript copy (with a full page and several phrases missing, in addition to the absence of quotation marks) did not distort the message of the review on the Milky Way. Thus, Hastie could confidently insist that Kant owed to Wright the disk theory of the Milky Way, but not the explanation of other nebulae as similar stellar disks, let alone their gradual evolution from nebulous matter and their gradual dissipation into it.[120]

If the professor of philosophy, E. Adickes, of the University of Tübingen, achieved somewhat more, it was only because he was also a most careful editor of Kant's unpublished manuscripts. As such, he could not help investigating the sources. While preparing his massive study of Kant's scientific views,[121] Adickes discovered, in addition to the shortcomings of the review itself, the deficiencies of the manuscript copy which Hastie had used. But his sharp eyes failed Adickes at the crucial point, in regard to the true meaning of Wright's words on the Milky Way as a vast ring of stars. Adickes quoted those words with puzzlement as Wright's "third theory," and asked whether they did not refer to other Milky Ways.[122]

Compared to Adickes' efforts, the treatment of Wright's ideas seems dispiritingly shabby in the famed *Source Book in Astronomy,*[123] published in 1929 under the co-editorship of Harlow Shapley. In the excerpts selected by Shapley from Letter VII of Wright's work, five lines, in which Wright explicitly identified his theory as a vast ring, were missing. One oversight often accompanies another. The only illustration Shapley reproduced from Wright's book was Plate XXII (Illustration X), on the orbital motion of the sun and of the earth around the center of the Milky Way. Although the quoted passages contained this minor point, their substance concerned the two parallel planes, and the ring and spherical-shell models of the Milky Way.[124]

Long accepted traditions about Wright must also have been at work in the case of Vera Gushee, a historian of science and the author of a well-researched article on Wright.[125] Since this was published in the influential *Isis,* the article might once and for all have opened the eyes of the scientific public, or at least of the historians of science. But Miss Gushee perceived only that Wright's analogy of the two parallel planes could not be considered a disk theory.[126] She felt doubtful that Wright had definitely committed himself to any particular theory. Nevertheless, Miss Gushee's warning that the analogy of the two parallel planes did not in itself suggest a disk theory was no small accomplishment, especially when viewed in contrast to the well-meaning but mistaken efforts of the prominent radiochemist, F. A. Paneth. A professor at Königsberg before moving to the University of Durham, Paneth's avocation was the study of the relation of Wright and Kant, the most renowned son of Königsberg. Upon arriving in Durham, he was immediately struck by the almost complete lack of interest in Wright in Wright's own homeland. Trying to remedy the situation, Paneth published several essays on Wright, of which the first appeared in 1941.[127] There Paneth claimed that Wright was the first to propose that the stars were not distributed at random, but "mainly concentrated in a flat disk," and that Wright's explanation revealed a "disk-like distribution of the visible stars."[128] He also praised the German abstractor of Wright's *Original Theory* for being "very careful," so that "Wright's convincing explanation stands out all the clearer in the translator's account."[129] He also gave unqualified praise to Hastie's translation of the manuscript copy of the German abstract. In the same paper, Paneth took the view that W. Herschel's idea about "the disk-like form of the stellar systems constituting our galaxy" is "just the picture of the universe first proposed by Thomas Wright in his *Original Theory*."[130] The most astonishing statement in Paneth's paper was, however, a little footnote—a classic case of the blinding force of long-cherished conclusions. In that footnote Paneth said that the anonymous (S.L.) reviewer only criticized "several minor points" in the *Original Theory*.[131] Paneth, who seems to be the only author until then to mention that review, failed to perceive its enormous significance for a correct appraisal of Wright. He was not to suspect the weakness in his proposal that a marble plaque be placed on Wright's house with the inscription: "He was the first to explain the Milky Way."[132]

Paneth was a major force behind the bicentennial celebration of the publication of the *Original Theory* on June 2, 1950, at the University of Durham. He was also one of the chief speakers and, needless to say, he heaped encomium on Wright's "disk theory" of the Milky Way.[133] A year later, he did the same at the Royal Institution, in a lecture whose printed form contained a bibliographical notice on Wright. Paneth listed Miss Gushee's article with the remark: "Fails to understand Wright's Original Theory."[134] The failure was Paneth's, whom neither the words of Wright, nor the anonymous criticism of 1751, nor Miss Gushee's provocative remark, could bring to see that a ring was a ring and a disk, a disk, and that, of the two, only the disk explained the Milky Way. The other speaker was Herbert Dingle, Professor of History and Philosophy of Science at the University of London, and later President of the Royal Astronomical Society. His lecture on "Thomas Wright's Astronomical Heritage" meant to vindicate Wright's work "for its originality, its power of imagination, and its critical judgment," especially when measured against his indebtedness to the astronomical tradition, ancient and recent.[135] Originality and imagination were indeed evident in Wright, but his theory of the Milky Way was as faulty as his account of past opinions on the Milky Way was fragmentary. This latter point should have impressed Dingle, a historian of science, but he found nothing strange, for instance, in Wright's omission of Aristotle's view. Nor was the lack of interest between Galileo and Wright due to the fact that, as Dingle claimed, the starry realm could not really be investigated by telescope. The reasons, as the previous chapters showed, lay elsewhere, but this could not be seen in the absence of a thorough study of the question before and during the relevant period.

Modern historians of astronomy are not to be commended for their reports on Wright. E. Zinner quoted both Miss Gushee's article and long excerpts from Letter VII of the *Original Theory* without perceiving the anomaly in the customary interpretation of Wright's theory of the Milky Way.[136] Zinner further asserted that Lambert developed his own ideas under the influence of Wright's work.[137] Perhaps G. Abetti should not, therefore, be judged too harshly for saying not a word about Wright in his *History of Astronomy*.[138] A. Pannekoek's reference to Wright is as short as it is vague.[139] Much the same variety of misstatements and half-truths or plain errors about Wright can be found in the host of essays and books on cosmology, scholarly and popular, that have appeared during the last quarter century. A typical example is the manner in which Milton K. Munitz, editor of the *Theories of the Universe*,[140] handled the matter. He claimed that to Wright "astronomy owes a prophetically correct analysis of the Milky Way."[141] The "proof" of this was the reproduction in full of the first half of Letter VII, but the complete omission of its second half in which Wright proposed his two models, the spherical shell and the ring.[142] All this showed that Munitz' reference, six years earlier, in connection with Miss Gushee's article, was of a decorative, not an instructive, character.[143]

The break in this pathetic perpetuation of a two-century-old myth came only with the public auction of a large number of Wright's manuscripts in 1966. These notes contained the text of Wright's *Second or Singular Thoughts,* which was published within two years. The Introduction by the editor, M. A. Hoskin, provided a wholly new picture of the development of Wright's cosmological ideas, leading to a badly needed warning on the fallacy of attributing to Wright the disk theory of the Milky Way.[144] Hoskin also called attention to Kant's mesmerizing influence on practically everyone who discussed Wright's theory. Contrary to Hoskin's observation, it is doubtful that Wright could justly be proud of his two explanations of the Milky Way.[145] They are faulty explanations, as was pointed out in 1751 by the anonymous reviewer in *The Gentleman's Magazine*.[146] The first detailed, correct explanation came from William Herschel, who in all likelihood stumbled on the truth without knowledge of either the ideas of Kant and Lambert, or those of Wright.[147] Wright was still alive in his stately hermitage at Byers Green, near Durham, when on June 17, 1784, and on February 3, 1785, Herschel read before the Royal Society his two epoch-making papers on the large-scale arrangement of stars, or on "the construction of the heavens."[148] The contents of his papers prompted no one in the audience or among the many readers of the *Philosophical Magazine* to recall the *Original Theory*. It was so completely forgotten that only a short phrase was devoted to it when, in 1793, *The Gentleman's Magazine* published an account in three installments of Thomas Wright's life, craftsmanship, hobbies, and scientific accomplishments.[149] Rarely was a truly original work more wronged. After all, unlike the great giants of astronomy before him, Wright dared to look at the Milky Way with the penetrating vision of the mind. That such a feat was reserved for an eccentric amateur bespeaks the often "wrong" routes of scientific history.

References

[1] Printed for the Author and sold by H. Chapelle, in Grosvenor Street MDCCL. The subtitle reads: "Founded upon the Laws of Nature and Solving by Mathematical Principles the General Phaenomena of the Visible Creation; and particularly the Via Lactea. Compris'd in Nine Familiar LETTERS from the AUTHOR to his FRIEND. And Illustrated with upwards of Thirty Graven and Mezzotinto Plates, By the Best Masters." The work was sponsored by 118 subscribers and soon became a rarity. It was recently published as a facsimile reprint by Macdonald in London and American Elsevier in New York in March, 1971, with an Introduction by M. A. Hoskin, who edited in the same volume the text of Wright's short cosmological essay of 1734, the "Elements of Existence, or a Theory of the Universe."

[2] Edited from the manuscript, with an Introduction by M. A. Hoskin (London: Dawsons of Pall Mall, 1968).

[3] It covers 15 pages in the reprint edition.

[4] *Ibid.*, p. 9.

[5] *Original Theory*, p. 1, note.

[6] *Ibid.*, p. 37.

[7] *Ibid.*, p. iii.

[8] *Ibid.*

[9] *Ibid.*, pp. 46–47.

[10] *Ibid.*, p. iv.

[11] *Ibid.*, p. 35.

[12] *Ibid.*, p. 33.

[13] As already suggested by the reference to the innumerability of worlds in the title of the Sixth Letter, p. 48.

[14] *Ibid.*, p. 58.

[15] *Ibid.*, p. 37.

[16] *Ibid.*, p. 48.

[17] *Ibid.*, p. 62.

[18] *Ibid.*

[19] There are two Plates marked XXI. One faces p. 62, the other faces p. 64. Wright obviously meant the latter, which is XXIII in order. There is no Plate actually marked XXIII.

[20] *Ibid.*, pp. 62–63.

[21] *Ibid.*, p. 63.

[22] *Ibid.*

[23] *Ibid.*, p. 64.

[24] Again, Wright's reference in the text was a misprint. He referred to Plates XXII and XXIII, but he must have meant Plate XXI (facing p. 62) and Plate XXII. These Plates should be studied together with Plates XXVIII and XXIX (illustrations XIII and XIV).

[25] *Ibid.*, pp. 65–66.

[26] *Ibid.*, p. 65.

[27] *Ibid.*

[28] *Ibid.*

[29] *Ibid.*, p. 73.

[30] *Ibid.*, pp. 168, 169, 171.

[31] *Second or Singular Thoughts*, p. 27 (Letter I).

[32] *Ibid.*, p. 39.

[33] *Ibid.*, p. 29.

[34] *Ibid.*, p. 39.

[35] *Ibid.*, p. 79 (Epigomena).

[36] *Ibid.*, p. 79.

[37] *Original Theory*, pp. 65–66.

[38] 21 (July 1751), pp. 315–17.

[39] *Ibid.*, p. 315.

[40] *Ibid.*, p. 316. According to Hoskin (see his Introduction to the reprint edition, p. xxxviii) the review by S. L. contained only "negligible" criticisms since it gave only "hints" of the difficulties of Wright's theory. These difficulties were not spelled out by Hoskin, perhaps because he took the view that a sufficiently thin spherical shell or ring could lead to the desired optical effect. But it seems doubtful that even a width considerably less than 1/10th of the radius (the ratio implied by Wright) would dispose of the difficulties raised by S. L.

[41] *Freye Urtheile und Nachrichten zum Aufnehmen der Wissenschaften und der Historie überhaupt* 8 (1751): 1–5, 9–14, 17–22; the three issues were dated Jan. 1, 5, and 8, respectively. They carried in turn the contents of Letters I–III, IV, and V–IX.

[42] See especially pp. 1–2 and 21–22 of the review.

[43] *Ibid.*, p. 20.

[44] The rest of the title reads: *und Theorie des Himmels oder Versuch von der Verfassung und dem mechanischen Ursprunge des ganzen Weltgebäudes nach Newtonischen Grundsätzen abgehandelt* (Königsberg und Leipzig: bey Johann Friedrich Petersen, 1755). An able translation of its first two parts by W. Hastie forms the main section of his *Kant's Cosmogony* (Glasgow: James Maclehose and Sons, 1900). Quotations are from this translation.

[45] *Ibid.*, p. 32.

[46] *Ibid.*, pp. 54–55.

[47] *Ibid.*, p. 62.

[48] *Ibid.*, pp. 136–37. Of these two points, the former was emphasized by K. G. Jones in his article, "The Observational Basis for Kant's *Cosmogony:* A Critical Analysis," *Journal for the History of Astronomy* 2 (1971): 29–34. Jones mentioned Wren, but not Gassendi, as a forerunner of Kant in considering other nebulae as other Milky Ways. Most probably, Kant did not depend on either, but he made no secret of his debt to Maupertuis.

[49] *Ibid.*, p. 57. Kant presented Sirius as the most probable central body in the Milky Way; see *ibid.*, p. 164.

[50] *Ibid.*, p. 148.

[51] *Ibid.*, pp. 152–53. This point, usually ignored in summaries of Kant's cosmological ideas, contains more visionary than scientific elements, and should cast a curious light on Kant's eagerness to decry Wright's "fanatical enthusiasm." See *ibid.*, p. 165.

[52] See the biography of Kant by his former student and trusted friend, Ernst L. Borowski, *Darstellung des Lebens und Characters Immanuel Kants,* a work originally published in 1804 with Kant's authorization, in *Immanuel Kant: Sein Leben in Darstellungen von Zeitgenossen: Die Biographien von L. E. Borowski, R. B. Jachmann und A. Ch. Wasianski,* edited by Felix Gross (Berlin: Deutsche Bibliothek, 1912), p. 89.

[53] *Freye Urtheile* 8 (1755): 429–32.

[54] *Ibid.*, p. 432.

[55] *Ibid.*

[56] *Ibid.*, p. 429.

[57] *Der einzig mögliche Beweisgrund zu einer Demonstration des Daseins Gottes* (1763), in Kant's *Gesammelte Schriften,* vol. 2, (Berlin: G. Reimer, 1905), pp. 63-163.

[58] *Ibid.*, Abtheilung II, Betrachtung 7, "Kosmogonie" pp. 137–41.

[59] *Ibid.*, p. 68.

[60] *Immanuel Kant: Briefwechsel,* with introduction, notes and indices by Otto Schöndörffer (Leipzig: Felix Meiner, 1924), vol. 1, pp. 36–40. The letter was dated November 13, 1765.

[61] *Cosmologische Briefe über die Einrichtung des Weltbaues* (Augsburg: Eberhard Kletts Wittib., 1761).

[62] *Briefwechsel,* vol. 1, pp. 38–39.

[63] *Cosmologische Briefe,* p. 150.

[64] *Photometria, sive de mensura et gradibus luminis, colorum et umbrae* (Augsburg: sumptibus viduae Eberhardi Klett, 1760). See Part VI, chap. 3, "De lumine Fixarum earumque distantia," pp. 504–11.

[65] For further details, see my *The Paradox of Olbers' Paradox* (New York: Herder & Herder, 1969), pp. 84-93 and 123–24.

[66] *Photometria,* p. 505.

[67] *Kurzgefasste Erläuterung der Sternkunde und den* [sic] *dazu gehörigen Wissenschaften* (Berlin: bey Christian Friedrich Himburg, 1778). See the sec-

tion, "Von den Fixsternen und erweiterte Aussichten in das Reich der Schöpfung," pp. 491–512.

[68] *Ibid.,* p. 512. Bode said much the same in the third and fourth editions of his popular astronomy, *Anleitung zur Kenntniss des gestirnten Himmels* (Berlin: bey Christian Friedrich Himburg, 1777 and 1778, pp. 658–59). In the second edition of that work (Hamburg: bey Dieterich Anton Harmsen, 1772), only Lambert was mentioned in connection with the Milky Way (p. 486 note).

[69] *Urania, a Compleat View of the Heavens: containing the Antient and Modern Astronomy, in Form of a Dictionary* (London: printed for T. Gardner, 1754). The pages of the book are not numbered. Articles are arranged in alphabetical order.

[70] *Conspectus praelectionum academicarum, continens fundamenta astronomiae* (Holmiae: in officinis librar. M. Swederi, 1779), vol. I, p. 137. Sixteen years later he repeated the same statements in the Swedish version of the book, *Astronomie* (Stockholm: tryckt hos J. P. Lindh, 1795), vol. I, p. 116.

[71] "Catalogue intéressant des Nébuleuses et des amas d'Etoiles observés par M. Messier à l'Observatoire de la Marine," in *Connoissance des Temps pour l'Année commune 1783* (Paris: de l'Imprimérie Royale, 1780), pp. 225–54. Messier ignored the Milky Way in a smaller list of nebulae published in 1771, and in the sequel to his big list published in 1781.

[72] *Astronomy explained upon Sir Isaac Newton's Principles and made easy to those who have not studied mathematics* (London: printed for W. Strahan etc., 1782), pp. 339–40.

[73] In 1797, 1798 and 1807. The second (Zeiss: beim Wilhelm Webel) contained notes that showed the influence of Bode's works, quoted above.

[74] Rafinesque even changed the title into *The Universe and the Stars: Being an Original Theory of the Visible Creation, founded on the Laws of Nature* by Thomas Wright. First American Edition from the London edition of 1750, with notes by C. S. Rafinesque, Prof. of historical and natural sciences. Philadelphia, 1837. Printed for Charles Wetherhill by H. Probasco, 119 North Fourth-street, below Callowhill. And sold by the principal Booksellers.—Rafinesque changed the orthography, inserted some of his remarks into the text, and, because of financial difficulties, omitted the Plates.

[75] *Ibid.,* p. 7, Preface by Rafinesque.

[76] *Ibid.,* p. 150, Notes to Letter V.

[77] *Ibid.,* pp. 4 and 5, Preface by Rafinesque.

[78] *Ibid.,* p. 152, Notes to Letter VII.

[79] *Ibid.,* p. 155, Notes to Letter IX.

[80] "Analyse historique et critique de la Vie et des Travaux de sir [sic] William Herschel," in *Annuaire pour l'an 1842 présenté au Roi par le Bureau des Longitudes* (2d ed.; augmented by the scientific notices of Arago; Paris: Bachelier, 1842), pp. 249–608.

[81] *Ibid.,* pp. 445–51.

[82] *Ibid.,* p. 446.

[83] *Ibid.,* p. 449 note. The *Clavis coelestis,* an elementary introduction to astronomy, was published in 1742 and was recently reprinted with an introduction by M. A. Hoskin (London: Dawsons of Pall Mall, 1967).

[84] This may be a harsh stricture on Arago, but a quick look at the history of the Milky Way in his posthumous, four-volume *Astronomie populaire* (edited by J. A. Barral; Paris: Gide, 1858; vol. 2, pp. 3–7), provides sufficient justification for it. There Arago hailed as a major advance Kepler's dictum in the *Epitome* that the Milky Way was a ring of stars, and poured ridicule on Gassendi for ascribing the origin of the Milky Way to God. Had Arago checked the context of Gassendi's statement, he would have seen that the real advance was made by Gassendi, not by Kepler. As for Wright, Arago sought refuge in the vaguest generalities, and excused himself in a note: "There exist three works of Wright. After some research I have succeeded in securing them, but due to some confusion they got lodged in the Library of Pulkovo before I could read them" (*op. cit.,* vol. 2, p. 8). F. G. W. Struve wrote, years before Arago complained: "I believed I have acted in accordance with the intention of the donor by depositing this rare gift in the Library of the Observatory." See his "Rapport fait à l'Académie Impériale des Sciences par M. W. Struve, sur une mission scientifique dont il fut chargé en 1847," in *Recueil des actes des séances publiques de l'Académie Impériale des Sciences de Saint-Pétersbourg, tenues le 28 décembre 1847 et le 29 décembre 1848* (St.- Pétersbourg, 1849), p. 80.

[85] *Etudes d'astronomie stellaire sur la voie lactée et sur la distance des étoiles fixes* (St. Pétersbourg: Imprimerie de l'Académie Impériale des Sciences, 1847), pp. 8 and 3 of the notes.

[86] "An Account of the Speculations of Thomas Wright of Durham," *Philosophical Magazine* 32 (1848): 241–52.

[87] First published in 1872.

[88] "An Account . . .", p. 244.

[89] See Wright's *Original Theory,* p. 40.

[90] The subtitle of the work reads, *From the Earliest Ages to the Middle of the Nineteenth Century,* reprinted from the London edition of 1852, with a new introduction by H. Wolf (New York: Johnson Reprint Corporation, 1966); see pp. 573–74.

[91] A volume on which Proctor worked for over two decades, and which was prepared for printing after his death by A. Cowper Ranyard (London: Longmans, Green and Co., 1892).

[92] *Ibid.*, p. 700 note 2.

[93] *Ibid.*, p. 699.

[94] *Ibid.*, p. 700. Years earlier, Proctor actually claimed that Wright's work "had been reprinted in a Hamburg journal of the year 1751," in his article, "The Construction of the Heavens," published in 1873 and reprinted in his *The Universe and the Coming Transits* (London: Longmans, Green & Co., 1874), p. 178.

[95] 3d ed.; London: Adam and Charles Black, 1893.

[96] *Ibid.*, p. 16.

[97] *The System of the Stars* (2d ed.; London: Adam and Charles Black, 1905), p. 339.

[98] *The Visible Universe: Chapters on the Origin and Construction of the Heavens* (London: Crosby, Lockwood and Son, 1893), p. 220.

[99] *Ibid.*, p. 221.

[100] *Ibid.*, p. 226.

[101] *Kosmos: Entwurf einer physischen Weltbeschreibung,* vol. 3 (Stuttgart: J. G. Cotta'scher Verlag, 1850), p. 213 note 92.

[102] *Ibid.*, p. 187.

[103] *Der Sternenhimmel,* translated from the second Dutch edition by F. Schlegel (Berlin: G. Reimer, 1850), pp. 375–78.

[104] *Ibid.*, p. 386.

[105] *Geschichte der Astronomie* (Munich: R. Oldenburg, 1877).

[106] *Handbuch der Astronomie ihrer Geschichte und Litteratur* (Zurich: F. Schulthess, 1890), vol. 1, p. 595.

[107] "Ueber die von Emanuel Swedenborg aufgestellte Kosmogonie," *Vierteljahrschrift der Astronomischen Gesellschaft* 14 (1879): 80–91. Curiously enough, Wolf omitted De Morgan's article, which he had mentioned ten years earlier in a short communication [*Vierteljahrschrift* 15 (1880): 367–71] on Wright's life and publications, prompted by Nyrén's article.

[108] *Op. cit.*, p. 87.

[109] *Geschichte der Himmelskunde von der ältesten bis auf die neueste Zeit* (Braunschweig: George Westermann, 1873), vol. 2, p. 216.

[110] Co-authored by A. Lancaster (Brussels: Xavier Havermans, 1882).

[111] "La voie lactée," in *Annuaire de l'Observatoire Royal de Bruxelles, XLVII Année* [1880], (Brussels: F. Hayez, 1879), pp. 233–42.

[112] *Ibid.*, p. 238.

[113] The paper was read at the meeting of July 31, 1895, and its text published in the *Proceedings of the Society of Antiquaries of Newcastle upon Tyne* 7 (1895): 99–104.

[114] *Ibid.*, p. 102.

[115] *Ibid.*, p. 104.

[116] Little need be said about the amateurish and rhetorical efforts made in Germany at the turn of the century to vindicate Kant's originality against that of Wright. See, for instance, M. Jacobi's essay, "Ein Vorläufer der Kant-Laplacischen Theorie von der Weltentstehung," published in the rather chauvinistic *Preussische Jahrbücher* 117 (1904): 244–54.

[117] See note 44, above.

[118] *Ibid.*, 192–205.

[119] *Ibid.*, p. LXVI.

[120] *Ibid.*, pp. LXVII–LXVIII.

[121] *Kant als Naturforscher* (Berlin: De Gruyter, 1924–25).

[122] *Ibid.*, vol. 2, p. 231.

[123] Edited by Harlow Shapley and Helen E. Howarth (New York: McGraw-Hill, 1929), pp. 113–16.

[124] *Ibid.*, p. 114.

[125] "Thomas Wright of Durham, Astronomer," *Isis* 33 (1941–42): 197–218.

[126] *Ibid.*, p. 215.

[127] "Thomas Wright of Durham and Immanuel Kant," *Durham University Journal* 32 (1941): 111–25.

[128] *Ibid.*, p. 116.

[129] *Ibid.*

[130] *Ibid.*, p. 121.

[131] *Ibid.*

[132] *Ibid.*, p. 125.

[133] See Paneth's own report, "Thomas Wright's 'Original Theory' of the Milky Way," in *Nature* 166 (1950): 49–50.

[134] "Thomas Wright and Immanuel Kant, Pioneers in Stellar Astronomy," *Proceedings of the Royal Institution of Great Britain* 35 (1951): 123. Paneth discussed mainly the biographical aspects in his "Thomas Wright of Durham," *Endeavour* 9 (1950): 117–25.

[135] Originally published in *Annals of Science* 6 (1948–50): 404–15. When the same paper was reprinted in his *The Scientific Adventure: Essays in the History and Philosophy of Science* (London: Sir Isaac Pitman and Sons, 1952), Dingle claimed in a footnote to the title that Wright was the originator of the disk theory (p. 113).

[136] *Astronomie: Geschichte ihrer Probleme* (Munich: Karl Alber, 1951), pp. 314–16 and 394.

[137] *Ibid.,* p. 316.

[138] Translated from the Italian by Betty Burr Abetti (London: Abelard-Schuman, 1952).

[139] *A History of Astronomy* (London: George Allen and Unwin, 1961), p. 316.

[140] New York: Free Press, 1957.

[141] *Ibid.,* p. 144.

[142] *Ibid.,* pp. 225–30. With rather dubious scholarship, Munitz used Rafinesque's edition as his source for the excerpts.

[143] "One Universe or Many?" *Journal of the History of Ideas* 12 (1951): 231–55; see p. 249 for reference to V. Gushee.

[144] See note 2 above; pp. 8–9.

[145] *Ibid.,* p. 9.

[146] The incisiveness of that review, referred to by Hoskin (see note 55 in his Introduction to the reprint edition), seems to have been taken somewhat lightly by him.

[147] Compare M. A. Hoskin, *William Herschel and the Construction of the Heavens* (New York: W. W. Norton, 1964), pp. 115–16.

[148] These articles wiil be discussed in detail in the next chapter.

[149] 63 (1793): 9–12, 126–27, 213–16. For reference to the *Original Theory,* see p. 127.

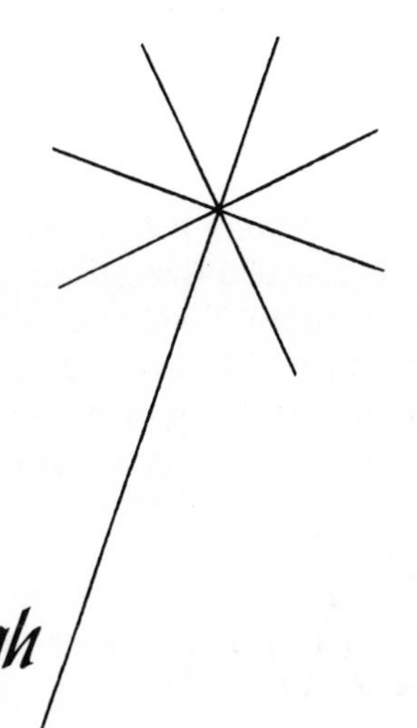

CHAPTER SEVEN

The Silent Breakthrough

ON March 19, 1784, William Herschel sent to J. J. de Lalande, the *doyen* of French astronomers, a copy of his paper on the motion of the solar system,[1] and expressed his appreciation[2] for the inspiration which he had received from a Mémoire of Lalande published in 1776.[3] April 7, the date of Lalande's reply,[4] clearly indicates his eagerness to communicate. The content of his letter reveals in turn his thinly disguised awe for the potentialities of Herschel's observations. Lalande addressed Herschel as "monsieur et illustre confrère," words that contrast sharply with the footnote in which J. E. Bode, the leader of German astronomy, just three years earlier revealed his total lack of information about the identity of the discoverer of Uranus: "In the June 1781 issue of the *Gazette litteraire* this elusive man is called Mersthel; in the *Journal encycloped.*, of July, Hertschel; in a note from Mr. Maskelyne to Mr. Messier in Mannheim, Herrschel; Mr. Darquier calls him Hermstel. What is his name really?—He must be a German by birth!"[5]

Herschel was possibly the finest gift of Germany to England. Bode, and the whole world of science, were to find that the discovery of Uranus was only the prelude to an amazing personal success in observational astronomy. Herschel's skill in building new telescopes, which dwarfed all others by comparison, became by 1783 the talk of the astronomical world. So was the news which began to spread, about an unsuspected wealth of stars and nebulae which he claimed to have sighted with his instruments. Lalande could not have been more interested in this aspect. He was then preparing his famous *Histoire céleste française*[6] which was to contain the positions of 50,000 stars. It was, therefore, natural for him to ask Herschel whether it was true that he had already counted 44,000 stars in a small, eight-by-three-degree area. In inquiring about its position, Lalande almost implied the answer with his next question: "Do you believe that the whiteness of the Milky Way comes from the small stars, or that it is a celestial substance that produces the nebulae?"[7]

In the history of science, few private inquiries have received a more monumental answer than that in the paper[8] which Herschel read before the Royal Society on June 17, 1784, of which he sent a copy to Lalande. To an attentive listener, the introductory paragraphs must have seemed a series of carefully phrased understatements. First came a short description of a new telescope even more powerful than those with which Herschel had already dazzled his contemporaries. There followed a transparent apology for the premature character of the conclusion which he was now to submit on the "interior construction" of the heavens. He also used an indirect method of calling attention to the radical novelty of the paper's message. Light could far better be appreciated when set against a dark background, and Herschel was painting it with broad and quick strokes: "Hitherto the sidereal heavens have, not inadequately for the purpose designed, been represented by the concave surface of a sphere, in the center of which the eye of an observer might be supposed to be placed."[9] Herschel knew, of course, that for more than a century the stellar universe was believed to be infinite in three dimensions. But his self-tutoring in astronomy gave him the impression that nothing of consequence had previously been said about the patterns differentiating that immensity. He was not so correct as he had thought. Nobody in an audience of professionals, however, felt impelled to take exception to this particular point. Apparently they all shared the same illusion.

At any rate, even if they had wished to challenge him on that score, the immigrant musician-turned-astronomer overwhelmed all of them in a very short time. Herschel, reading his paper, might have been likened to a magic peddler emptying a sack full of a great variety of gems of unsuspected brilliance, while making astonishing comments. The starry sky should henceforth be considered, he remarked, as "a naturalist regards a rich extent of ground or chain of mountains, containing strata variously inclined and directed, as well as consisting of very different materials."[10] The richness of the starry soil was mentioned first. In Herschel's estimate, the number of stars in a 15-degree length of the Milky Way was at least 50,000. These stars were the ones large enough to be distinctly counted. "Besides these," he added, "I suspected at least twice as many more, which, for want of light, I could only see now and then by faint glittering and interrupted glimpses."[11]

Herschel gave more than a glimpse of the Milky Way. He boldly went on to unfold its very structure and shape. Again, using the language of a naturalist, he described it as that great stratum of stars in which our sun was probably placed, perhaps not far from the center of its thickness. The proof of this latter point was the appearance of the galaxy, which, as Herschel noted, could only be derived if one supposed that two parallel planes formed the boundaries of a large group of stars: "For, suppose a number of stars arranged between two parallel planes, indefinitely extended every way, but at a given considerable distance from each other; and, calling this a sidereal stratum, an eye placed somewhere within it will see all the stars in the direction of the planes of the stratum projected into a great circle, which will appear lucid on account of the accumulation of the stars; while the rest of the heavens, at the sides, will only seem to be scattered over with constellations, more or less crowded, according to the distance of the planes or number of stars contained in the thickness or sides of the stratum."[12]

Herschel might have borrowed the analogy from conversations with others. The dicta of Wright, Kant, and Lambert could by then have filtered into astronomical parlance. But Herschel might very possibly have discovered it himself, for independent discoveries are not rare in science. At any rate, Herschel's development of the analogy had nothing of the vagaries of Wright's style nor of the imprecision of his reasoning. The diagram (Illustration XVII) which Herschel offered showed this all too well. It was a masterly rendering of a spherical feature in two dimensions, and an explanation for that branching in the Milky Way, which, as Wright's anonymous critic had already remarked, must be explained in any satisfactory theory of the Milky Way: "Let us now suppose, that a branch, or smaller stratum, should run out from the former, in a certain direction, and let it also be contained between two parallel planes extended indefinitely onwards, but so that the eye may be placed in the great stratum somewhere before the separation, and not far from the place where the strata are still united. Then will this second stratum not be projected into a bright circle like the former, but will be seen as a lucid branch proceeding from the first, and returning to it again at a certain distance less than a semicircle."[13] Such a supposition, he added, "will satisfactorily, and with great simplicity, account for all the phaenomena of the milky way, which, according to this hypothesis, is no other than the appearance of the projection of the stars contained in this stratum and its secondary branch."[14]

The solution also eliminated the necessity of supposing the sun to be a special star, occupying the center around which the Milky Way was placed as a ring of stars. That Herschel made a special reference to the ring model of the Milky Way should not suggest that he spoke with an eye on Wright, for the ring model was much older than the *Original Theory* and was also a most natural hypothesis. The reference not only indicated Herschel's uneasiness about a center of the universe, but also the fact that he looked far beyond the Milky Way, proud though he might be of his model. He certainly was not its prisoner. Its parallel planes could readily be replaced by any surface to account for the irregular contours of the Milky Way: "And thus any kind of curvatures, as well as various different degrees of brightness, may be produced in the projections."[15] Moreover, he carefully considered the visual effects produced when the model was viewed from a distance. Once looked upon from the outside, the stratum would appear "as one of the less circles of the sphere," and this circle would shrink gradually into a lucid spot as the observer moved away from the boundary plane of the stratum. The spot itself could be of any shape "according to the position, length, and height of the stratum."[16] In other words, the Milky Way could appear to a very distant observer like any one of the various nebulae, depending on the position of the observer relative to its principal plane.

Throughout the paper, the Milky Way was indeed considered one of the nebulae about which Herschel boldly announced that they should "open a new view of the heavens."[17] A principal reason for this was his belief that his new telescope had resolved into stars all the nebulae listed in Messier's catalogue.[18] He eagerly contrasted his detailed description of two of them with the vague vignettes given by Messier. Equally important were the 466 new nebulae and clusters of stars which he added to Messier's 103. Herschel described with obvious delight the rather fantastic shapes shown by some newly discovered nebulae. Some had the shape of a fan, others resembled an electric brush, as he put it, and others reminded him of the shape of a comet. His most significant remark was that nebulae seemed to form strata, in much the same manner as stars formed the Milky Way. Did he suggest that nebulae could form supersystems of their own? He certainly seemed to be excited by the new vistas, and made no secret of his hope that many more nebulae remained to be detected. With disarming assurance, he promised the Royal Society additional catalogues of nebulae and star clusters, each comprising two or three hundred items.

With these statements, Herschel had already inscribed his name forever in the history of astronomy. But he was only halfway through the first of a dozen or so historic papers on the subject. The second half of the paper was filled with the first results of his famous star-gauging. Although based on patently uncertain assumptions, the technique signified Herschel's firm resolve to measure the distance to the limits of the Milky Way. The star-gauges were deceptively simple formulae derived from the geometry of a cone, whose vertex was the telescope. If the stars were at roughly equal distances from one another in the Milky Way, and if practically all stars in that finite cone could be seen, then from their number the length of the cone, or the distance from the telescope to the border of the galaxy, could readily be estimated. In addition, Herschel took the view that all stars were equally bright, and that the magnitudes traditionally assigned to them provided a clue to their distance.

In more than one sense, Herschel moved on some "nebulous ground."[19] (He used the expression to denote areas of the sky very rich in nebulae.) He speculated on the gravitation of the solar system toward the convergence of the two strata of the Milky Way, where stars were more numerous. Before long he voiced caution: "The subject is new, and we must attend to observations, and be guided by them, before we form general opinions."[20] That by observations one should succeed was a hope which he based on the prospect of improving his telescopes. Herschel called attention to the fact that only the possession of better telescopes could, for instance, have revealed to Lalande that the Milky Way consisted only of stars and not of stars plus some nebulous material.[21] (Needless to say, Lalande would have also needed a correct picture of the Milky Way.)[22] It was with a restatement of his intention to improve his already astonishing telescopes that Herschel concluded his paper: "by applying ourselves with all our powers to the improvement of telescopes, which I look upon as yet in their infant state, and turning them with assiduity to the study of the heavens, we shall in time obtain some faint knowledge of, and perhaps be able partly to delineate, the *Interior Construction of the Universe*."[23]

Observations made with an even better telescope formed the basis on which Herschel built a rich and startlingly original fabric of hypotheses about the Milky Way and other nebulae, in the paper[24] which he read eight months later, on February 3, 1785. Not that Herschel wanted to "indulge a fanciful imagination and build worlds of our own."[25] He resolutely stated that a chief aim of observations was to make conjectures possible. His confidence must have been all the stronger since he felt justified in stating that he had by then "viewed and gaged this shining zone in almost every direction."[26] The shining zone was the Milky Way, but even more important were the 900 or so nebulae which Herschel could already claim to have observed. With astonishing daring, Herschel went on to examine a totally new topic, the formation of various forms of large groups of stars (nebulae) on the basis of gravitational attraction. The first on his list was the class of globular clusters *(i)* which had their origin in the attraction exercised by a particularly large star. Similar globular clusters could also form, Herschel argued, if some stars happened by chance to be close to one another *(ii)*. If such chance agglomerations were not roughly symmetrical at the start of the process, strange forms of nebulae, resembling rows, hooks, and branches, could be the result *(iii)*. For a fourth case, Herschel considered the mutual attraction of two clusters of stars resulting in binary nebulae, evoking systems of binary stars *(iv)*. The fifth case was a consequence of the fourth, namely, the development of "great cavities or vacancies" *(v)* by the drift of stars and clusters from their original position, which Herschel assumed to be an essentially homogeneous distribution.

Herschel skirted the question of the gravitational paradox of an infinite number of stars in an infinite universe. The objection raised by Bentley against the infinite, Newtonian universe never received any public consideration from Herschel, although he mentioned the possible destruction of stars by collision due to gravitational attraction. To preserve the "whole" structure, Herschel referred to some unknown means provided by the "great Author of it,"[27] in addition to two physical factors. The first was the "indefinite extent of the sidereal heavens," which, to recall Herschel's rather superficial reasoning, "must produce a balance that will effectually secure all the great parts of the whole from approaching to each other."[28] Such would have been the case only if the indefinite was truly infinite, but then further and even graver problems would have arisen, of which Herschel had apparently no inkling. The second factor was a "projectile force" imparted to every star, a point on which Herschel went into no detail. In a sense he always remained an amateur.

Herschel might not have been aware of the gravitational paradox of an infinite universe, but he certainly read of its optical counterpart. He dodged the issue all his life. In 1781, in discussing the parallax of the fixed stars,[29] he stated as his second postulate that the magnitudes of stars were the results of their distance, so that a "star of the second, third, or fourth magnitude is two, three, or four times as far off as one of the first."[30] To enhance the reasonableness of this assumption, Herschel referred to a paper by Halley, dated 1720, in which Halley built, on the identical assumption, his concentric shell model of an infinite universe of stars and its optical paradox.[31] Still, as was the case with the gravitational paradox, Herschel was willing to consider only the optical appearances arising from indefinitely but not infinitely large congeries of stars. Even stranger, he again took Halley's assumption as his starting point for discussion of the visual appearance of distant clusters, and strata of stars and nebulae. "We will at present suppose that those of the second magnitude are at double, and those of the third at treble the distance, and so forth."[32] The quantity of light sent from stars located in each shell was of no concern to Herschel: "It is not necessary critically to examine what quantity of light or magnitude of a star intitles it to be estimated of such or such a proportional distance."[33] At any rate, he was not willing to proceed by logic to strict infinity but only to the indefinitely great. He merely spoke of the light of single stars, and that could be handled by what he himself called a "coarse estimation." A star of the second magnitude, in the seventh shell, was the limit to the naked eye, and no observer was to be bothered by the light of a star located fifty times farther than the nearest or first-magnitude star, or at the outer edge of a globular cluster, of such size.

On the basis of such patently questionable considerations, Herschel then described, inconsistently enough, the appearance of the "whole universe" to an observer located in the middle of a cluster of stars: "The whole universe, therefore, to him will be comprised in a set of constellations, richly ornamented with scattered stars of all sizes. Or if the united brightness of a neighbouring cluster of stars should, in a remarkable clear night, reach his sight, it will put on the appearance of a small, faint, whitish, nebulous cloud, not to be perceived without the greatest attention."[34] To an observer placed in the third type of cluster, "the heavens will not only be richly scattered over with brilliant constellations, but a shining zone or milky way will be perceived to surround the whole sphere of the heavens, owing to the combined light of those stars which are too small, that is, too remote to be seen."[35] By making use of the telescope, the Milky Way would prove to be a stratum of stars and all the nebular patches would reveal themselves as composed of stars. The conclusion then presented itself that the actual appearance of the heavens, even when viewed through a powerful telescope, was a function of "the confined situation in which we are placed,"[36] and that our Milky Way, "a crookedly branched nebula," was not even "the most considerable of those numberless clusters that enter into the construction of the heavens."[37]

The second main part of the paper was devoted to the actual determination of the true location and extent of the "crooked branches" that constituted the boundaries of the Milky Way. Herschel sounded a beautifully phrased *caveat* on the finality of his conclusions based on a great number of gaugings, describing the technique in ample detail. The drawing (Illustration XVIII) he made of "our sidereal system," "will not be called a bad one," he remarked, "when it shall be considered how very limited must be the pencil of an inhabitant of so small and retired a portion of an indefinite system in attempting the picture of so unbounded an extent."[38] What he really meant was that the drawing accurately mirrored the real situation. Otherwise he would hardly have added that one could derive from the drawing a definite estimate about the extent of the Milky Way: "Now, to have some idea of the wonderful extent of this system, I must observe that this section of it is drawn upon a scale where the distance of Sirius is no more than the 80th part of an inch; so that probably all the stars, which in the finest nights we are able to distinguish with the naked eye, may be comprehended within a sphere, drawn round the large star near the middle, representing our situation in the nebula, of less than half a quarter of an inch radius."[39] Although he missed the true value by more than one order of magnitude,[40] the vistas he opened represent a most extraordinary achievement in the history of science. The drawing itself has become the most frequently printed diagram of the Milky Way.

From Herschel's gaugings it followed that the Milky Way was, in all probability, as Herschel termed it, a "detached nebula."[41] His conviction on this point was firmer than suggested by his pleasing rhetoric: "It is true that it would not be consistent confidently to affirm that we were on an island unless we had actually found ourselves every where bounded by the ocean, and therefore I shall go no farther than the gages will authorise; but considering the little depth of the stratum in all those places which have been actually gaged, to which must be added all the intermediate parts that have been viewed and found to be much like the rest, there is but little room to expect a connection between our nebula and any of the neighbouring ones."[42] Clearly, the pivotal point of the first synthesis he reached about the construction of the heavens was that our Milky Way *was* an island. But the evidence was ambiguous. His lifelong struggling with it had set the stage for controversies that raged for a hundred years after his death concerning the idea of island universes. Herschel was also prophetic in pointing out the crucial role which telescopes were to play in that battle: "I ought also to add, that a telescope with a much larger aperture than my present one, grasping together a greater quantity of light, and thereby enabling us to see farther into space, will be the surest means of compleating and establishing the arguments that have been used: for if our nebula is not absolutely a detached one, I am firmly persuaded, that an instrument may be made large enough to discover the places where the stars continue onwards."[43]

Herschel's remained, however, a lonely voice for several generations so far as the contents of the third part of his paper were concerned. There he had ventured into a wholly unexplored field, the evolution of nebulae. Primarily he wanted to substantiate his statement that the Milky Way had "fewer marks of profound antiquity upon it than the rest."[44] The burden of proof forced him into conjectures with an extraordinarily modern ring. The Milky Way and all nebulae of type *(iii)* had to be relatively young because otherwise, Herschel argued, they would not have many sections with a markedly even distribution of stars, and many rather loosely packed clusters. In view of this, Herschel forecast the gradual breaking up of the Milky Way into clusters of stars; that is, "to a stratum of two or three hundred nebulae."[45] He noted, in fact, that that number of initial groupings of stars could already be distinguished within it.[46]

The rest of Herschel's paper was devoted to the description of some peculiar features of our nebula and with comments on some other nebulae of extraordinary shape or brilliance. Many of these details proved to be the starting point of research that still attracts interest today. The great dark opening, later called the "coal-sack," was such a feature, together with the "purity or clearness" toward the poles of our galaxy. There followed the description of compound nebulae and the assignment of distances to several nebulae. Herschel spoke of distances as large as 600 to 6,000 times the distance of Sirius. To the Andromeda nebula, Herschel assigned a distance of 2,000 times the distance of Sirius, an estimate which was short of the mark by a factor of about 150. He also believed that he had almost resolved stars in it, a feat that was not achieved until the twentieth century. Among the nebulae he found a ring-shaped one and, most important, some which he called planetary nebulae. The apparently single, very luminous star in their centers, surrounded by "milky whiteness," greatly excited his imagination. He saw in them the great "laboratories of the universe." "The stars forming these extraordinary nebulae, by some decay or waste of nature, being no longer fit for their former purposes, and having their projectile forces, if any such they had, retarded in each others' atmosphere, may rush at last together, and either in succession, or by one general tremendous shock, unite into a new body."[47] The new star of 1572 in Cassiopeia seemed to him to exemplify this situation. He was certain that a less dramatic but equally important role could be played by these planetary nebulae: "To have a fixed point somewhere in the heavens, to which the motions of the rest may be referred, is certainly of considerable consequence in Astronomy; and . . . [they] are bright and small enough to answer that end."[48]

Herschel must have been elated and rightly so. His two papers were a dazzling display of original data and theories in which his brilliantly reasoned analysis of the appearance of the Milky Way, based on overwhelming observational evidence, held the central position. When, in 1798, F. T. Schubert's three-volume *Theoretische Astronomie* saw print in St. Petersburg, Herschel's two great papers on the construction of the heavens were accorded a detailed summary.[49] As a result of his papers, the Milky Way received, for the first time in a manual of astronomy, an up-to-date treatment commensurate with its importance. Schubert's work also carried Herschel's famous diagram (Illustration XVIII) and a detailed explanation of it, together with Herschel's estimate of the dimensions of the Milky Way and his speculations on its dynamics. Schubert insisted that, before Herschel, telescopes could not wholly dissipate the nebulosity in the Milky Way, and that the number of stars seen by the telescope outside the Milky Way was practically the same as that within it. Thus, the explanation that the Milky Way as a congeries of stars "remained a probable opinion until it was raised to the rank of certainty by Herschel with the aid of telescopes constructed by him."[50] In claiming this, Schubert overstated the case, but this was certainly more pardonable than the opposite attitude.

The baffling fact is that the scientific world remained singularly unimpressed by what Herschel said of the Milky Way. The first sign of this came when J. D. Cassini, director (1784–1793) of the Paris Observatory, read, in November, 1784, a report on the verification of new discoveries made in England about the fixed stars before the French Academy. This was printed in full in 1787 and carried in a footnote references to Herschel's 1785 catalogue of double stars. "Since no one had, nor could anyone hope to have, instruments [telescopes] powerful enough to resolve these marvels [double and triple stars], one had to take it all on his words," remarked Cassini,[51] who also hastened to reassure the world of science, on the basis of his own observations in the late summer and early fall of 1784, that Herschel was trustworthy. Curiously enough, Cassini made no mention, even in a footnote, of the great objective of Herschel's observations—the construction of the heavens and, in particular, the structure and shape of the Milky Way.

Lalande, who had asked Herschel the question about the Milky Way, was not moved to any comment of consequence. So far as the Milky Way was concerned, the first two editions (1761, 1771) of his massive, three-volume *Astronomie* ignored not only Wright and Kant, which was understandable, but also Lambert. It defies explanation, however, that, when the third, revised edition was published in 1792,[52] Lalande's one-page account of the Milky Way increased by only four lines despite the chain of Herschel's publications. Those four lines were an indirect defense of Lalande's belief that the Milky Way was not merely a congeries of stars, but that it contained real nebulosity too, although as Lalande put it, tongue in cheek: "But Mr. Herschel sees there stars in such a large number that he is convinced that they are the cause of that whiteness."[53] Lalande's *Astronomie* was not to contain information about the Milky Way of Herschel. Nine years later, in 1801, the Preface of Lalande's *Histoire céleste française*[54] included no mention of Herschel's star catalogues, although ample references were made to more than a dozen catalogues published by others. Was this a slighting of Herschel, or a grudging hint that Herschel already belonged in a class of his own? At any rate, Lalande, a friend of the Herschels, noted that with Herschel's telescopes one would see not 80,000 stars (Lalande's *Histoire céleste* was to contain 50,000), but sixty million. Herschel was duly mentioned as the discoverer of Uranus, but was silently ignored in connection with the Milky Way, the richest realm of stars. "I do not speak of the Milky Way," wrote Lalande, and took issue with Herschel on the significance of empty (starless) spaces in the sky.[55] Lalande hinted at the structure of the Milky Way only indirectly when he offered the conjecture that "the situation of our system among all the neighboring systems must naturally keep us closer to those which have much mass, and farther away from those that have less; perhaps this is the reason why we see so many stars in certain parts of the sky and so few in others."[56] The crowning touch of the baffling performance came in Lalande's 200-page history of astronomy from 1781 to 1802, which he attached to his *Bibliographie astronomique,* published in 1803.[57] There in a year-by-year account, the astronomical events of the years 1784 and 1785 received less than two pages each, with Herschel's share limited to three lines. In 1784, wrote Lalande, "Herschel continues to scan with his telescope of 20 feet the whole sky. He counts 44,000 stars in a space of 8 by 3 degrees."[58] In 1785 "announcement is made of 1300 nebulae observed by Herschel, of planetary nebulae and of his observation of the rotation of Mars."[59]

J. B. J. Delambre, who succeeded Lalande in 1807 as professor of astronomy at the Collège de France, devoted a long chapter to the methods of cataloguing stars in his three-volume *Astronomie théorique et pratique,*[60] but Herschel's catalogues were not mentioned, nor was his explanation of the Milky Way. All that Delambre said of the Milky Way in almost two thousand pages was that the ancients considered it a great circle, although it was rather a zone and a great nebula.[61] Such was scarcely a creditable performance in 1814, but Delambre could surprise one with his highhandedness. His history of astronomy during the eighteenth century[62] considered only those astronomers whose careers did not continue into the nineteenth century. It was, therefore, a history not only without Herschel but without Lagrange, and, above all, Laplace. Laplace himself had a surprise or two concerning Herschel's work. Between 1796 and 1824, Laplace saw through press five editions of his famous *Exposition du système du monde,* but in none of them did Herschel receive credit for his effective discovery of the real structure of the Milky Way. With respect to the Milky Way, Laplace merely stated that our sun belonged to the stellar group that forms the Milky Way, and that many other nebulae are similar star systems which, when viewed at closer range, would appear similar to our own galaxy. Laplace's interest in nebulae originated in his "nebular" hypothesis of the solar system. To strengthen the theory, he referred, from the fourth edition (1813) on, to Herschel's description of some nebulae as good examples of that diluted original form of the sun, from which the system of the planets was supposed to develop.[63] In that celebrated "system of the world" Laplace did not even trouble to inform his reader that the appearance of the Milky Way presupposed the confinement of most stars between two parallel layers. Kant, whose ideas both on the Milky Way and on the formation of planetary systems were soon to be mentioned with increasing frequency, remained a nonentity to Laplace, and so did Lambert—to say nothing of Wright.

Insensitivity to details of scientific history is not infrequent among cultivators of pure science, but can hardly be condoned in a work such as the *Histoire de l'astronomie depuis 1781 jusqu' à 1811,* and published, surprisingly, in 1810.[64] In this work, the subject was treated topically, with the first of the three main divisions covering the discoveries made by observations. Of this, almost one-third, or thirty pages, were devoted to Herschel's discoveries,[65] but not one word to his studies of the Milky Way, although Herschel's lists and speculations of the nebulae and "planetary nebulae" were discussed extensively. Ironically, the work was what it purported to be, a sequel to J. S. Bailly's monumental history of astronomy,[66] whose third or final volume came to a close with discoveries and observations from the year of 1781. Bailly mentioned the discoverer of Uranus as Mr. Hartchell.[67] He did not seem to know anything about the explanation of the Milky Way given by Kant, Lambert, and Wright. Bailly rather complained that Democritus' idea of the Milky Way as a congeries of stars did not fit all the observations.[68] Herschel's great papers of 1784 and 1785, with which Bailly undoubtedly had become familiar, could have given him food for thought. But political events had already begun to dominate Bailly's pursuits completely. They also brought about the tragic finale to his life on the guillotine in 1793.

That passage of time had also extinguished the flame of interest which Herschel's explanation of the Milky Way lighted in the pages of the *Astronomisches Jahrbuch.*[69] The first major flicker of that flame came when F. X. von Zach, who was soon to become first director of the observatory at Gotha and editor of the astronomical review, *Monatliche Correspondenz,* sent, while visiting in London in 1785, a fresh report[70] to the *Jahrbuch* describing Herschel's latest papers on the construction of the heavens. The report accurately conveyed in 24 paragraphs all the points made by Herschel. But there was no hint in von Zach's words about the striking novelty and value of Herschel's explanation of the Milky Way. The next year, J. H. Schroeter, in Lilienthal, near Bremen, who emulated Herschel with a reflecting telescope of 18 inches in aperture, made a perplexing reference to Herschel in the pages of the *Jahrbuch.* "The Germans should be proud of him," he wrote, but the reason given was the telescopes Herschel had built and not the explanation he had given of the Milky Way.[71]

In general, the *Jahrbuch*'s reports on Herschel matched the preferences of the astronomical fraternity. It was eager to hear of such news as the rotation of the rings of Saturn, the sighting of its moons, and those of Uranus, and the like. Typical in this respect were the reports sent in 1786, 1787, 1788, and 1789 to the *Jahrbuch* by Count von Brühl, the Saxon ambassador in London. Thus, the important news in 1789 was the discovery by Herschel of the seventh and eighth moons of Saturn. This was followed by the puzzling remark: "The diligence and the fruits which he [Herschel] harvests with it, should excuse this untiring explorer of the heavens for his all too long silence."[72] What really required an excuse was the prosaic and brief treatment, in the same issue of the *Jahrbuch,* of Herschel's list of another thousand nebulae.[73]

This strange pattern continued in 1791 when the *Jahrbuch* carried two detailed summaries, by E. G. Fischer, of Herschel's ideas on the construction of the heavens. In the first,[74] Fischer's approach to his topic could not have been more auspicious. He deftly remarked that, although Herschel's discoveries within our planetary system created a greater stir than did his exploration of the Milky Way, the latter achievement would become, at least in retrospect, of far greater importance.[75] However, Fischer offered not one concrete detail on the structure and shape of the Milky Way as outlined by Herschel. Typically, in his second paper,[76] which was devoted to a comparison between Herschel and Kant, Fischer argued that Kant's views on the Milky Way and on the structure of the cosmos were more encompassing than those of Herschel. If he praised Herschel, it was in part because Herschel contributed to the glory of the German genius. The field was thus open for Fischer to vindicate Herschel's independence of Kant. He did this by a rather surprising claim. According to it, Herschel had long ago grown unfamiliar not only with German publications but even with his native tongue.[77] When, in the same year, a full translation of Herschel's three great papers (1784, 1785, 1789) was published in Königsberg, including the extensive list of stars and the diagrams, the translator, G. M. Sommer, an assistant director of the local library, not only insisted on the priority of Kant over Lambert, but also praised the superior quality of Kant's ideas on the Milky Way.[78] The occasion for such a display of local pride has rarely been more ill-chosen.

The only sign of enthusiastic excitement at Herschel's explanation of the Milky Way remained a private matter for a long time. It came from G. F. Lichtenberg, professor of physics at Göttingen. Lichtenberg had far-ranging interests, a bold imagination, and a strong attraction to astronomical and cosmological topics; he became one of Herschel's first correspondents among German men of science. Unique in many respects, Lichtenberg was rather alone as he noted in his letter to Herschel of August 12, 1784, that Lambert's speculations on the Milky Way were but vague anticipations of Herschel's explanation.[79] Lichtenberg stood equally alone in his reaction to Herschel's next paper on the Milky Way which contained the famous drawing of its outlines:

> Highly honoured Sir,
>
> Our mutual friend, Herr Planta, [Secretary of the Royal Society] has brought me your priceless letter accompanying your present [Herschel's paper], for which I render you most obliging thanks; it was as if I received a new pair of eyes to see the heavens. When I first looked at the engraving, I took the star-cluster for a strange nebula which you had perhaps seen outside our system; when, however, I read on and found that this showed a cross section of the nebula of which our sun is but a point—I cannot lie —I was so enraptured at the grand conception and so filled with that admiration for the immense Architect, which let me feel the delight I had experienced when I got my first correct idea about the structure of our own world-system. In one word, I owe to the reading of your paper one of the pleasantest hours of my life for the last twenty years, for which I have to thank you. If I ever visit England again, certainly it will be only and alone with the intention of seeing you and your instruments before I die.[80]

The emotions of the letter can easily be recognized in the essay which Lichtenberg wrote, probably within a few days, so that it might be printed in the *Göttingisches Taschenbuch zum Nutzen und Vergnügen* for 1786, a pocket calendar which also carried literary and scientific reports, written in a popular style.[81] Lichtenberg's enthusiasm comes through clearly in that gem of analogy by which he tried to bring to the level of the motley readership of the *Taschenbuch* Herschel's explanation of the Milky Way: "These considerations put him in the position to grasp and outline a section of our nebula; he did so just as someone in a forest, the trees of which stood evenly but deeply distributed, would be in the position to draw its boundaries without moving from his spot, as his sight towards the open fields [surrounding the forest] would not be completely blocked in any direction, so that it [the edge of the forest] would appear to him closer here and farther there."[82] Lichtenberg also showed a rare grasp of the importance of Herschel's famous diagram of the contours of the Milky Way, which he described in detail. His enthusiasm for Herschel continued unabated as may be seen from his cosmological contributions to subsequent volumes of the *Taschenbuch.* There he spoke of Herschel as "the great expander of astronomy,"[83] as "the extraordinary genius,[84] and also as "the one who alone possesses the keys to the innermost sanctuary of the great temple of nature."[85]

Lichtenberg's enthusiasm and instinctive grasp of the unusual importance of Herschel's explanation of the Milky Way had, however, failed to penetrate into manuals of astronomy written around 1800 in Germany or elsewhere. When, in 1808, Bode, still the center of German astronomy, saw through press the third edition of his famous textbook on astronomy, he felt no need to rewrite its thirty-year-old, first-edition version of the Milky Way and nebulae.[86] This meant that, so far as Bode's textbook was concerned, Herschel still had as little connection with the Milky Way as he had had in 1778. Herschel received no credit for his work on the Milky Way in the four-volume text by Abel Bürja, preacher to the French colony in Berlin before taking the post of professor of mathematics at the Military Academy of Berlin.[87] However, he did ascribe to Herschel the idea that the universe consisted of stellar systems similar to the Milky Way.[88] At the University of Tübingen, J. G. F. Bohnenberger, professor of astronomy, also failed to do Herschel justice in relation to the Milky Way. Bohnenberger's textbook on astronomy, numbering more than 700 pages, belabored the point that our Milky Way, when viewed from a great distance, should look like other small nebular patches in the sky.[89] He said not a word in regard to its structure, and merely credited Herschel in a footnote with the resolution of the Milky Way's nebulosity into an astonishingly large number of stars by means of his telescopes.[90] Herschel was once more passed over when Bohnenberger brought his work to a close with remarks on the possible motion of the solar system toward the constellation Hercules.[91]

Herschel's telescopes were recalled with awe and admiration by Giuseppe Piazzi, the discoverer of the first asteroid, in his textbook on astronomy, where he described his visit with Herschel at Slough in 1787. He referred to Herschel as the discoverer of many nebulae, but all that Piazzi's two-volume textbook offered on the Milky Way was: "According to the now generally shared opinion it is nothing else but the fusion of the light of an immense number of stars. Such an opinion becomes factual evidence when one has access to a telescope of great strength and light, as are Herschel's 20-footers."[92] The real usefulness of Herschel's giant telescopes could not have been concealed more effectively. Both the facts and the history of the question suffered when, in 1806, in the *Dictionnaire des sciences et des arts*[93] Herschel was described as the one who finally proved that the fusion of the light of many stars was the sole cause of the whiteness of the Milky Way. This might perhaps have been acceptable if the anonymous author of the article had been justified in mentioning no one between Manilius and Herschel. The most ironical slighting of Herschel probably came in 1813, in Delambre's *Abrégé d'astronomie*.[94] There Bode was praised for his catalogue of 17,240 nebulae and clusters, published in 1801, while Herschel's name turned up only in a list of names of new constellations, one of which received the name, "Le télescope de Herschel."

In Herschel's adopted homeland, his findings on the Milky Way found no acceptance in the leading textbooks on astronomy of the early 1800's. What Olinthus Gregory wrote in 1803 in his *Treatise on Astronomy* was not only bewilderingly short, but also revealed an undercurrent of misgiving as to the validity of Herschel's inferences: "The milky way he [Herschel] supposes to be that particular nebula in which our sun is placed; and, in order to account for the appearance it exhibits, he supposes its figure to be much more extended towards the apparent zone of illumination, than in any other direction; which is a supposition that he thinks allowable, from the observations he has made on the figures of other nebulae."[95] This was certainly an extremely poor rendering of a magnificent feat, to say nothing of the fact that Herschel did not form his idea of the Milky Way on the basis of other nebulae. There was nothing hypothetical, however, in Gregory's own statement about the nature of the Milky Way. His claim that the Milky Way was a "collection of innumerable little stars"[96] had value only as an illustration of the barriers which great scientific discoveries meet even in scientific circles.

Gregory, professor of mathematics at the Royal Military Academy in Woolwich, had formerly been a member of the faculty at Cambridge, although not its most prominent one. Yet, even the Plumian Professor of Astronomy there, S. Vince, showed no greater appreciation for Herschel's ideas on the Milky Way. In fact, all he said on the subject was that it had been shown by Herschel that the number of stars was immense, especially in the Milky Way. He did so in a massive, three-volume work, proudly entitled *A Complete System of Astronomy,* of which he published a second, revised edition in 1814.[97] Not that Vince deliberately ignored Herschel—he spoke with admiration of Herschel's work on nebulae, and gave ample credit to Herschel for his conclusion that the solar system appeared to be moving in the direction of Hercules. But there was not a hint in Vince's work of the fact that for Herschel stellar motions had a far-reaching cosmological signifiance, and that Herschel had already begun to develop that idea a quarter of a century earlier.

Evidences of stellar motion were to provide Herschel with a glimpse into the actual working of the laboratories of the heavens, the stellar clusters and nebulae. He seized on the topic with an intensity that set the tone of his paper,[98] read on June 11, 1789. In his notebooks there were now data on a thousand more nebulae for which he excitedly tried to find some common denominator. His first synthesis had begun four years earlier, with a significant allusion to the predicament of a naturalist surrounded by an incredible wealth of specimens. The richness of his own findings had now taken on startling proportions. Should he merely enumerate the gems in his bag, or, to quote his words, should the investigator of the skies be "less inquisitive than the natural philosopher, who sometimes, even from an inconsiderable number of specimens of a plant, or an animal, is enabled to present us with the history of its rise, progress, and decay?"[99] What Herschel clearly wanted was the clue to a sound comparison of Milky Ways, and he believed he had found it in the phases of the motion of the stars caused by gravitational attraction. The various phases were indicated for Herschel by the degree of condensation of stars into clusters. Dense clusters were old, and close to their final demise in spectacular collisions of stars. Large, and relatively empty, clusters were specimens at a youthful phase of their development.

Herschel knew, of course, that his analogies had to be clothed in mathematical language. He made much of the idea that overwhelming probability supported the assumption that all clusters of stars mirrored the same central power at work. In other words, they were the logical products of the fundamental laws of nature, instead of the formations of blind chance. From this it followed that all clusters were as much alike as were the various individuals of one species. Herschel's mathematical analysis of the situation was markedly generic, implying great oversimplification of a patently complex phenomenon. But even in the absence of rigorous procedures, of which investigators of nebulae could avail themselves only a century later, there was something contagious in his claim that the evolutionary view of galaxies was "no longer an unfounded hypothesis, but is fully established on grounds which cannot be overturned."[100]

His approach brought, as he proudly put it, a "new kind of light."[101] The thousands of galaxies, whose existence was not even suspected just ten years earlier, appeared in that new light as thousands of distinct, individual, island universes (to anticipate a more modern expression), each living its own life, and each at a different stage of development, some ripening, some aging, some already turning into ashes, others just being reborn from the cosmic dust. A vast variety of members of one species, going through their life-cycles in virtual isolation from one another—such was the vision which the manifold probings into the composition of the Milky Way inspired in the mind of Herschel. It was a fitting conclusion to the first phase of his heroic efforts to fathom the depths not of a universe of stars, but of a universe of Milky Ways.

One point that remained obviously unexplained in that grandiose cosmic portrait was the cosmic ashes or dust into which the stars declined, and out of which they were to rise anew. Herschel's phrasing could readily create the impression that stars were the basic entities in the universe. Only on occasion did he hint that certain phenomena, especially the peculiar surroundings of planetary nebulae, seemed to resist his classifications.[102] Observations forced him, before long, to recognize that, besides stars, there also existed true nebulous matter in interstellar space. The momentous discovery came on the night of November 13, 1790. His record of it starts with the exclamation: "A most singular phenomenon!"[103] It was an eighth-magnitude star surrounded by a luminous atmosphere, of the same hue throughout, without even the slightest evidence of gradation. In this last circumstance lay the crux of the matter. Herschel had seen hundreds of nebulous patches, parts of which could not be resolved into stars. But in every case there was a stepwise transition from areas of great brilliance and distinct stars to areas of irresolvable faint nebulosity. In the Milky Way, for instance, he invariably found a chain of "well-connected steps . . . from the most evident congeries of stars to other groups in which the lucid points were smaller, but still very plainly to be seen; and from them to such wherein they could but barely be suspected, till I arrived at last to spots in which no trace of a star was to be discerned."[104] It was on that invariable pattern that he based his belief that all nebulous areas would be resolved into stars with better telescopes.

But now something had been sighted which suggested that the inference was not at all conclusive. Until then, Herschel explained, he felt like a naturalist whom the almost imperceptible transitions between different species prompted to believe that there was probably no boundary between the animal and the vegetable kingdoms. Should, however, the naturalist consider a man and a tree side by side, he could hardly avoid being struck by the difference. As Herschel deftly noted: "a glance like that of the naturalist, who casts his eye from the perfect animal to the perfect vegetable, is wanting to remove the veil from the mind of the astronomer."[105] He did not hesitate to cast off the veil which, as he now realized, blocked his vision: "Our judgment, I may venture to say, will be, that the nebulosity about the star is not of a starry nature."[106]

If the new conclusion had a bearing on anything, it was the question of the extent of the Milky Way, whose depths were invariably lost in some misty nebulosity. If nebulous matter constituted a species of its own, distinct from stars, were not the gaugings of the Milky Way subject to serious uncertainties and ambiguities? Could the Milky Way still be considered as an entity with well-defined limits, when the astronomer's line of sight ended in a nebulous background, that could in itself be a congeries of stars, but might also be a strictly nebulous domain, with unfathomable depths? And could not those depths be imagined as so far-reaching as to include all the other nebulous patches, or island universes? None of these questions appeared on Herschel's mental horizon. Not that this newly established substance of the heavens had not given him food for thought. If it awakened anything in his mind, it was his irrepressible penchant for speculating on the birth, growth, decay, and rebirth of stars and galaxies. In a complete reversal of his previous conviction, he now derived stars from nebulae: "If, therefore, this matter is self-luminous, it seems more fit to produce a star by its condensation than to depend on the star for its existence."[107] Clearly, Herschel was not a celestial geometer, engrossed with the strict implications of postulates about geometrical structure, concerning finiteness or infinity. He always remained a celestial naturalist, unexcelled, to be sure, concentrating on evolutionary patterns that could be rigorously investigated only a century or so afterward.

Herschel investigated them anyhow. With no thermodynamics, kinetic theory, spectroscopy, or atomic physics at his disposal, he could not do much more than open up important, though somewhat poetical, perspectives. His evolutionary process could now be extended beyond the stars into their "ashes," that is, into the nebulous matter: "The surmise of the regeneration of stars, by means of planetary nebulae, expressed in a former Paper, will become more probable, as all the luminous matter contained in one of them, when gathered together into a body of the size of a star, would have nearly such a quantity of light as we find the planetary nebulae to give."[108] He obviously felt elated, for he suggested in the same breath a possible experimental verification of the new, star-generating process. He also conjectured boldly as to whether the nebulous material could not be caused by the flux of light from the stars. True, the subtility of light particles was beyond imagination, or as he put it, "unconceivable."[109] But equally inconceivable was the number of stars providing that flux. Here Herschel betrayed himself, saying that the number of the emitting bodies was "almost infinitely great."[110] This was the imprecision of the naturalist unconcerned with or shying away from strict infinity. His overwhelming preoccupation was not with extent but with transformation, a fact which rapidly began to produce a one-sidedness in his accounts of the Milky Way, the nebulae, and the universe.

The Preface, written for his third and last catalogue of nebulae and star clusters,[111] offered further evidence that his mind was fixed on classification and evolution. In a most expressive passage, he called attention to the fact that distances of stars and nebulae were indications of an immense past.[112] There is indeed something slightly ironic in the initial and often-quoted phrase of the great paper of June 20, 1811,[113] in which Herschel restated that a knowledge of the construction of the heavens was always the ultimate object of his observations. Even at the start of his observations, the parameter of time seemed to fascinate him as much as the parameter of space. Now he showed little interest in the latter. The "new light," which he expected from a most painstaking and detailed classification of nebulae, was to inform about sequence in time, not correlation in space.[114] Three years later, a thorough reworking of the problem of classification ended with a conclusion which extolled the category of time. This emphasis was ultimately derived from the primacy which he had come to accord to the nebular over the stellar: "the state into which the incessant action of the clustering power has brought it at present, is a kind of chronometer that may be used to measure the time of its past and future existence; and although we do not know the rate of going of this mysterious chronometer, it is nevertheless certain, that since the breaking up of the parts of the milky way affords a proof that it cannot last for ever, it equally bears witness that its past duration cannot be admitted to be infinite."[115]

This is not to suggest that Herschel had completely abandoned consideration of the extent of the Milky Way. But the nebulous matter dampened interest in this respect as much as it effaced clear contours and limits. Characteristically, he now listed a group of nebulae under the heading of "objects of an ambiguous construction."[116] His sensitivity to the implications of assumptions and basic standpoints made all too clear to him that a reassessment of the value of his giant telescopes was inevitable. His spectacular career started with his unparalleled skill in improving the resolving power of telescopes by orders of magnitude. Formerly, he had believed with absolute assurance that all nebulous patches would one day reveal themselves as distinct groups of stars to his ever-improving telescopes. Now the nebulosity seemed to reveal inherent ambiguity never to be resolved by telescopes, however perfect: "We have indeed no reason to expect that an increase of light and distinctness of our telescopes would free us from ambiguous objects; for by improving our power of penetrating into space, and resolving those which we have at present, we should probably reach so many new objects that others, of an equally obscure construction, would obtrude themselves, even in greater number, on account of the increased space of the more distant regions of their situation."[117]

His original ambition to fathom the heavens remained to the end a factor to reckon with. He disclosed that he was making a series of observations "for ascertaining a scale whereby the extent of the universe, as far as it is possible for us to penetrate into space, may be fathomed." He still hoped, although in his mid-seventies, to make one "final investigation of the universal arrangement of all these celestial bodies in space."[118] Such disclosures came, ironically, in the paper which contained the finishing touches to his second synthesis of the construction of the heavens. In that synthesis the universe was "almost infinite" and the contours of the Milky Way evanesced in the midst of that newly found nebulosity. The only distinct and general feature of that synthesis was the evolutionary process which ruled the appearance of each of the innumerable species of stars and nebulae.

Herschel's report of his final fathoming of the heavens[119] started on a note that recalled the opening remark of his great paper of 1784, that until then the heavens had been considered a vast spherical surface. By this statement Herschel really meant that astronomers, as he now rephrased the idea, had until recently been largely concerned with the determination of the angular separation of stars. A grasp of the real construction of the heavens demanded, however, the exploration of stars in the third dimension; that is, to use a new expression of Herschel, their profundity. The choice of the word was appropriate in more than one respect. Herschel went about his business with profound dedication, speculation, and skill. All three were in full use as the near-octogenarian made his last attempt to clear up the ambiguities of the sky, two of which remained outstanding: the so-called ambiguous nebulosities, and the doubt they cast upon the true contours of the Milky Way. His conclusions reflected ambiguity itself. To begin with, he had to recognize that Halley's concentric shell model of stars did not match the count of stars of the first twelve magnitudes. The marked discrepancy between the computed and the observed numbers raised serious doubts in him about the value of "gaging" as a clue to estimating the depth of a given field of stars.

Even more upsetting had been the recognition that "the utmost stretch of the space-penetrating power of the 20 feet telescope could not fathom the Profundity of the milky way, and that the stars which were beyond its reach must have been farther from us than the 900dth order of distances."[120] Years earlier he would have immediately had recourse to the "space-penetrating power" of an even larger telescope to escape from the dilemma. Had he not constructed one telescope with a 40-foot focal length, and an aperture of 48 inches? It was the wonder of the age when mounted in 1788. Soon, however, it proved to be a mixed blessing, apart from its mechanical awkwardness. As Herschel had already remarked in 1799, 598 years would have been needed to scan just the Northern Hemisphere with it.[121] Herschel had never used it extensively, nor did his final findings strengthen his faith that it could clear up the principal ambiguities that vexed him. Although Herschel stated its space-penetrating power as the 2300th order of distances, or 2300 times the distance of Sirius, he could not help adding: "it would then probably leave us again in the same uncertainty as the 20 feet telescope."[122]

His mind, eager to find the true size of the Milky Way, instinctively recoiled from accepting such a dispiriting outcome. A year later, in his last words on the topic,[123] he spoke of the 40-foot telescope as an instrument that "may then truly be called a finder."[124] Immediately afterward there followed a section whose title gives a penetrating glimpse of Herschel's most cherished conception: "The Milky Way is not an ambiguous object."[125] The conclusion of that section had no room for nebulosity. The Milky Way was again pictured as a row of stars, clearly resolvable as far as telescopes could reach: "the application of a higher magnifying power evinced that the doubtful appearance was owing to an intermixture of many stars that were too minute to be distinctly perceived with the lower power."[126] The same firmness characterized his conclusion: "when our gages will no longer resolve the milky way into stars, it is not because its nature is ambiguous, but because it is fathomless."[127]

The Milky Way was not to be so easily disposed of. Herschel unwittingly played the prophet when he brought his paper to a close with a discussion of "ambiguous celestial objects."[128] Unfortunately, the ambiguity lay not only in the nebulosity of those celestial objects but also in Herschel's style. When he said that the Milky Way was "fathomless," he hardly intended to relinquish his long-standing belief that nebulae, by and large, represented systems of stars independent of it. But, by 1818, his first papers on the construction of the heavens were already gathering dust. The scientific public, always eager to regale itself with the last tidbits, took his latest words as quite separate from utterances spanning over more than thirty years. This set the stage for century-long and often illogical arguments on the question of island universes. The mysterious absence of logic in the long history of man's groping about the truth of the Milky Way was, it seems, to assert itself once again. It certainly evidenced itself in Herschel's obituaries and eulogies following his death in 1822. They contained no reference to Herschel's first work on the Milky Way, on his discovery of the true explanation of that whitish zone in the sky. His was indeed a "silent breakthrough," fittingly symbolized by the quiet of the little church of Upton, where his epitaph silently proclaims that he was the one who —*coelorum perrupit claustra*—broke through the vault of the skies.

Herschel's obituary notice in *The Gentleman's Magazine* not only enveloped in vagueness the views "of a daring sublimity" which Herschel had developed about "our own system" (the Milky Way was not mentioned as such), but erroneously ascribed them to the observations made by the "stupendous instrument" (meaning the 40-foot telescope) which Herschel completed in 1787.[129] Accuracy, which never had been the chief virtue of journalism, should certainly have been the hallmark of the eulogy delivered on Herschel by J. B. J. Fourier,[130] "perpetual secretary" of the French Academy, who is best remembered for his analysis of the differential equations governing the transfer of heat. But Fourier was an orator as well, free to indulge in pleasing generalities. In a lengthy oration of some twenty pages, he allotted a mere dozen lines to Herschel's work on the Milky Way. The only specific information included was a remark that Herschel had no knowledge of the ideas of Kant and Lambert. Fourier had no inkling of Wright, which suggests that what he said of Kant was secondhand information. Nor did Fourier care to specify the points on which Kant and Lambert had anticipated Herschel. Again, Fourier did not identify that "explanation [of the Milky Way] which Herschel derived from positive and manifold observations, and which had been anticipated by the famous philosopher of Königsberg and by the academician from Berlin."[131]

When, in 1832, Bessel outlined, in a long lecture[132] before the Physico-economical Society of Königsberg, the achievements of astronomy during the previous hundred years and the current prospects, nothing was said of Herschel's explanation of the Milky Way. Bessel brushed aside Herschel's "magnificent undertakings" as details belonging not to astronomy proper but to the physical description of the heavens![133] In Vienna, J. J. Littrow, director of the Observatory there, in 1835 explained the lenticular shape of the Milky Way without referring to Herschel.[134] H. W. Brandes, professor of physics at the University of Leipzig and one of the editors of a greatly revised edition of Gehler's *Physikalisches Wörterbuch,* gave no adequate picture of Herschel's incredible labors and achievements in his article, "Die Milchstrasse," which appeared there in 1837.[135] He failed to realize that Herschel was probably unaware of the publications of Kant and Lambert to whom, especially the former, the article gave undue space.

Bessel's style suddenly became replete with superlatives when in 1843 he devoted a long, special lecture to Herschel's achievements of which the greatest, Bessel declared, was ascertaining the boundaries of the Milky Way! This mistaken encomium was all the more belated since Bessel insisted that, from the very beginning of his career, Herschel's work had loomed before him in the splendor of extraordinary excellence.[136] Actually, Bessel's attention had been drawn to Herschel by the publication, a year earlier, of Arago's long essay on Herschel's life and work. There Arago made the much overdue remark about the "painstaking analysis [of the Milky Way] which Herschel substituted for the imperfect insights of his predecessors [Wright, Kant, and Lambert]."[137] But then Arago, who ended his essay with a list of Herschel's papers, gave an account of Herschel's ideas on the Milky Way which did not extend beyond Herschel's paper of 1785. Arago failed to so much as hint that Herschel had developed doubts about the possibility of sighting the true limits of the Milky Way. Nor did he seem to be aware of the imprecision of Herschel's later utterances, which led to the most vainly debated question in astronomy for the rest of the century. Herschel's vague question as to whether the Milky Way was unfathomable became the question of whether or not the whole universe was contained within an unfathomable Milky Way, or, in other words, whether there was one universe or many.

References

[1] "On the Proper Motion of the Sun and Solar System: with an account of several Changes that have happened among the Fixed Stars since the time of Mr. Flamsteed" (1783), in *The Scientific Papers of Sir William Herschel,* edited by J. L. E. Dreyer (London: The Royal Society and The Royal Astronomical Society, 1912), vol. 1, pp. 108–30. This edition of Herschel's papers will be quoted as Dreyer. The most important papers of Herschel on the Milky Way, and on the stellar and nebular realm in general, are reprinted almost in full in *William Herschel and the Construction of the Heavens* by M. A. Hoskin, with astrophysical notes by D. W. Dewhirst (New York: Norton and Norton, 1964), a work noted for its scholarship and the incisive analysis of its subject. There one also finds the basic bibliography on Herschel's life and work, on the milieu in which he pioneered, on his instruments and ideas.

[2] For the text of the letter accompanying Herschel's paper, see C. A. Lubbock, *The Herschel Chronicle: The Life Story of William Herschel and his Sister Caroline Herschel* (Cambridge: University Press, 1933), p. 201.

[3] "Sur les taches du soleil et sur sa rotation," *Histoire de l'Académie des Sciences avec les mémoires de mathématique et de physique, Année 1776,* pp. 457–514.

[4] Lubbock, *The Herschel Chronicle,* p. 201.

[5] "Ueber einen im gegenwärtigen 1781sten Jahre entdeckten beweglichen Stern . . ." *Astronomisches Jahrbuch für das Jahr 1784* (Berlin: G. J. Decker, 1781), p. 211 note.

[6] Paris: Imprimerie de la République, An IX [1801].

[7] Lubbock, *The Herschel Chronicle,* p. 201.

[8] "Account of some Observations tending to investigate the Construction of the Heavens," in Dreyer, vol. 1, pp. 157–66.

[9] *Ibid.,* p. 157.

[10] *Ibid.,* p. 158.

[11] *Ibid.*

[12] *Ibid.,* pp. 160–61.

[13] *Ibid.,* p. 161.

[14] *Ibid.*

[15] *Ibid.*

[16] *Ibid.*

[17] *Ibid.,* p. 158.

[18] "Catalogue intéressant des Nébuleuses et des amas d'Etoiles, observés par M. Messier à l'Observatoire de la Marine," in *Connoissance des Temps pour l'Année commune 1783* (Paris: de l'Imprimerie Royale, 1780), pp. 225–54.

[19] Dreyer, vol. 1, p. 165.

[20] *Ibid.*

[21] *Ibid.*, pp. 162–63.

[22] Lalande's views on the Milky Way will be discussed shortly.

[23] Dreyer, vol. 1, p. 166.

[24] "On the Construction of the Heavens," Dreyer, vol. 1, pp. 223–59.

[25] *Ibid.*, p. 223.

[26] *Ibid.*

[27] *Ibid.*, p. 225.

[28] *Ibid.*

[29] "On the Parallax of the Fixed Stars," read on December 6, 1781; in Dreyer, vol. 1, pp. 39–57.

[30] *Ibid.*, p. 52.

[31] Herschel's reference is: "Dr Halley on the Number, Order and Light, of the Fixed Stars," *Phil. Trans.* vol. XXXI." It contained an elaboration of the solution given by Halley to the optical paradox in a paper, "Of the Infinity of the Sphere of Fix'd Stars," immediately preceding the former. For further details, see notes 51 and 52 in Chapter V.

[32] Dreyer, vol. 1, p. 226.

[33] *Ibid.*

[34] *Ibid.*

[35] *Ibid.*

[36] *Ibid.*

[37] *Ibid.*, p. 227.

[38] *Ibid.*

[39] *Ibid.*, p. 252.

[40] His diagram suggested that the Milky Way was some six thousand light years in diameter, whereas the actual estimate is about fifteen times greater.

[41] Dreyer, vol. 1, p. 245.

[42] *Ibid.*, p. 247.

[43] *Ibid.*, pp. 247–48.

[44] *Ibid.*, p. 252.

[45] *Ibid.*, pp. 252–53.

[46] *Ibid.*, p. 253.

[47] *Ibid.*, p. 259.

[48] *Ibid.*

[49] *Zweiter Theil, Theoretische Astronomie,* pp. 53–59.

[50] *Ibid.*, p. 54. Curiously, Schubert added, on p. 57, that "already Kepler submitted thoughts similar to those of Lambert and Herschel on the position and shape of the Milky Way," and referred to Kepler's *Epitome*. Schubert did not mention Wright or Kant. Twenty-five years later the French edition of the work still retained the same remark *(Traité d'astronomie théorique* [St. Pétersbourg: Imprimerie de I'Académie Impériale des Sciences, 1822, vol. 2, p. 84]). Schubert's *Populäre Astronomie* (St. Petersburg: Kayserl. Akademie der Wissenschaften, 1804–10) reported much the same about Herschel's work both in its historical part (vol. 1, p. 139), and in its section on physical astronomy (vol. 3, pp. 60–74). In his last discussions of the Milky Way and of its relation to the whole of the universe, which appeared in his *Vermischte Schriften* (Leipzig: F. A. Brockhaus, 1840, vol. 6, pp. 89–116 and vol. 7, pp. 3–57), Schubert closely followed Lambert's hierarchial ordering of stars and galaxies. He identified some nebulae as independent star systems of equal rank with the Milky Way, and put their distance from us at 200 times the diameter of the Milky Way, which he gave as 500 times the distance of the nearest stars.

[51] "Vérification des nouvelles découvertes faites en Angleterre sur les Etoiles fixes," in *Histoire de l'Académie Royale des Sciences, Année MDCCLXXXIV* [1784], *avec les Mémoires de mathématique et de physique pour la même Année* (Paris: Imprimerie Royale, 1787), pp. 331–41. For quotation, see p. 333.

[52] Paris: Imprimerie de P. Didot l'Ainée, 1792.

[53] *Ibid.*, vol. 1, p. 271.

[54] Paris: Imprimerie de la République, An IX [1801].

[55] *Ibid.*, p. iv.

[56] *Ibid.*, p. v.

[57] Paris: Imprimerie de la République, An XI [1803], pp. 661–880.

[58] *Ibid.*, p. 666.

[59] *Ibid.*, p. 668.

[60] Paris: Mme Ve Courcier, 1814. See chapter 16, in vol. 1, pp. 409–80.

[61] *Ibid.*, pp. LVI and LXIV.

[62] *Histoire de l'astronomie au dix-huitième siècle* (Paris: Bachelier, 1827), a work published five years after Delambre's death.

[63] On this reference to Herschel, see 4th ed. (1813), p. 431; 5th ed. (1824), p. 395. On Laplace's references to the Milky Way, see 1st ed. (1796), vol. 2, pp. 306–07; 2d ed. (1799), pp. 348–49; 3d ed. (1808), vol. 2, pp. 393–94; 4th ed., p. 445; 5th ed., p. 395.

[64] Paris: chez Courcier, 1810. The title page has the following information about the rather elusive author, M. [Monsieur] Voiron: "Docteur de la Faculté des Sciences, Ancien Professeur de Belles-Lettres, actuellement Professeur de Mathématiques transcendantes au Prytanée Militaire."

[65] *Ibid.*, pp. 11–40.

[66] *Histoire de l'astronomie moderne, depuis la fondation de l'école d'Alexandrie jusqu' à l'époque de MDCCLXXXII* (Paris: Chez les Frères Debure, 1779–82).

[67] *Ibid.*, vol. 3, p. 85.

[68] *Ibid.*, pp. 261–62.

[69] This widely circulating Yearbook was published in Berlin, usually three years ahead of the year for which it contained the information about the position of planets, eclipses, and comets to be observed. In the period under question, Johann Elert von Bode was its editor, but he seemed to be much more interested in obtaining some original material for his *Jahrbuch* from Herschel than in the latter's ideas. Subsequent references will be given as *A. J. for 1788* [1785], the date in brackets indicating the year of publication.

[70] *A. J. for 1788* [1785], pp. 246–54.

[71] *A. J. for 1789* [1786], p. 153.

[72] *A. J. for 1793* [1790], p. 113.

[73] *Ibid.*, pp. 104–07.

[74] *A. J. for 1794* [1791], "Ueber die Anordnung des Weltgebäudes, ein freier Auszug aus Herrn Herschels Schriften über diese Materie von Herrn Professor Fischer in Berlin," pp. 213–26. [Ernst Gottfried] Fischer was at that time professor of mathematics in the Gymnasium "zum grauen Kloster" in Berlin, and from 1810 until his death in 1831, assistant professor of chemistry at the University of Berlin.

[75] *Ibid.*, p. 214.

[76] "Einige Anmerkungen zu dem Vorigen Aufsatz von ebendemselben," *ibid.*, pp. 226–33.

[77] *Ibid.*, pp. 227–28. This last remark was somewhat exaggerated.

[78] *William Herschel über den Bau des Himmels.* The subtitle reads: "Drey Abhandlungen aus dem englischen übersetzt nebst einem authentischen Auszug aus Kants allgemeiner Naturgeschichte und Theorie des Himmels." (Königsberg: bey Friedrich Nicolovius, 1791). See especially pp. 201–04. This translation of Herschel's three papers was referred to in a rather mediocre article by Prof. Ideler on Herschel's ideas on the construction of the heavens, in *A. J. for 1807* [1804], pp. 113–29.

[79] *Lichtenbergs Briefe,* edited by Albert Leitzmann and Carl Schüddekopf, (Leipzig: Dieterich'sche Verlagsbuchhandlung, 1902), vol. 2, p. 136. In the same letter, Lichtenberg also noted the excitement created in London by Herschel: "When you stayed in London last year, an English scientist, whose name I keep to myself, wrote [to me]: Your countryman, Mr. Herschel, is now in town to teach our astronomers how to see" (*ibid.,* p. 136).

[80] *Ibid.,* pp. 238–39.

[81] Etwas von Herschels neuesten Entdeckungen," in *Georg Christoph Lichtenbergs Vermischte Schriften* (Göttingen: Heinrich Dieterich, 1803), vol. 6, pp. 333–46.

[82] *Ibid.,* pp. 337–38. Only five years earlier, the Milky Way received a mere passing reference as a zone of light from Lichtenberg, in his lengthy essays on the structure of the world, published in the *Taschenbuch,* for the years 1779 and 1780. See *ibid.,* pp. 172–210 and 221–53; the short reference to the Milky Way is on p. 236. Typically, the astronomer mentioned there was J. Ferguson.

[83] In 1797, see *ibid.,* vol. 7, p. 19.

[84] In 1792, see *ibid.,* vol. 7, p. 6.

[85] In 1791, see *ibid.,* vol. 6, p. 417.

[86] *Erläuterung der Sternkunde und der dazu gehörigen Wissenschaften* (3d rev. ed.; Berlin: Himburg, 1808), see pp. 319, and 321. The first edition was discussed in the previous chapter.

[87] *Lehrbuch der Astronomie* (Berlin: bei Schöne, 1794–1806).

[88] *Ibid.,* vol. 3 (1798), p. 101.

[89] *Astronomie* (Tübingen: J. G. Cotta, 1811), p. 695.

[90] *Ibid.,* p. 695.

[91] *Ibid.,* p. 697.

[92] *Lezioni elementari di astronomia ad uso del real Osservatorio di Palermo* (Palermo: dalla Stamperia Reale, 1817), vol. 1, p. 239.

[93] Edited by M. Lunier (Paris: chez Le Normant, 1806), vol. 3, pp. 522–23.

[94] J. B. J. Delambre, *Abrégé d'astronomie ou leçons élémentaires d'astronomie théorique et pratique* (Paris: Mme Veuve Courcier, 1813), p. 193.

[95] London; printed for G. Kearsby, 1803, p. 481.

[96] *Ibid.*, p. 42.

[97] London: G. Woodfall, 1814; see especially chap. 27, "On the fixed stars," vol. 1, pp. 487–508.

[98] "Catalogue of a Second Thousand of new Nebulae and Clusters of Stars; with a few introductory Remarks on the Construction of the Heavens," in Dreyer, vol. 1, pp. 329–69.

[99] *Ibid.*, p. 330.

[100] *Ibid.*, p. 335.

[101] *Ibid.*, p. 337.

[102] See his papers of February 3, 1785 (Dreyer, vol. 1, p. 257) and of June 11, 1789 (Dreyer, vol. 1, p. 337).

[103] "On Nebulous Stars, properly so called," read on February 10, 1791; in Dreyer, vol. 1, pp. 415–25; for entry quoted, see pp. 421–22.

[104] *Ibid.*, p. 415.

[105] *Ibid.*, p. 416.

[106] *Ibid.*

[107] *Ibid.*, p. 423.

[108] *Ibid.*, p. 424.

[109] *Ibid.*

[110] *Ibid.*

[111] "Catalogue of 500 new Nebulae, nebulous Stars, planetary Nebulae, and Clusters of Stars; with Remarks on the Construction of the Heavens," read on July 1, 1802; in Dreyer, vol 2, pp. 199–234.

[112] *Ibid.*, p. 213.

[113] "Astronomical Observations relating to the Construction of the Heavens, arranged for the Purpose of a critical Examination, the Result of which appears to throw some new Light upon the Organization of the celestial Bodies," in Dreyer, vol. 2, pp. 459–97; for quote, see p. 459.

[114] *Ibid.*, p. 459. As Hoskin aptly noted: "The importance of the temporal dimension in his theory increases . . ., whereas the spatial distribution of

nebulae and clusters is now significant in so far as it offers clues to their development in time," *William Herschel and the Construction of the Heavens,* p. 130.

[115] "Astronomical Observations relating to the sideral part of the Heavens, and its Connection with the nebulous part: arranged for the purpose of a critical Examination," read on February 24, 1814; in Dreyer, vol. 2, pp. 520–41; for quote, see p. 541.

[116] *Ibid.*, pp. 526–28.

[117] *Ibid.*, p. 528.

[118] *Ibid.*, p. 540.

[119] "Astronomical observations and experiments tending to investigate the local arrangement of the celestial bodies in space, and to determine the extent and condition of the Milky Way," read on June 19, 1817; in Dreyer, vol. 2, pp. 575–91.

[120] *Ibid.*, pp. 588–89.

[121] "On the Power of penetrating into Space by Telescope . . .," read on November 21, 1799; in Dreyer, vol. 2, pp. 31–52; see p. 52.

[122] *Ibid.*, p. 589.

[123] "Astronomical observations and experiments, selected for the purpose of ascertaining the relative distances of clusters of stars, and of investigating how far the power of our telescopes may be expected to reach into space, when directed to ambiguous celestial objects," read on June 11, 1818; in Dreyer, vol. 2, pp. 592–613.

[124] *Ibid.*, p. 608.

[125] *Ibid.*, p. 609.

[126] *Ibid.*

[127] *Ibid.*

[128] *Ibid.*, pp. 611–13.

[129] *The Gentleman's Magazine and Historical Chronicle* 92 (1822): 275.

[130] "Eloge historique de Sir William Herschel prononcé dans la séance publique de l'Académie royale des sciences, le 7 Juin 1824," *Mémoires de l'Académie royale des sciences de l'Institut de France,* Année 1823, tome VI (Paris: chez Firmin Didot, 1827), pp. LXI–LXXXI.

[131] *Ibid.*, p. LXXVI.

[132] *Populäre Vorlesungen über wissenschaftliche Gegenstände,* edited by H. C. Schumacher (Hamburg: Perthes-Besser & Mauke, 1848), pp. 1–33.

[133] *Ibid.*, p. 15.

[134] *Die Wunder des Himmels oder gemeinfassliche Darstellung des Welt-systems* (Stuttgart: Hoffman, 1835), vol. 2, p. 358.

[135] Sechster Band. Dritte Abtheilung, Me–My. Leipzig: E. B. Schwickert, 1837, pp. 2281–88.

[136] "Sir William Herschel," in *Abhandlungen von Friedrich Wilhelm Bessel,* edited by Rudolf Engelmann (Leipzig: Verlag von Wilhelm Engelmann, 1876), vol. 3, pp. 458–78.

[137] See especially p. 451 in his essay, quoted in note 80 of the previous chapter.

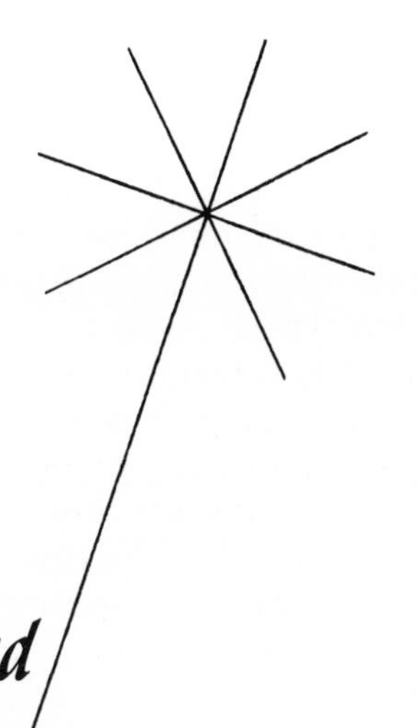

CHAPTER EIGHT

The Myth of One Island

"WE arrive, therefore, at the perhaps unexpected but incontestable result that the system of Herschel, which he proposed in 1785, collapses completely, by virtue of the subsequent investigations of its author; and that Herschel himself had entirely abandoned it." Such was the famous claim made in 1847 by F. G. W. Struve,[1] first director of the Pulkovo Observatory, and one of the giants among astronomers during the second quarter of the nineteenth century. Nevertheless, he had to admit that most astronomers held fast to the explanation of the Milky Way proposed by Herschel in 1785. The reasons for this, according to Struve, lay in two factors. First, Herschel's explanation of the Milky Way represented a compact whole, impressive in the boldness and geometrical precision of its construction. Second, Herschel had never renounced it in its entirety.[2]

Struve's claim, which was not without a touch of contradiction, rested on two supports. One was his interpretation of various statements of Herschel, although Struve himself acknowledged that, after 1802, one could find only partial conclusions on the subject of the Milky Way in Herschel's papers.[3] The other was Struve's meticulous analysis of the data in several new stellar catalogues, especially the ones compiled by Bessel in Königsberg, by Argelander in Bonn, the complete re-evaluation of Lalande's *Histoire céleste française* sponsored by the Royal Astronomical Society, and, last but not least, the data obtained by Struve himself with the aid of the famed Pulkovo refractor, a masterpiece of Fraunhofer. "It now remains to compare the results with the phenomenon of the Milky Way,"[4] he wrote as he summarized the significance of his material in six points. The six points were not much to boast about. Their most specific item concerned the alleged position of the sun as slightly removed from the densest layer of the stratum of stars. In consequence, Struve argued, the Milky Way was not really a great circle. As to its structure Struve could not have offered more generic statements: "It is, therefore, beyond doubt that the phenomenon of the condensation of stars towards a main line of the equatorial disk is most closely tied to the nature of the Milky Way, or rather, the condensation in question and the appearance of the Milky Way are identical phenomena."[5]

This was hardly a new revelation, but Struve was convinced that nothing specific could be revealed about the contours of the Milky Way. "Herschel proved in 1817 that the Milky Way was unfathomable even for his 40-foot telescope."[6] Struve's phrasing was slightly at fault. The culprit was the letter "t." Instead of writing "Herschel a prouvé," he should have written "Herschel a trouvé," that is, "Herschel found it." Herschel, as noted in the previous chapter, crowned that discovery of his not with a rigorous conclusion, but with a comment pervaded with the resignation of old age, barring him from any further exploration of the question. Struve, however, clearly sought in Herschel's words a confirmation of his own cherished conviction that the Milky Way was strictly unfathomable. This was, in fact, the principal conclusion which he deduced from his own analysis of the star catalogues. "The same uncertainty, which Herschel found with his 40-foot telescope," wrote Struve, "obtains, concerning the limits of the visible stars, in all other directions of the celestial vault, including also the directions toward the poles of the Milky Way. Nowhere are we in a position to distinguish the very last [row of] stars."[7] Struve could not have been more categorical in stating his grand conclusion: "It follows that *if we consider all the fixed stars which* surround the sun, as forming a great system, that of the Milky Way, we are in a *perfect ignorance about its extent and that we have not the slightest idea* of the external form of that immense system."[8]

Whatever the merit of Struve's claim that his conclusions perfectly matched Herschel's final ideas about the Milky Way, he certainly did not follow Herschel in the bold speculations on which

Herschel had put so much emphasis.[9] Compared with Herschel's boldness, Struve's approach was distinctly timid, and contrasted oddly with the directives he formulated. In his opinion, the man of science should never retreat when confronted by an obscure phenomenon, or a difficult investigation. Study of previous work, acquisition of new data, would always give him the assurance of making progress, provided he used "a calm speculation, without yielding to the influences of an excited and preoccupied imagination." Persistence, coupled with mathematics, would then lead him to "unexpected results which might, however, possess a high degree of certainty."[10] The "unexpected" was to elude Struve.

The results of what he called his "further researches on the Milky Way," undoubtedly possessed the glitter of mathematical certainty, but their ultimate effect was to wrap the Milky Way in a studied vagueness. Struve found that counts of stars of the first nine magnitudes provided him with a specific law about the density of stars in the successive first nine spherical shells, each the width of the average distance of stars from one another. The law indicated that "the radii of the spheres which separate the stars of different class [magnitude] form a geometrical progression very close to the ratio of $1/\sqrt{2}$."[11] This could mean that stars were not homogeneously distributed; that is, their density diminished with distance, an inference which Struve plainly did not wish to entertain. To explain the difference between the theoretical power of telescopes and the actual star counts obtained with them at various distances which they could theoretically fathom, Struve sought refuge in the theory of the absorption of starlight by interstellar ether.

Such an escape from the dilemma held, however, an implication of potentially disastrous consequence. If the depths of the Milky Way were to be conceived of as unlimited, then those depths might also be regarded as infinite. Struve was so conscious of this possibility that, in the next chapter of his book, entitled "On the extinction of the light of fixed stars in its passage through the celestial spaces,"[12] he began wih a detailed discussion of Olbers' famous paper on the optical paradox of an infinite number of stars. Olbers, as is well known, proposed as a solution the absorption of light, without being able to provide an independent proof of it.[13] Struve believed that his star counts gave him precisely that proof, for he concluded, on the basis of those counts, that light lost

1/107th of its intensity in passing through unit distance—that is, the distance of a first-magnitude star.[14] In his words, "the intensity of light decreased in a proportion greater than the inverse square of distances."[15] Thus, the threatening specter of the paradox of infinity seemed to be removed, or, to quote Struve, (1) "by admitting the extinction [of starlight], the brilliance of the Milky-Way-without-limit will no longer be infinite."[16]

Once this problem was disposed of, five specific and confidently quantitative conclusions could be drawn: (2) the stars situated beyond the reach of the Herschelian telescope of 20 feet, contribute only 12 percent to the total brilliance of the Milky Way; (3) in the direction of the central regions of the Milky Way the stars invisible to the naked eye contribute 0.92007th to its brightness; (4) the luminosity of the sky in the direction of the poles is only 0.22 percent of that in the central regions of the Milky Way; (5) there is a distinct variation in the portion which stars of different magnitudes contribute to the total luminosity in the given directions; (6) the luminosity of the Milky Way is six times stronger than the luminosity of the background sky elsewhere. "All those conclusions are in admirable conformity with the general appearance of the celestial vault," so Struve claimed.[17]

On closer inspection, the actual situation was not nearly so admirable. First, there was a strange little phrase added to the conclusion: (3) "We have neglected here the stars situated beyond the reach of the Herschelian telescope, in the direction of the pole; but we know that their number is extremely small, and that their contribution to the total luminosity is entirely negligible."[18] But could one be categorically sure that the number in question was small? Were not inferences of this type always somewhat hypothetical, and were they not particularly so in the framework of Struve's overcautious positivism? The inability to reach the limits of the Milky Way with Herschel's telescopes presented no logical basis for Struve's peremptory conclusion that the Milky Way was strictly unfathomable. By the same token, the small count of stars toward the galactic poles provided no rigorous justification for Struve's statement that we *do* know that the number there remains very small, even beyond the reach of the best telescopes. One could say at most that the density of stars was much less in those directions, but then, in assuming an infinite universe, one still had to face the problem of inhomogeneity in addition to that of Olbers' Paradox. If the density of stars in the polar directions was much less than that toward the plane of the Milky Way, but remained homogeneous, their combined light was still infinite.

Struve might have recognized that the absorbing medium was certain, sooner or later, to heat up to the incandescence of stars. For this he needed nothing more than the knowledge of thermal processes, as known in 1847. In fact, it was not a thermodynamicist but an astronomer, John F. W. Herschel, worthy successor of his father, William Herschel, who first pointed out, only a year later, the unsatisfactory character of the absorption of starlight as a solution to the paradox.[19] A different matter was Struve's silence about the long-standing efforts of William Parsons, the third Earl of Rosse, to build telescopes even larger than those of the elder Herschel. By 1840, Rosse's findings, using a telescope three feet in diameter, strongly supported the view that better telescopes will indeed resolve into stars those nebulae which were formerly classed as "irresolvable." The same pattern was borne out when, in 1842, Rosse's famous six-foot-diameter telescope was put into operation. The first findings, as reported in 1844 to the Royal Irish Academy by T. R. Robinson, a friend and collaborator of Rosse, were recorded in its *Proceedings* as follows:

> He [Robinson] could not leave this part of his subject without calling attention to the fact that no REAL nebula seemed to exist among so many of these objects chosen without any bias: all appeared to be clusters of stars, and every additional one which shall be resolved will be an additional argument against the existence of any such. There must always be a very great number of clusters, which from mere distance will be irresolvable in any instrument; and if it prove to be the case that *all* the brighter nebulae yield to this telescope, it appears unphilosophical not to make universal Sir J. Herschel's proposition, that "a nebula, at least in the generality of cases, is nothing more than a cluster of discrete stars."[20]

Rosse himself, in the short paper he read before the Royal Society on June 13, 1844, struck a most exemplary note of caution, which would have stood Struve in good stead. According to Rosse, two extremes in the interpretation of data were to be avoided. One was to assume "that the absence of all symptoms of resolvability [in a given nebula] was evidence conclusive that the object was not a cluster."[21] The other extreme was optimism based on the higher resolving power of greater and better telescopes: "it would be very unsafe to conclude that such will always be the case, and thence to draw the obvious inference that all nebulosity is but the glare of stars too remote to be separated by the utmost power of our instruments."[22] The principal result of further observations with Rosse's giant telescope consisted in the detection of spiral structure in a dozen or so nebulae. In his famous paper of 1850,[23] Rosse was prompted to remark that the subject of the dynamical laws governing the evolution of galaxies had thereby become "more mysterious and more inapproachable" than ever.[24] He was firmly convinced that spiral nebulae represented a phenomenon basically different from the shape of the Milky Way, a conjecture which was to be proven wrong. Rosse said nothing about the general question of resolvability versus irresolvability. Since, however, his paper contained an impressive list of nebulae previously listed as irresolvable, faith in ultimate resolvability was inevitably strengthened.

The man who gave a most authoritative support to such an interpretation of Rosse's novel findings was the younger Herschel. In his Presidential Address to the British Association meeting at Cambridge in June, 1845, he characterized the previous year as "an epoch in astronomy."[25] The reason for this was the recent completion of Rosse's six-foot reflector and, of course, its resolution of hitherto unresolved nebulae. "A not unnatural or unfair induction would therefore seem to be, that those which resist such resolution do so only in consequence of the smallness and closeness of the stars of which they consist; that, in short, they are only optically and not physically nebulous."[26] Herschel even strengthened his interpretation by calling attention to the almost insensible gradations that formed a bridge between the shapes of spherical nebulae (almost invariably resolvable) and of ellipsoidal nebulae which frequently were not: "though there is no doubt a great number of elliptic nebulae in which stars have *not* yet been noticed, yet there are so many in which they *have,* and the gradation is so insensible from the most perfectly spherical to the most elongated elliptic form, that the force of the general induction is hardly weakened by this peculiarity; and for my own part I should have little hesitation in admitting all nebulae of this class to be, in fact, congeries of stars."[27]

Herschel strongly felt that in view of Rosse's observations "the idea of a *nebulous matter,* in the nature of a shining fluid, or condensible gas, must, of course, cease to rest on any support derived from actual observation in the sidereal heavens, whatever countenance it may still receive in the minds of cosmogonists from the tails and atmospheres of comets, and the zodiacal light in our own system."[28] By this Herschel meant that the actually observed nebulosities should not be considered evidences of the "primaeval chaos."[29] At the same time, he emphasized that the features of the Magellanic Clouds clearly indicated the dynamics of the formation of nebulae. This was all the more curious in that he did not specify the quality of material on which those dynamics operated. As will be seen shortly, it was his description of the Magellanic Clouds which was before long to be seized upon as proof of the existence of a truly nebulous matter.

Another clear indication of the direction in which Rosse's findings influenced speculation on the status of nebulae may be found in the *Thoughts on Some Important Points relating to the System of the World* by J. P. Nichol, professor of astronomy at the University of Glasgow.[30] The preface to its first edition, in 1846, stated that the "supposed distribution of a self-luminous fluid, in separate patches, through the Heavens, has, beyond all doubt, been proved fallacious by that most remarkable of telescopic achievements—the resolution of the great Nebula in Orion into a superb cluster of Stars: and this discovery necessitates important changes in previous speculations in Cosmogony."[31] It was with obvious pride that Nichol reproduced a letter from Rosse, dated March 10, 1846, on the complete resolution of the great nebula in Orion, although Rosse's words did not suggest the general resolvability of nebulae.[32]

For Nichol, the real import of Rosse's impressive findings lay deeper. The demise of a nebulous matter extending to infinity implied, according to Nichol, a revision which is best stated in his own words: "The Infinite we had built up after the fashion of what had become familiar, was yet, with all its greatness, only an IDOLA, and could fill neither Space nor Time. It was indeed a grand and noble Temple, but yet not the Temple of the Universe."[33] The true temple or architecture of the universe which began to emerge in Nichol's mind was a vast, co-ordinated, and presumably finite system of stars; he even suggested that the magnificent, now resolved, great nebula in Orion was so immense as to outclass all other nebulae taken together. He was inclined to view the "nebulous heavens as a whole,"[34] without, however, providing specifics about the co-ordination of nebulae in that wholeness. Of the fact of co-ordination he had no doubts. *"Aggregation* or *association* is the law of the Stellar Universe," and he dismissed the probability that "any Orb" (nebula) was "resting alone, amidst a dark desert in Space." The nebulae either resembled "a cluster of islets in the Pacific," or "an ocean peopled with subservient islands."[35]

Nichol used the same simile in his lectures before the Mercantile Library Association in New York early in 1848.[36] He compared the impact of the new telescopes to the report which brought word about "large groups of islands and vast continents" to a man who had never before left his own small island in the South Pacific.[37] Vast as their array could be, Nichol did not characterize them as infinite in extent. He had already explained the pale whiteness of the Milky Way as the effect of the combined light of a finite layer of stars. "It may be shown that if these stars were diffused through all space, our heavens would on every side blaze with a splendor so bright that the Sun would not be visible to us at noonday."[38] This was a fine application of Olbers' Paradox to the Milky Way, although Nichol did not mention Olbers by name. Nor did he explicitly inform his audience that, because of the same paradox, the universe of stars could not be infinite. His baffling vagueness at crucial junctures can best be seen in the words he attributed to the elder Herschel concerning the Milky Way: "It is clear that this great multitude of fixed stars is not what it was formerly supposed—the *infinite* heavens. This great mass of stars is simply a vast cluster or congeries, and nothing more."[39] Herschel did not say this; he simply evaded the question of the strict infinity of the universe. In the same manner, Nichol avoided giving a tentative outline of the co-ordination of all the stellar systems. About the existence of such an overall and implicitly finite system, he spoke with firm assurance. From "looking over the whole stellar Universe," he felt impelled to conclude that "those mighty firmaments, those gorgeous systems lying apart in Space," were only "special groups in regard of the great all."[40] With the younger Herschel's great surveys of both stellar hemispheres at his disposal, Nichol could have improved on his often poetical vagueness with some definite arguments against what he clearly tried to discredit: the idea of strictly isolated "island universes." Yet, those arguments went unperceived even by the author of those great surveys.

Impressive as Rosse's observations were, they were not the decisive factor which unsettled Struve's meticulously shaped interpretation of the Milky Way. They did not even become the crucial step in ushering in a new conception of the system of nebulae which was to dominate astronomical thought well into the twentieth century. The magic stroke was almost a trivial ruse if a painstaking tabulation of all known nebulae might be so termed. The younger Herschel's famous catalogue of the nebulae of the Northern Hemisphere, published in 1833,[41] contained a most momentous piece of information which should have been derived from it by "mere inspection," as he noted fifteen years later in his great survey of the southern sky.[42] The "obvious" concerned the sudden increase in the number of nebulae at high latitudes, that is, toward the northern galactic pole. The crowding of the nebulae toward the southern galactic pole appeared to be almost as striking. However, the only inference he drew from this was that "the nebulous system is distinct from the sidereal, though involving, and perhaps, to a certain extent, intermixed with the latter."[43]

Herschel's mind was clearly dominated by the debate on resolvability versus irresolvability, and unmistakably leaned towards the former. "Such a medium," he wrote of strictly gaseous nebulae, "is purely hypothetical. We see, after all, only the luminous portions of a nebula, and can have no other knowledge of their aggregation or segregation than what our telescopes afford us. The distinction between nebulae properly so called, and those which we are to consider as certainly or very probably clusters of stars, resting, as it must do, on the merely temporary and conventional ground of the capacity or incapacity of our telescopes, wholly or partially to resolve them, can never become a permanent ground of classification, since every new improvement in the powers of the telescope will cause more and more nebulae to pass into the class of clusters."[44] He remained strangely silent concerning the relevance of the crowding of nebulae toward the galactic poles to the shape of the "sidereal creation." This is not to suggest that he had not entertained curious conjectures. While at the Cape of Good Hope, he wrote to William Rowan Hamilton that "it is impossible to resist the conviction that the Milky Way is not a stratum but a ring."[45] In the eleven editions of his *Outlines of Astronomy* from 1849 to 1871, he confessed to ignorance about the true shape of the Milky Way: "Beyond the obvious conclusion that its form must be, generally speaking, *flat,* and of a thickness small in comparison with its area in length and breadth, the laws of perspective afford us little further assistance in the inquiry."[46] Not even so much was said on the question of whether some or many nebulae were independent systems, or formed part of our galaxy. He retained to the end, however, a glowing admiration for Rosse's achievements through which the "probability has almost been converted into a certainty," that ultimately all nebulae could be resolved into clusters of stars.[47]

This was a rather puzzling attitude, not only because of a discovery made by Huggins in 1864, of which more will be said later, but also in view of Herschel's own description of the Magellanic Clouds in his survey of the southern sky. There the evidence was unmistakable that, within a huge system, clearly distinct from the tract of the Milky Way, there existed a complexity of celestial formations which defied any attempt, as Herschel put it,[48] to make an exact drawing of it: "The Nubecula Major like the Minor, consists partly of large tracts and ill-defined patches of irresolvable nebula, and of nebulosity in every stage of resolution, up to perfectly resolved stars like the Milky Way, as also of regular and irregular nebulae properly so called, of globular clusters in every stage of resolvability, and of clustering groups sufficiently insulated and condensed to come under the designation of 'clusters of stars'."[49]

It was this description, which became, in the hands of two distinguished British thinkers, decisive evidence for their own interpretations of the universe. One of them was William Whewell, the renowned Master of Trinity College, deeply committed to vindicating God and man in the world of science. In his *The Plurality of Worlds,*[50] he took the presence of nebulosity among clusters of stars as proof that distant nebulae could not necessarily be considered as congeries of stars, each with a retinue of planets, or abodes where manlike beings could exist.[51] What Whewell wanted to prove was that terrestrial life was a most extraordinary phenomenon, clearly indicative of a purposeful design in the universe. To him, the universe could not be a set of island universes and remain purposeful. He might have done a greater service to the cause of natural theology had he seized upon the actual design in the clustering of possible island universes around the galactic poles. Whewell sensed no relevance there, just as he carefully avoided the most portentous question, whether the number of nebulae was infinite or finite, and, if finite, whether or not they all formed one huge system with the Milky Way as its principal plane.[52]

This question was also ignored in the lengthy and spirited rebuttal which David Brewster penned[53] to Whewell's claim that life, especially higher forms of it, was restricted to our earth. Brewster made the most of Rosse's results, and considered it an inevitable conclusion, "though the evidence of demonstration is wanting, that all nebulae are clusters of stars."[54] He dubbed Whewell a "star-dust philosopher,"[55] and in the same style he declared that questions about the substance of nebulae were as immaterial as if it were "granite or greywackle."[56] With the same bravado, he claimed that no specific shape indicated a purely nebular status for any nebula, since stars were observed even in the annular nebula between β and γ Lyrae.[57] According to Brewster, planetary systems as abodes of life were innumerable, and he quoted Newton's words from his First Letter to Bentley on the homogeneous distribution of stars in the universe.[58] In his overzealous admiration for Newton, Brewster not only failed to suspect that Bentley was not so wrong after all, but he also failed to realize that one could, by 1854, speak of a homogeneous distribution of nebulae only if all of them were indeed resolvable into stars.

Whewell's real antagonist was not Brewster, a devout Christian, but Herbert Spencer, chief spokesman of mechanistic evolution. Spencer was, however, at one with Whewell and Brewster in his failure to consider the problem of an infinite, homogeneous universe of stars or nebulae, when in 1858 he wrote his famous essay, "The Nebular Hypothesis."[59] In the cosmic presence of the "nebulous" substance Spencer saw the basic substratum from which all formations, non-living as well as living, would evolve and into which they would eventually return. He claimed that "the notion, of late years idly repeated and uncritically received, that the nebulae are extremely remote galaxies of stars like those which make up our own milky-way, is totally irreconcileable with the facts—involves us in sundry absurdities."[60] The opposite view, which he championed, he submitted in an equally sweeping fashion: "What, then, is the conclusion that remains? This only:—that the nebulae are not further off from us than parts of our own sideral system, of which they must be considered members; and that when they are resolvable into discrete masses, these masses cannot be considered as stars in anything like the ordinary sense of that word."[61]

Phrases, however mighty, should derive their true strength from facts, while Spencer could only outline generalities. Nevertheless, he seemed to be the first to spell out explicitly the statistical significance of the higher frequency of nebulae in areas of the skies most devoid of stars, and in particular in areas around the galactic poles: "If there were but one nebula, it would be a curious coincidence were this one nebula so placed in the distant regions of space, as to agree in direction with a starless spot in our own sideral system. If there were but two nebulae, and both were so placed, the coincidence would be excessively strange. What then shall we say on finding that there are thousands of nebulae so placed? Shall we believe that these far-removed galaxies, dispersed through infinite space have in thousands of cases happened to agree in visible position with the thin places in our own galaxy? Such a belief is next to impossible."[62] Then, after pointing out that "scarcely any nebulae lie near the galactic circle (or plane of the Milky Way); and the great mass of them lie round the galactic poles," he asked: "When to the fact that the general mass of nebulae are antithetical in position to the general mass of stars, we add the fact that local regions of nebulae are regions where stars are scarce, and the further fact that single nebulae are habitually found in comparatively starless spots; does not the proof of a physical connexion become overwhelming? Should it not require an infinity of evidence to show that nebulae are not parts of our sideral system?"[63]

When new evidences were recognized, they all seemed to give firm support to Spencer's position. The first great blow to the upholders of "resolvability in principle" came in 1864, with William Huggins' observations of the spectra of some nebulae which seemed to offer a test case for decision. The first account of his findings could not have concealed their significance more effectively in technical jargon. "I then found that the light of this nebula, unlike any other ex-terrestrial light which had yet been subjected by me to prismatic analysis, was not composed of light of different refrangibilities, and therefore could not form a spectrum. A great part of the light from this nebula is monochromatic."[64] What he meant was that the nebulae in question could not consist of stars, which invariably produced a broad, continuous spectrum. The next year Huggins disclosed more of the implications of his findings on the status of the nebulae. First, the presence of minute, closely associated points of light in a nebula could no longer be taken as a trustworthy proof that the object consisted of stars. Second, one had to reconsider the intrinsic value of inferences on the enormous distances assigned to many nebulae. Huggins made no secret of his willingness to view nebulae possessing a gaseous spectrum as an independent and major component in "that great group of cosmical bodies to which our sun and the fixed stars belong."[65] The elemental impact which the discovery made on Huggins was only revealed a quarter of a century later, when he recalled how "nebula after nebula yielded" to Rosse's giant telescopes and "the opinion began to gain ground that all nebulae may be capable of resolution into stars."[66] What was truly at stake was whether nebulae were external independent systems, island universes, "cosmical sandheaps too remote to be separated into their component stars,"[67] or merely the early stages of a universal evolutionary process. With his carefully constructed telescope-spectrograph combination, the case could at long last be tested. The tension of the moment stood out vividly in his memory even after a lapse of twenty-five years:

> On the evening of the 29th of August, 1864, I directed the telescope for the first time to a planetary nebula in Draco. The reader may now be able to picture to himself to some extent the feeling of excited suspense, mingled with a degree of awe, with which, after a few moments of hesitation, I put my eye to the spectroscope. Was I not about to look into a secret place of creation?
>
> I looked into the spectroscope. No spectrum such as I expected! A single bright line only! At first, I suspected some displacement of the prism, and that I was looking at a reflection of the illuminated slit from one of its faces. This thought was scarcely more than momentary; then the true interpretation flashed upon me. The light of the nebula was monochromatic, and so, unlike any other light I had as yet subjected to prismatic examination, could not be extended out to form a complete spectrum . . .
>
> The riddle of the nebulae was solved. The answer, which had come to us in the light itself read: Not an aggregation of stars, but a luminous gas. Stars after the order of our own sun, and of the

> brighter stars, would give a different spectrum; the light of this nebula had clearly been emitted by a luminous gas . . . There remained no room for doubt that the nebulae, which our telescopes reveal to us, are the early stages of long processions of cosmical events, which correspond broadly to those required by the nebular hypothesis in one or other of its forms.[68]

The second salvo opened in 1867 with a paper by Cleveland Abbe of the Cincinnati Observatory.[69] He counted the number of nebulae and clusters in every rectangular area of the sky 30 minutes wide and 10 degrees high. The conclusions he derived from his Tables indicated (1) that clusters were members of the Milky Way, (2) that most nebulae, resolved and unresolved, lay outside it and (3) that "the visible universe is composed of systems, of which the *Via Lactea,* the two *Nubeculae,* and the Nebulae, are the individuals, and which are themselves composed of stars (either simple, multiple, or in clusters) and of gaseous bodies of both regular and irregular outlines."[70] The expression, "visible universe," is worth noting here, since, by the end of the century, it had become customary to picture the universe as consisting of two parts: one visible and confined to the Milky Way, and another, truly infinite, which was believed to be forever beyond the reach of observations. If, however, the Milky Way was "the visible universe," it could only form the principal plane. Or, as Abbe had proposed, the disk of the Milky Way was to be pictured as cutting "nearly at right angle the axis of a prolate ellipsoid, within whose surface all the visible nebulae are uniformly distributed."[71] Abbe spoke in fact of the "north and south poles of the universe," by which he meant the end points of the prolate ellipsoid to which all nebulae seemed to be confined. He even conjectured that the sun's distance from the center of the visible universe was one-third of the shorter semiaxis of its shape, the prolate ellipsoid.[72]

Such a brave calculation did not long survive. Two years later, Richard A. Proctor, a prolific author of informative books popularizing astronomical topics, produced a map-form analysis of Abbe's Tables. It showed that the prolate ellipsoid shape of the "visible universe" did not entirely correspond to the data. Still, Proctor hastened to add that Abbe's considerations clearly proved that "no extension of telescopic observation can appreciably affect our views respecting the distribution of the nebulae."[73] In stating this, he did not wish to deny absolutely the possibility that some nebulae could perhaps be truly "external"—that is, systems of stars wholly independent of the Milky Way. Interestingly enough, he singled out the Great Nebula in Andromeda as a possible case, emphasizing at the same time that perhaps only one in several hundred nebulae could really be external.[74] To Proctor, the starry and nebular sections of the sky were "parts of a single scheme,"[75] and were "members of the same scheme."[76] As proof of this, he noted not only the clustering of the majority of nebulae around the galactic poles, but also a peculiarity inherent in Abbe's Tables. These indicated that practically all unresolvable nebulae were rather close to the galactic poles. To Proctor's mind, honed on statistical analysis, this forcefully suggested a pattern which he spelled out with characteristic impetuosity: "I believe that the irresolvable and scarcely resolvable nebulae are composed of stars really smaller and really closer than those forming the clusters and easily resolvable nebulae."[77]

In addition to his impetuosity, Proctor was also indefatigable in pursuing a worthy target, which in this case consisted of the potentialities of a thorough analysis of the statistics of stellar and nebular distribution. In fact, Abbe's Tables became for Proctor an incentive to reopen the whole question of stellar distribution which, until then, had been more or less explicitly based on the principle of homogeneous distribution according to brightness and size. Independent of Abbe, Proctor had developed strong misgivings about the higher frequency of bright stars in the belt of the Milky Way. Such could not be the case, he argued in 1867, if stars were indeed homogeneously distributed according to size and brightness within a flat disk, with the sun located near the center plane.[78] Three years later, Proctor decided to take an exceedingly thorough look at the matter by plotting all the 324,198 stars listed in Argelander's *Durchmusterung* on equal-surface projection charts. In the presentation of his charts to the Royal Astronomical Society, he merely stated that the disk theory should therefore be abandoned.[79] A year later, he not only disclosed that the work had taken 400 hours, but he explained in detail the main implications of his charts. These involved two points. One was the claim that the whiteness of the Milky Way was largely due to many small and relatively nearby stars—clear defiance of the principle of homogeneity. The other was an emphatic reassertion of the idea of a one-island universe. These two points formed a closely integrated whole which is best presented in his own words: "The sidereal system is altogether more complicated, altogether more varied in structure, than has hitherto been supposed. Within one and the same region co-exist stars of many orders of real magnitude, the greatest being thousands of times larger than the least. All the nebulae hitherto discovered, whether gaseous or stellar, irregular, planetary, ring-formed, or elliptic, exist within the limits of the sidereal system. They all form part and parcel of that wonderful system whose nearer and brighter parts constitute the glories of our nocturnal heavens."[80]

From this it followed that the sidereal system consisted of strata or streams of stars, and of nebular zones, often in the form of sections of a ring, and lying basically in the same plane and intermingled with one another. Proctor saw a classic case of this in the position and composition of the Magellanic Clouds: "The evidence afforded by the Nubeculae seems to me decisive in favor of the intimate association between the stellar and nebular systems."[81] This was not without an irony which did not escape him. Years earlier, Sir John Herschel in part had based on a similar analysis of the Magellanic Clouds his somewhat ambivalent opinion that the nebulae formed a system "distinct from the sidereal system, though involving, and perhaps to a certain extent intermixed with the latter."[82] Herschel's lifelong unwillingness to draw overly definite cosmological conclusions was something to contend with, especially in view of his status as the undisputed grand old man of British science in the 1860's. Proctor praised the "tentative and philosophical note in which Sir John Herschel reasoned respecting the universe," and the manner in which he offered his views "rather as suggestions than as theories."[83] Proctor indeed did his best to let his conclusions appear at least in partial agreement with the views of the two Herschels, and he eagerly seized on some of their pronouncements. This could be done the more readily since, to both Herschels, theories were tools and not idols, and they modified them as the observational evidence seemed to demand. Proctor could not have been more accurate when he described "as characteristic of the Herschel family," the power "of weighing evidence without reference to preconceived opinions; of indicating the theories suggested by this or that portion of the evidence, and as readily indicating reasons which must lead to the abandonment of such theories."[84]

Proctor should, in fact, have italicized the moral which he then recommended to all: "It would be well for the cause of scientific progress if the power were not so uncommon as it is."[85] He noticeably failed to administer the medicine to himself. He certainly deserves much credit for calling attention to the complexities of the Milky Way. At the same time, he guided speculation in the wrong direction by reinforcing the conviction that nebulae were parts of a one-island universe.

A good illustration of this conviction is the discussion of the topic by Quirico Filopanti (pen name for Giuseppe Barilli), professor of mechanics at the University of Bologna and author of many popularizing works on physics and astronomy. Never very profound, Filopanti ably interpreted the intellectual fashions of his day and spelled them out with impressive directness. What he said about the Milky Way and galaxies in his *Lezioni di astronomia,*[86] published in 1877, was superficial as to the history of the question, but incisive as to the status of nebulae. Thus, he credited Dante, somewhat chauvinistically, with divining the starry composition of the Milky Way. In the introductory chapter he warned his reader that he was ultimately to show that "our cosmos is surrounded from every side by a desert that cannot be traversed by light."[87] That there were innumerable similar worlds completely severed from one another in an infinite space, was a fact which, he wrote, "I positively believe."[88] He elaborated on this theme in a chapter specially devoted to the Milky Way and nebulae, in which he stated that the question of resolvability was of secondary importance.[89] What really mattered, he argued in a succeeding chapter, was the optical paradox (he did not mention Olbers by name) of an infinite universe of stars, and he saw only one escape from the dilemma.[90] It consisted in assuming that, between galaxies, there was an absolutely empty abyss, void even of the ether, the vehicle of light propagation. All visible nebulae could, therefore, belong only to our Milky Way whose size he put at 3000 light-years, and claimed that the diameters of the other nebulae were very much smaller.[91]

Filopanti was in distinct disagreement with the view of those who preferred to estimate the size of the Milky Way at between five and ten times that value. But even the most authoritative astronomers were voicing the same opinion about the status of nebulae. Convincing proof of this is the manner in which Simon Newcomb, the leading American astronomer, expressed himself on the subject. He found, in the clustering of nebulae around the galactic poles, overwhelming evidence against the view which considered them systems similar in size to the Milky Way and separated from it by "great vacant space."[92] To him speculations about invisible galaxies and systems constituted an entirely idle topic. The "whole visible universe" could, therefore, be pictured as a sphere cut in half by the thin disk of the Milky Way, the two resulting hemispheres being filled with clusters of stars and nebulae of small size in comparison with it (Illustration XIX). In his final discussion of the topic, in 1902, he combined the question of island universes with Olbers' Paradox. Since he found no evidence for a hierarchical distribution of an infinite number of stars, nor for absorption of starlight—two solutions of the paradox which he considered possible—he took the view that a finite universe of stars was the only reasonable assumption. He did not profess to know the size of that "universe," but he felt convinced that "the great mass of stars is included within a limited space, the extent of which we have as yet no evidence [of]. Outside of this space there may be scattered stars or invisible systems. But if these systems exist, they are distinct from our own."[93] This meant that "island universes," if they existed, were invisible and beyond the ken of science.

One should not, therefore, be unduly surprised at the unhesitating manner in which Agnes M. Clerke, the renowned historiographer of nineteenth-century astronomy, characterized as "unanswerable" the "argument from co-existence in nearly the same region of space" of nebulae and clusters.[94] "There is no maintaining," she wrote confidently, "nebulae to be simply remote worlds of stars in the face of an agglomeration like the Nubecula Major, containing in its (certainly capacious) bosom *both* stars and nebulae. Add the evidence of the spectroscope to the effect that a large proportion of these perplexing objects are gaseous, with the facts of their distribution telling of an intimate relation between the mode of their scattering and the lie of the Milky Way, and it becomes impossible to resist the conclusion that both nebular and stellar systems are parts of a single scheme."[95]

In her great monograph, *The System of the Stars,*[96] she started her discussion of the status of the nebulae in a tone explicable only if consensus on the subject had really been as firm as her style suggested: "The question whether nebulae are external galaxies hardly any longer needs discussion. It has been answered by the progress of research. No competent thinker, with the whole of the available evidence before him, can now, it is safe to say, maintain any single nebula to be a star system of coordinate rank with the Milky Way."[97] The vision of a one-island universe, the co-ordination of all ascertainable heavenly objects, stars, nebulae, clusters, and star-streams into "one mighty aggregation," was for her a "practical certainty."[98] The edifice seemed to her so compact that she instinctively looked for its capstone. This she believed to coincide with the "hoodlike accumulation of nebulae in Coma Berenices." In a poetic vein, she set its distance equal to the extent of the main galactic stream, and added that it "may thus be said to constitute the polar cap of a sphere equatorially girdled by the Milky Way."[99]

There were two main faults in this pleasing flight of fancy which presented the "observable universe" as two hemispheres of nebulae joined by the disk of the Milky Way, as a principal plane. First, it set somewhat arbitrary limits to cosmological inquiry. "With the infinite possibilities beyond science has no concern,"[100] declared Miss Clerke, which sounded all the more unscientific since, during the previous decade, the problem of the gravitational and optical paradox of an infinite universe had received renewed attention. What really betrayed the frailty of the one-island-universe theory was the fact which Miss Clerke unabashedly admitted without perceiving the seriousness of the position: "As to the distances of the nebulae, we know nothing positive; they no doubt vary extensively, nor can either fineness of grain, or faintness of light (both of which may be inherent qualities) serve to distinguish between those nearest to, and those further away from us."[101] The lack of any solid information on the distances of nebulae should in itself have been a powerful warning against theorizing about the shape of a structure which presumably reached deep into space. But thinking on the question allowed no doubts. Indeed, when re-editing his "Nebular Hypothesis" in 1897, Spencer no longer claimed, as he had forty years earlier, that belief in the existence of truly extragalactic nebulae was "next to impossible," but that such a belief was simply "impossible."[102]

More often than not explicit reference to this problem was lacking in studies on the detailed construction of the Milky Way, prompted, in large part, by Proctor's publications. Good illustrations of this are the monographs and articles by C. Easton[103] and G. Celoria,[104] to mention only two outstanding investigators. Their studies are remembered because both tended to see the Milky Way as a ring or double ring of stars. They and their critics equally lacked the perspicacity of S. L., the reviewer of Thomas Wright's *Original Theory,* as to the true merit of a ring model of the Milky Way. There were, however, some notable cases in which the idea of the Milky Way as an all-inclusive star system was clearly discernible, although the studies in question did not have the Milky Way as a direct objective. The first such study was the pioneering effort of K. Schwarzschild to account for the stellar universe in terms of a four-dimensional geometry.[105] His special target was the determination of the numerical value of the curvature of space. A finite positive value, implying a closed space, came logically from a finite total mass. Schwarzschild was only too ready to see this value in the total number of stars then estimated in the Milky Way. Eight years later, in 1908, he again resorted to the curvature of a four-dimensional space to resolve the paradox of an absolute empty space surrounding the Milky Way. He spoke of it as the all-encompassing system of all observable celestial objects, and attributed to it "not a local but a universal significance."[106]

A similar conviction seemed to guide Lord Kelvin, otherwise a most spirited defender of the infinity of space, when he first discussed the gravitation of stars on the basis of the kinetic theory of gases and treated stars as molecules. In the process he faced the "old and celebrated hypothesis that if we could see far enough into space the whole sky would be seen occupied with discs of stars all of perhaps the same brightness as our own sun, and that the reason why the whole of the night-sky and day-sky is not as bright as the sun's disc, is that light suffers absorption in travelling through space."[107] This was an objectionable formulation of the paradox, lacking reference either to Olbers, or, far more important, to the idea of an infinite universe of homogeneously distributed stars. But Kelvin had already gone to great length to show that the distribution of stars was not uniform, that it gradually diminished in density from the inner to the outer shells of a sphere with a total radius of 3×10^{16} km. His choice of this as a distance at which a star would show a parallax of one-thousandth of a second of arc was probably dictated by its correspondence to a galactic radius of about three thousand light-years, a figure rather close to the lower limit of the then generally accepted value of the size of the Milky Way. Kelvin could then readily show that a thousand million suns within a sphere of that radius would occupy only four percent of the heavenly vault. In a similar sense he exploited the point that not all stars were at peak brilliance at the same time. Curiously, he spoke in the same context of vibrations sent through the infinite ether by the close interaction of molecules, and considered the star, 1830 Groombridge, a visitor from another sphere of stars because of its high velocity. However, he ignored all such spheres except ours, notwithstanding the explicit mention of infinite space in the title of his paper.

This dichotomy in thinking was clearly evident in major statements on the topic by leading astronomers, as the twentieth century began. In 1906, Hermann Kobold of Kiel Observatory offered, in his monograph[108] on the structure of the stellar realm, conclusions which were equivalent to the theory of the one-island universe. To be sure, Kobold voiced his views on a note of resignation. He did not wish to be trapped by "formulas" which "formed only very fragile bridges over the abyss which surrounds the already explored area of our knowledge" about the universe.[109] But he did not, on the other hand, wish to face the full import of the optical paradox of an infinite universe.[110] In such perplexity what could have been a more welcome panacea than the vision of a vast, spherical system of stars enclosing the Milky Way and all other nebulae?[111] The inclusion of nebulae in the one-island universe received a rather weighty endorsement the following year in a major address on the Milky Way. It was delivered by Max Wolf, director of the new Heidelberg Observatory, before the Verein Deutscher Naturforscher und Ärzte (Association of German Scientists and Physicians), meeting in Dresden in September, 1907.[112] The forty-one-year-old Wolf was already world-famous for his work in stellar photography, and he made the most of the 52 photographs which he projected during his lengthy lecture. They ultimately served to buttress two of his major conclusions. The first concerned the spherical shape of the visible stellar universe, with the Milky Way as its main cross section. The other rejected the possibility that some nebulae were island universes. In this context, Wolf not only mentioned the grouping of spirals around the galactic poles, but also argued that no extragalactic nebulae could appear as distinct as did the spirals. Then he added: "By taking everything into account, we seem to be justified in considering it likely that the nebulae and clusters of stars represent an essential part of our star-island and perhaps lie relatively close to us. They all form, together with the stars of the Milky Way, an organic whole. Distant, isolated Milky Ways [*Milchstrasseninseln*], have never been sighted by man."[113]

Wolf spoke with unqualified approval of a special claim by the dean of German astronomy, Hugo von Seeliger, of the Munich Observatory, who had worked for years to establish an exact basis for estimating the dimensions of the Milky Way. Seeliger's qualifications for the task could not have been more outstanding. By completing a painstaking reduction of the luminosity data of the famed *Bonner Durchmusterung,* he proved himself to be in full control of the technical part of his enterprise. He was a master at the mathematical analysis of a staggeringly complex problem. He organized the statistics of star counts on a mathematical basis far superior to anything used before. Most important, the one-island universe must have particularly appealed to him, for he did not conceal his misgivings about infinity. To him, an infinite universe uniformly filled with matter savored of metaphysics and was inherently not representable (*vorstellbar*).[114] Not that he wished to abandon the idea of infinity. In fact, he sacrificed the universal validity of the inverse-square law of gravitation in order to get rid of the gravitational paradox of an infinite universe.[115] Such mental preconditioning was subtly present in his half dozen papers on the distribution of stars in space. His convictions were categorically stated in a lengthy essay, published in 1911.[116] There he not only restated his claim, warmly seconded by Wolf, that the density of stars rapidly decreased to zero at a given distance, but he also submitted that such a zero-point density, marking the boundaries of the Milky Way, occurred at 910 times the distance of Sirius, or about 8,000 light-years.[117]

Strong support for Seeliger's conclusions seemed to come from the work of J. C. Kapteyn, who, in the early 1900's, became one of the world's leading astronomers, although lacking the use of an observatory at the University of Groningen. Kapteyn's principal aim was that of the older Herschel, to fathom the construction of the sidereal universe. His work was based on the analysis of stellar photographs made available to him by observatories all over the world. From those photographs he derived hundreds of new stellar parallaxes, together with a better understanding of the wide distribution of star luminosities, on which any inference on the contours of the "visible universe" ultimately rested. Kapteyn was not a man of spectacular utterances. In two of his major addresses on the structure of the stellar universe, one in London in 1908,[118] the other in Washington in 1913,[119] one looks in vain for grand conclusions. Even at the end of his long and distinguished career, he called the final résumé of his ideas a "First Attempt at a Theory of the Arrangement and Motion of the Sidereal System."[120] Nevertheless, he seemed to share in full the conviction that the visible universe was embodied in one single system centered around the Milky Way. It was not without reason that astronomers began to speak of the Kapteyn-universe.[121]

The only well-argued disclaimer against the one-island universe came from the pen of Joseph Plassmann, professor of astronomy at the University of Freiburg i. B., but his massive textbook of astronomy, first published in 1898, never received the attention it deserved.[122] He discussed the question of an all-inclusive Milky Way in terms of the optical paradox of an infinite universe, and also pointed out that a finite universe presupposed a non-Euclidean geometrical structure.[123] With remarkable boldness he claimed that the Great Nebula in Andromeda—and many "irresolvable" nebulae—was separated from the Milky Way by distances a hundred or a thousand times greater than the galactic diameter.[124] While the estimate turned out to be exaggerated in the case of the Andromeda nebula, Plassmann's view of the universe as a four-dimensional system of Milky Ways had lasting merit. There was also food for thought in his remark on the deeper aspects of cosmic four-dimensionality: "The fourth dimension is a limit-concept which justifies the finiteness of the universe, and at the same time saves us from picturing it as enclosed by a fence; such a universe, precisely because of its most shining appearances, carries on its forehead the mark of finiteness and contingency."[125]

Plassmann never came to be counted as a "notable" man of science. Curiously, the "notable" dissent on the question of the one-island universe came from those whose principal aim was vindication at any price of the correctness of the idea of an infinite universe.

The efforts of S. Arrhenius were too weighted with "metaphysics in reverse," and offered nothing noteworthy so far as the Milky Way and the nebulae were concerned.[126] More interesting were the speculations of C. V. Charlier, the first twentieth-century champion of a hierarchically ordered infinite universe. This concept meant not only that galaxies were not uniformly distributed in space, but that, except for a relatively few nebulae, galaxies had to be assigned distances which put them forever beyond the ken of observation.[127] Thus Charlier "saved" infinity for the universe by incorporating a dichotomy between its observable and unobservable parts. The former was, to all intents and purposes, a one-island universe. Observability was one matter, the explanation of the total effect of all galaxies in the universe, another. For, if galaxies were the building blocks of the universe, at an infinite distance their hierarchical "supergroup" had to contain an infinite number of them, again posing, for anyone mindful of logic, the optical and gravitational paradox of an infinite universe. Infinity often produces indefinite problems, and the former objection to Charlier's idea might be of that type. But there was nothing indefinite in the fact that Charlier could present no evidence whatever of a hierarchical grouping of galaxies. Such evidence would have implied that the distances of some visible galaxies were larger by several orders of magnitude than the dimensions of the Milky Way. The meager data then available seemed to indicate the contrary. In fact, the measurements made by K. Bohlin[128] of a parallax of 0″.17 for GC4373 placed it less than 20 light years from us.

One who examined Bohlin's measurements with patent misgivings was Henri Poincaré. In his lectures on cosmogonical hypotheses he cautioned suspension of judgment on the matter.[129] But in that work he was concerned with the evolution of galaxies, not with their coordination. The major clue to that evolution, he believed, lay in the study of spiral nebulae: "A fact which strikes everybody," went the concluding remark of the Preface of his work, "is the spiral form of certain nebulae; it occurs too often to be thought of as a product of chance. One should realize how incomplete are all cosmogonical theories in which this point is overlooked."[130] Poincaré did not seem to sense that the question which he posed a little earlier about the distance of those spirals was no less momentous: "Are they at enormous distances outside the Milky Way and are themselves Milky Ways seen from afar?"[131] Spiral nebulae were no longer "the still less useful" celestial objects that Maxwell dubbed them half a century earlier.[132]

Poincaré was not the only towering figure of early twentieth-century astronomy who failed to consider in detail the deeper problems raised by the concept of a one-island universe. Eddington, the rising young giant, presents us with the same problem, all the more keenly felt since Eddington burst onto the scene, in 1914, with a monograph[133] giving a masterful analysis and summary of the latest research on the stellar and nebular universe. Eddington's perspicacity may be credited with the forceful suggestion that spirals were a very special class of nebulae. "We have no reason to believe that the arguments which convince us that the irregular and planetary nebulae are within the stellar system apply to the spirals."[134] Yet, he admitted in the same breath that the grouping of spiral nebulae around the galactic poles might equally be viewed as evidence of their independence of the Milky Way as of its influence on them. What posed the problem for Eddington was the dynamical aspect, or evolution, of galaxies. If the spirals were part of *the* system, then it was impossible to make a sound conjecture, he believed, about their real nature. "If, however, it is assumed," Eddington argued, "that these nebulae are external to the stellar system, that they are in fact systems coequal with our own, we have at least an hypothesis, which can be followed up, and may throw some light on the problems that have been before us. For this reason the 'island universe' theory is much to be preferred as a working hypothesis; and its consequences are so helpful as to suggest a distinct probability of its truth."[135]

It was one thing to speculate about island universes, another to pursue rigorously the implications of a universe consisting of an infinite number of island universes. This latter path was never followed by Eddington, the most philosophical of twentieth-century astronomers and cosmologists. During a long and brilliant career, he never touched on the paradoxes of an infinite universe. This was all the more curious since Eddington became one of the earliest and most perceptive advocates of the General Theory of Relativity, which provided a conceptually consistent picture of a finite universe and, with that as a basis, a satisfactory explanation of Olbers' Paradox. Eddington was to take delight in estimating and re-estimating the size of the universe which, on the basis of General Relativity, had to be finite. But, in 1914, General Relativity was still a few years away, and, with it, a consistent explanation of the possibility of a finite universe. Thus, it should be all the more noteworthy from a psychological viewpoint that, in 1914, Eddington could construct a magisterial synthesis of the latest research on the "visible universe" without mentioning any difficulty but that of the spiral nebulae. For, impressive as Eddington's provisos could be about the possible revision of man's picture of the universe, he felt that his presentation of the universe contained "something that will last, notwithstanding its faulty expression."[136] That "something" was the inclusion of all stars in one single system. It was no accident that the very first figure in Eddington's book was not called "hypothetical Section of the Milky Way," but "Hypothetical Section of the Stellar System"[137] (Illustration XXa). The very last sentence of the book carried the same message: "If we have still to leave the stellar universe a region of hidden mystery, yet it seems as though, in our exploration we have been able to glimpse the outline of some vast combination which unites even the farthest stars into an organised system."[138] Ultimately, the "Stellar System" proved too puny to contain most nebulae. The culprit was not the "nebulous" matter but some of the very stars in the nebulae. By the time Eddington's book came off press in 1914, novae had been observed in several nebulae, and some other stars out there were to signal an even more momentous message.

References

[1] *Etudes d'astronomie stellaire. Sur la voie lactée et sur la distance des étoiles fixes* (St. Pétersbourg: Imprimerie de l'Académie Impériale des Sciences, 1847), p. 34.

[2] *Ibid.,* p. 50.

[3] *Ibid.*

[4] *Ibid.,* p. 61.

[5] *Ibid.,* p. 62.

[6] *Ibid.,* p. 63.

[7] *Ibid.*

[8] *Ibid.*

[9] See the introductory part of his great paper of 1784, quoted in note 8 of the previous chapter.

[10] *Etudes d'astronomie stellaire,* p. 67.

[11] *Ibid.,* p. 79.

[12] *Ibid.,* pp. 83–93.

[13] Olbers himself evaluated the chief merit of his paper in the following words, in his letter to Bessel of June 22, 1823: "I have recently sent to Bode's *Jahrbuch* a small essay on the transparency of cosmic spaces, in which, in my opinion, even if I have not demonstrated, but at least I have made it very probable, that the cosmic spaces are not absolutely transparent, and that precisely because of the absence of absolute transparency, astronomical observations are possible." *Briefwechsel zwischen W. Olbers und F. W. Bessel,* edited by A. Erman (Leipzig: Avenarius & Mendelssohn, 1852), vol. 2, p. 244.

[14] *Etudes d'astronomie stellaire,* p. 87.

[15] *Ibid.,* p. 86.

[16] *Ibid.,* p. 91.

[17] *Ibid.,* p. 92.

[18] *Ibid.*

[19] In his "Humboldt's *Kosmos,*" *The Edinburgh Review* 87 (1848): 185.

[20] "On Lord Rosse's Telescope," *Proceedings of the Royal Irish Academy* 3 (1844): 130.

[21] "Observations on Some of the Nebulae," *Philosophical Transactions* 133–34 (1843–44): 323.

[22] *Ibid.,* p. 324.

[23] "Observations on the Nebulae," *Philosophical Transactions* 140 (1850): 499–514.

[24] *Ibid.,* p. 503.

[25] *Report of the Meeting of the British Association for the Advancement of Science* (London: John Murray, 1846), p. xxxvi.

[26] *Ibid.,* p. xxxvii.

[27] *Ibid.*

[28] *Ibid.,* p. xxxviii.

[29] *Ibid.*

[30] Edinburgh: John Johnstone, 1848.

[31] *Ibid.*, p. vi.

[32] *Ibid.*, p. 110.

[33] *Ibid.*, p. 111.

[34] *Ibid.*, p. 113.

[35] *Ibid.*, p. 66. The expression "island universe" first appeared prominently in the German form "Weltinsel" in Humboldt's *Kosmos: Entwurf einer physischen Weltbeschreibung,* vol. I (Stuttgart: J. G. Cotta'scher Verlag, 1845), p. 92.

[36] Published under the title, *Views of Astronomy* (New York: Greeley and McElrath, Tribune Buildings, 1848). The text originally appeared in *The New-York Tribune* as reported by its "phonographic writer," Oliver Dyer.

[37] *Ibid.*, p. 8.

[38] *Ibid.*, p. 7.

[39] *Ibid.*

[40] *Ibid.*, p. 11.

[41] "Observations of Nebulae and Clusters of Stars, made at Slough, with a Twenty-feet Reflector, between the years 1825 and 1833," *Philosophical Transactions* 123 (1833): 359–505.

[42] *Results of Astronomical Observations made during the Years 1834, 5, 6, 7, 8 at the Cape of Good Hope, being a completion of a telescopic survey of the whole surface of the visible heavens commenced in 1825* (London: Smith, Elder & Co., 1847), p. 133.

[43] *Ibid.*, p. 136.

[44] *Ibid.*, p. 139.

[45] See Günther Buttmann, *The Shadow of the Telescope: A Biography of John Herschel,* translated by B. E. J. Pagel and edited by D. S. Evans (New York, Charles Scribner's Sons, 1970), p. 93.

[46] *Outlines of Astronomy* (11th ed.; London: Longmans, Green and Co., 1871), p. 574. For the same passage, see p. 532 in the 1st edition (London: Longman, Brown, Green and Longmans, 1849).

[47] *Ibid.*, p. 639.

[48] *Results of Astronomical Observations,* p. 147.

[49] *Ibid.*, p. 146.

[50] First published anonymously in London in 1853, under the title, *Of the Plurality of Worlds: An Essay*. References are to the American edition, with an Introduction by E. Hitchcock (Boston: D. Lothrop, 1854).

[51] *Ibid.*, p. 161.

[52] Whewell's silence about the question of the infinity of the world, and about its gravitational and optical paradoxes, is puzzling to say the least. His massive treatises on the philosophy and history of the sciences had provided him with more than one logical opportunity.

[53] *More Worlds than One: The Creed of the Philosopher and the Hope of the Christian* (London: John Murray, 1854).

[54] *Ibid.*, p. 171.

[55] *Ibid.*, p. 175.

[56] *Ibid.*, p. 215.

[57] *Ibid.*, p. 176.

[58] *Ibid.*, pp. 249–50.

[59] Such at least was the title under which the essay is remembered and was re-edited by Spencer in 1899, with more than a few changes and amplifications, in *Essays Scientific, Political and Speculative* (New York: D. Appleton, 1899), vol. 1, pp. 108–81. The original appeared "anonymously" as a review of six books, under the title, "Art. VII.—Recent Astronomy and the Nebular Hypothesis," *Westminster Review* 70 (1858): 185–225. Subsequent references are to the latter.

[60] *Ibid.*, p. 195.

[61] *Ibid.*, p. 191.

[62] *Ibid.*, p. 187.

[63] *Ibid.*

[64] "On the Spectra of some of the Nebulae," *Philosophical Transactions* 154 (1864): 438.

[65] "On the Spectrum of the Great Nebula in the Sword-handle of Orion," *Proceedings of the Royal Society of London* 14 (1865): 42.

[66] "The New Astronomy: A Personal Retrospect," *The Nineteenth Century* 41 (1897): 916.

[67] *Ibid.*

[68] *Ibid.*, pp. 916–17.

[69] "On the Distribution of Nebulae in Space," *Monthly Notices of the Royal Astronomical Society* 27 (1867): 257–64.

[70] *Ibid.*, p. 262.

[71] *Ibid.*

[72] *Ibid.*, p. 263.

[73] "Distribution of the Nebulae," *Monthly Notices of the Royal Astronomical Society* 29 (1869): 338.

[74] *Ibid.*, p. 342.

[75] *Ibid.*, p. 341.

[76] *Ibid.*, p. 342.

[77] *Ibid.*

[78] "Star-streams," reprinted in *The Universe and the Coming Transits* (London: Longmans, Green & Co., 1874), pp. 1–16.

[79] "On a Chart of 324,198 Stars," reprinted in *The Universe and the Coming Transits,* pp. 157–71.

[80] "The Construction of the Heavens," reprinted in *The Universe and the Coming Transits,* pp. 172–205: quotation is from pp. 201–02.

[81] "Distribution of the Nebulae," p. 342.

[82] *Results of Astronomical Observations,* p. 136.

[83] "Distribution of the Nebulae," p. 342 note.

[84] *Ibid.*

[85] *Ibid.*

[86] Milano: L. Bortolotti, 1877.

[87] *Ibid.*, p. 30.

[88] *Ibid.*

[89] *Ibid.*, p. 486.

[90] *Ibid.*, pp. 536–38.

[91] *Ibid.*, p. 498.

[92] *Popular Astronomy* (4th rev. ed.: New York: Harper & Brothers, 1882), p. 494.

[93] *The Stars: A Study of the Universe* (London: John Murray, 1902), p. 233.

[94] *A Popular History of Astronomy during the Nineteenth Century* (2d ed.; London: Adam & Charles Black, 1887), p. 456.

[95] *Ibid.*, pp. 456–57. For the same statements, see p. 505 in the 3rd edition (1893) and p. 422 in the 4th (1902).

[96] 2d ed.; London: Adam & Charles Black, 1905; first published in 1890.

[97] *Ibid.*, p. 349.

[98] *Ibid.*

[99] *Ibid.*, p. 359.

[100] *Ibid.*, p. 349.

[101] *Ibid.*, p. 359.

[102] See note 59 above, *op. cit.*, p. 112.

[103] *La voie lactée dans l'hemisphère boréal. Cinq planches litographiées, description détaillée, catalogue et notice historique.* Avec un préface par H. G. van de Sande Bakhuyzen (Dordrecht: Blussé et Cie; Paris: Gauthier-Villars et Fils, 1893). In striking evidence of the status of historical research on the Milky Way, Easton believed that his few references contained "almost all the works of that kind." (See p. 9.) Easton's idea of a ring model of the Milky Way was illustrated in his "A New Theory of the Milky Way," *The Astrophysical Journal* 12 (1900): 136–58; see also his *La distribution de la lumière galactique, comparée à la distribution des étoiles cataloguées dans la Voie lactée boréale* (Amsterdam: J. Müller, 1903).

[104] "Sopra alcuni scandagli del cielo eseguiti all' Osservatorio Reale di Milano et sulla distribuzione generale delle stelle nello spazio," *Pubblicazioni del Reale Osservatorio di Brera in Milano,* No. XIII. (Milano: Ulrico Hoepli, 1877).

[105] "Ueber das zulässige Krümmungsmaass des Raumes," *Vierteljahrschrift der Astronomischen Gesellschaft* 35 (1900): 337–47.

[106] See his collection of four addresses, *Über das System der Fixsterne* (Leipzig: B. G. Teubner, 1909), pp. 42 and 22.

[107] "On Ether and Gravitational Matter through Infinite Space," *Philosophical Magazine* 2 (1901): 161–77. For quotation see p. 175.

[108] *Der Bau des Fixsternsystems* (Braunschweig: F. Vieweg, 1906).

[109] *Ibid.*, p. 215.

[110] *Ibid.*, pp. 25–26 and 186.

[111] *Ibid.*, pp. 181–83 and 227–28.

[112] "Die Milchstrasse," in *Verhandlungen der Gesellschaft Deutscher Naturforscher und Ärzte: 79. Versammlung in Dresden 15–21 September 1907,* edited by Albert Wangern (Leipzig: F. C. W. Vogel, 1908), pp. 178–93.

[113] *Ibid.*, p. 188.

[114] "Ueber das Newton'sche Gravitationsgesetz," *Astronomische Nachrichten* 137 (1894): col. 133.

[115] "Ueber das Newton'sche Gravitationsgesetz," *Sitzungsberichte der mathematisch-physikalischen Classe der K. b. Akademie der Wissenschaften zu München* 26 (1896): 373–400.

[116] "Ueber die räumliche Verteilung der Sterne im schematischen Sternsystem," *Sitzungsberichte* . . . 1911, pp. 413–61.

[117] *Ibid.*, p. 423.

[118] "Recent Researches in the Structure of the Universe," *Proceedings of the Royal Institution of Great Britain* 19 (1908): 300–15.

[119] "The Structure of the Universe," *Science* 38 (1913): 717–24.

[120] *The Astrophysical Journal* 55 (1922): 302–28.

[121] See, for instance, the article by J. Jeans, "The Motions of the Stars in a Kapteyn-Universe," *Monthly Notices of the Royal Astronomical Society* 82 (1922): 122–32.

[122] *Himmelskunde: Versuch einer methodischen Einführung in die Hauptlehren der Astronomie* (Freiburg im Breisgau: Herdersche Verlagsbuchhandlung, 1898). A second edition was published in 1913.

[123] *Ibid.*, p. 519.

[124] *Ibid.*, pp. 518–19.

[125] *Ibid.*, p. 520.

[126] A typical example of this is the cursory treatment of the Milky Way in Arrhenius' *Die Vorstellung vom Weltgebäude im Wandel der Zeiten,* translated from the Swedish by L. Bamberger (Leipzig: Akademische Verlagsgesellschaft, 1908), pp. 153–54.

[127] "Wie eine unendliche Welt aufgebaut sein kann," *Arkiv för Matematik, Astronomi och Fysik* 4 (1908); Nr. 4.

[128] "Die Parallaxe des planetarischen Nebels GC4373," *Astronomische Nachrichten* 184 (1910): cols. 231–35.

[129] *Leçons sur les hypothèses cosmogoniques* (Paris: A. Hermann, 1911), p. 269. Contrary to Poincaré, Bohlin did not measure the parallax of Andromeda.

[130] *Ibid.*, pp. xxiv–xxv.

[131] *Ibid.*, p. xxiv.

[132] In his Adams-Prize-winning essay on Saturn's rings (1857); see *The Scientific Papers of James Clerk Maxwell,* edited by W. D. Niven (Cambridge: Cambridge University Press, 1890), vol. 1, p. 291.

[133] *Stellar Movements and the Structure of the Universe* (London: Macmillan, 1914).

[134] *Ibid.,* p. 242.

[135] *Ibid.,* p. 243.

[136] *Ibid.,* p. 31.

[137] *Ibid.*

[138] *Ibid.,* p. 261.

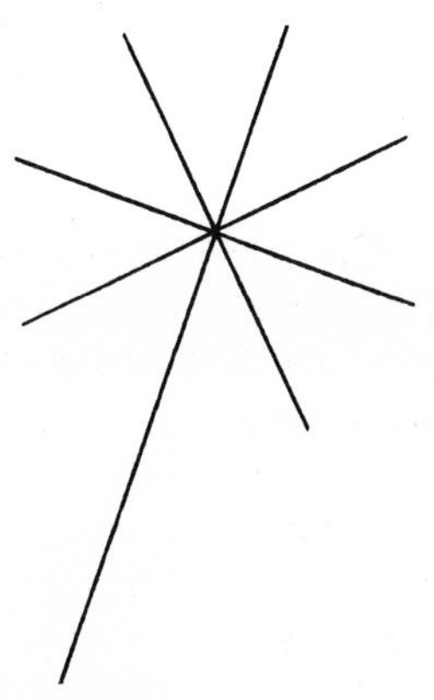

CHAPTER NINE

The Beacon from Andromeda

SYSTEMS constructed by science resemble imposing edifices which, on closer inspection, reveal a crack or two. The flaw of Eddington's impressive synthesis lay in his remark concerning the new estimation of the distance of the Lesser Magellanic Cloud. This distance, some 32,000 light-years, was, Eddington noted, a new record.[1] It was obtained in 1913 by Ejnar Hertzsprung,[2] who found that the period-luminosity law for Cepheid variable stars in that Cloud, established in 1912 by Henrietta S. Leavitt, could be used for determining its distance. Enormous as 32,000 light-years might be, they represented a distance not unlike the dimensions then assigned to the Milky Way. Hertzsprung himself did not suggest that the same law might be the instrument for verifying signals from much greater depths of the universe. His major objective actually concerned the distribution of 68 Cepheid variables, and he found it highly significant that none of these was more than a few thousand light-years away from the plane of the Milky Way.[3] The spark that kindled a new and epoch-making study of the nebulae came, in keeping with the erratic story of the Milky Way, through sheer chance, and not through immediate exploitation of what should have been the logical source—the period-luminosity law of the Cepheid variables.

Since chance usually favors the strongest and best, fortune descended upon Mount Wilson Observatory. There, George W. Ritchey found a nova on a photograph of the spiral nebula NGC6946 taken July 19, 1917. Word of the discovery was sent by telegram to various observatories, among them Lick Observatory on Mount Hamilton, south of San Francisco Bay. One of its staff, Heber D. Curtis, promptly set himself the task of re-examining all the photographs taken with the 36-inch Crossley refractor, and found three novae in two spirals, NGC4527 and NGC4321. Meanwhile, Ritchey performed a similar study of the plates at Mount Wilson, and established the flare-up of a nova in the Great Nebula of Andromeda and of another, Z Centauri, in NGC5253 during August, 1885. With a flash of intuition Curtis realized the significance of the individual data, and, eleven days after Ritchey's discovery, he went on record in a note to the *Publications of the Astronomical Society of the Pacific* that "the occurrence of these new stars in spirals must be regarded as having very definite bearing on the 'island universe' theory of the constitution of spiral nebulae."[4]

Interest suddenly reached fever pitch. Two months after Ritchey's discovery, the list of novae in spirals already contained eleven entries—which promptly divided the astronomers into two major camps. The issue was whether or not the new evidence of novae supported the island-universe theory. The respective leaders were Curtis and Harlow Shapley of Mount Wilson Observatory; both published notes in the October, 1917, issue of the *Publications*. A debate which became a classic in the history of science ensued. As A. Sandage of Palomar Observatory wrote: "Both men strongly believed their own position . . . Everyone, once his belief is set, will rationalize the facts to suit himself. The debate in question is an excellent example of the process . . . The arguments, pro and con, as they were advanced in 1917 to 1921, constitute a psychological study of the first order."[5]

The note by Curtis[6] was a reply to an argument, based on analogy, against the theory of island universes. By that argument, if the spirals were systems similar to our galaxy—in which twenty-six novae were listed—more novae should have been seen in the spirals. Curtis' rejoinder was based upon the assumption that the brightness of novae could vary by as much as sixteen magnitudes. Thus, if a nova of the fifteenth magnitude in our galaxy was at a distance of 20,000 light-years, the distance would be twenty million light-years if it were in a spiral. At such a distance, an island universe ten minutes of arc in apparent diameter would have a diameter of about 60,000 light-years: "not an impossible dimension," added Curtis, "so far as we may judge from our very imperfect knowledge of the size of our own Galaxy."[7] Curtis was not nearly so attentive to the imperfect knowledge concerning a crucial point in his argument: the change of magnitude in novae. Nor was he reluctant to make the most of analogies when they served his purpose; hence, he assumed that the novae newly discovered in some spirals corresponded to the brightest ones in our galaxy. He set their difference at ten magnitudes and their respective distances as one to one hundred. This meant that the spirals were far outside the Milky Way.

Shapley's paper[8] was, typically, full of references to the "uncertainties" in the various aspects of the data obtained on novae in spirals. It was all too apparent that he wished to weaken the arguments in support of the theory of island universes. His mind seemed already made up on the subject, and in this respect, the manner in which he exploited A. van Maanen's measurements of the rate of rotation of some spirals was particularly revealing. The results, Shapley argued, implied that the arms of a spiral nebula should rotate at the speed of light, if it was as big and as distant as an island universe had to be. The observed rotation rate of spirals seemed, therefore, to imply that they were parts of the Milky Way. Shapley wasted not a moment in delivering his *coup de grâce* to island universes. Spirals were not only relatively close objects, he insisted, but they were not even systems of stars, since no stars could be seen in their arms. He was in fact ready, as will shortly be seen, to dismiss novae as real parts of the spirals. Still, for the time being, he rested content with restating his two-pronged attack in emphatic terms: "We are not prepared to accept velocities of rotation of the order of the velocity of light . . . Measurable internal proper motions, therefore, can not well be harmonized with 'island universes' of whatever size, if they are composed of normal stars."[9]

One can only guess the full extent of Shapley's motives, but his eagerness to systematize was all too evident. He proceeded in a truly methodical manner, as may be seen from his work on redetermining the distances of globular clusters which was already in progress. Through this work he also meant to ascertain the true limits of the Milky Way, as an all-embracing system of stars, and, last but not least, to discredit the theory of island universes. With the haste of a born systematizer, he felt, early in 1918, that the data at hand entitled him to sketch the major features of the system of stars.[10]

The new system was the old one-island universe on a far more grandiose scale. In arriving at the new synthesis, Shapley relied upon three methods of measuring the distances of globular clusters. The three dealt, respectively, with novae, B-type stars, and Cepheid variables, but it was the last which apparently provided the most convincing evidence of the need to attribute to the galaxy a diameter of more than 200,000 light-years. The most distant globular clusters appeared to be at least that far away and, to Shapley, they also marked "the extent and arrangement of the total galactic organization."[11] From this, he concluded that "all known sidereal objects become a part of a single enormous unit."[12] In that single unit all globular clusters and the two Magellanic Clouds, "extensive and massive systems" though they might be, were "clearly subordinate factors." Shapley's satisfaction with the new picture could also be felt in his meticulous attention to such details as the hundred-thousandfold increase in the volume "commonly assigned to the stellar universe."[13]

The new size of the galaxy, or stellar universe, was one of the four reasons advanced by Shapley in support of his contention that the spiral nebulae were not "separate galaxies of stars," or island universes. It was, in fact, with the question of spirals as other island universes that his report came to a close. "The cluster work," as Shapley referred to his pioneering observations, "strongly suggests the hypothesis that spiral nebulae, while not closely related in history or dynamical development to the average star, are, however, members of the galactic organization, appearing to avoid the regions of enormous masses and forces more widely than do the globular clusters."[14] This was, of course, the old argument of Herbert Spencer. It was not without an unusual weight of its own, but it also acted as a blindfold. Shapley's case amply illustrated that point by the manner in which he tried to defend the purely "nebular" or gaseous character of spirals, in spite of the fact that novae could be seen in a number of them. Around 1918, the generally accepted explanation of relatively nearby or "galactic" novae was that they were "the penetration of nebulosity by a star with considerable velocity."[15] With a quick sleight of hand Shapley turned the situation inside out. In the case of novae in spirals, it was not the star that penetrated a "nebulous" mass but the latter which enveloped a star by rapid motion. From this there followed the deceivingly simple explanation of the frequency of novae in spirals: it was a "function of the dimensions of the nebula, its velocity in space, and the stellar density of its neighborhood—that is, the distance from the galactic plane."[16]

The weakness of such reasoning is usually apparent to all except those already convinced of the theory it was meant to prove. If island universes were to be distrusted, spirals could only be gaseous in nature. Thus, they became a conceptual filter which effectively colored the interpretation of ingeniously assembled data. Success or failure with these data depended heavily on the original intuition, whose role was thoroughly elaborated in Curtis' lecture[17] of March 15, 1918, in Washington, only a few weeks after Shapley's discussion saw print. Curtis spoke, of course, of the intuition of others, of long-deceased men of science, who "with some strange gift of intuition have looked ahead from meager data, and have glimpsed or guessed truths which have been fully verified only after the lapse of decades or centuries."[18] He certainly did not seem to want to create the impression that he himself argued a cause largely on the basis of intuition supported by a meager body of data.

The increase in the estimated number of spirals was anything but meager. It grew from less than 10,000 to about 700,000 through a re-examination by Curtis of photographs "on all available regions taken at Lick Observatory during the past twenty years."[19] In addition to his paper, "On the Number of Spiral Nebulae,"[20] Curtis could boast of another paper in press, which had a more direct bearing on the status of spirals. To him, a champion of island universes, the apparent absence of stars in spirals demanded an explanation diametrically opposed to that which suited Shapley's purpose. Naturally, Curtis searched for what he called "occulting regions" in the spirals. His study of seventy-nine spirals[21] provided him with the evidence he sought. After all, in 1917 the spirals looked, even through the best telescopes, mere systems of nebulous patches. The potential fallacy of his argument did not dawn on him as he listed evidences of that "occulting" matter outside as well as between the whorls of spirals, in spirals seen edgewise, and in those lying at an appreciable angle to our line of sight. The "occulting" matter was in spirals everywhere. It manifested itself, to quote Curtis' descriptive passage, "in such appearances as 'lanes' more prominent on one side of the major axis of the elongated elliptical projection, in a greater brightness of the nebular matter on one side of this major axis, in a fan-shaped nuclear portion, or in various combinations of these effects."[22] His eagerness prompted him to unremitting search, lest the slightest possible occurrence of that "occulting" matter should be overlooked. Thus, he quickly seized upon E. E. Barnard's studies on dark gaseous clouds in the Milky Way,[23] which became, in his reading, something closely analogous to the "occulting" areas in spirals.

Clearly, none of these fresh observational data could be marshalled in direct evidence of the island-universe status of the spirals. It was still a plausible belief, not a cogently argued conclusion, that the brightness of novae in spirals implied a distance of millions of light-years and a size comparable to the Milky Way. But once this point was reached, Curtis could confidently claim that there was no need to be concerned about the large average radial velocity, 500 miles per second, of some spirals. Curtis felt that this put them "in a class apart."[24] The apparent grouping of spirals around the galactic poles had for a long time been interpreted in the opposite sense; namely, that they were a "subclass" of the one, inclusive star system centered on the Milky Way. With the existence of "occulting" matter granted, the absence of spirals along the plane of the Milky Way and at small angles to it seemed satisfactorily explained.

Curtis was not alone in considering the spirals as island universes, but he was then undoubtedly the most articulate spokesman of that view. Shapley, in a paper published on the heels of Curtis' essay, appeared to overstate the case, however, when he wrote that the island-universe hypothesis of spirals "has many adherents and appears to be growing in general acceptance."[25] Shapley based his counter-arguments on "recent research," but he belabored results that were somewhat "old" in view of the feverish activity that then characterized work at Mount Wilson and Mount Hamilton. Shapley patently ignored the fact that Curtis spoke of galactic novae as being 100,000 light-years away. While Shapley was only too eager to find fault with arguments based on the high radial velocities of spirals and on their spectra, he unhesitatingly assigned a diameter of 300,000 light-years to the galaxy. He described his reasons as ones which "all seem definitely to oppose the 'island universe' hypothesis of the spiral nebulae."[26] Such a style contrasted not only with Curtis' less categorical statements, but also with the wide cosmological horizons opened up by Curtis' interpretation of the spirals. "The field of research is, like our subject matter, practically infinite," wrote Curtis—words which naturally fitted the theory of island universes.[27] Shapley's contentions implied, moreover, a dichotomy between the Milky Way, as the observable part of the universe, and the infinitely wide and unobservable part lying outside it. Or, as Shapley somewhat lamely remarked: "We have no evidence that somewhere in space there are not other galaxies; we can only conclude that the most distant sidereal organizations now recognized—globular clusters, Magellanic clouds, spiral nebulae—cannot successfully maintain their claims to galactic structure and dimensions."[28]

The interest aroused in the status of nebulae by the observations and reflections of Curtis and Shapley may be measured by the invitation extended them to present their positions together on April 26, 1920, before the National Academy of Sciences. The immediate issue was the true distance of many globular clusters which seemed to determine "The Scale of the Universe," as the Shapley-Curtis encounter was called.[29] The position Shapley defended was that "the diameter of the whole system of globular clusters was about 300,000 light years."[30] The main difficulty of this position arose from uncertainties about the zero-point of the distance scale of Cepheid variables. Readiness to minimize this revealed Shapley's thinking as much as did his dicta on the dethroning of man's "earthly gods" by the Copernican revolution, to say nothing of the "Shapleyan revolution," which displaced the solar system from the center of the galaxy.[31] Metaphysics aside, Shapley did scant justice to scientific objectivity when he argued that "there appears as yet no reason for modifying the tentative hypothesis that the spirals are not composed of typical stars at all, but are truly nebulous objects."[32] The tone of his argument suggested that his convictions on this point were anything but tentative or recent. He also echoed some of his earlier statements when he hinted broadly at a distinction between the observable and non-observable part of the universe: "But even if spirals fail as galactic systems, there may be elsewhere in space stellar systems equal to or greater than ours—as yet unrecognized and possibly quite beyond the power of existing optical devices and present measuring scales."[33] He did not seem to anticipate that his conclusion would be wholly upset by the "modern telescope" which, in his words, was "destined to extend the inquiries relative to the size of the universe much deeper into space, and contribute further to the problem of other galaxies."[34]

The most valuable part of Shapley's discussion of the problem concerned the high radial velocities of spirals. Since V. M. Slipher at Lowell Observatory had begun, in 1912, systematic study of the velocity of spirals, it had become increasingly clear that most spirals were receding, relative to the solar system. The inference could, therefore, plausibly be drawn that the spirals were being expelled, by some unknown mechanism, from the center of the galaxy. But this was a point to which Shapley did not then refer. It was, however, mentioned by Curtis, who phrased Shapley's position as follows: "The spirals are probably of nebulous constitution and possibly not members of our own galaxy, driven away in some manner from the regions of the greatest star density."[35] Curtis himself was so perplexed by the recessional velocities of spirals that he felt the theory of island universes "must be definitely abandoned" should those velocities be proven real,[36] or, in his own words, "should the results of the next quarter-century show close agreement among different observers."[37] Actually, he did his best to emphasize the disagreement between various observers, and listed considerations which seemingly weakened the evidence: "With regard to the observed excess of velocities of recession, additional observations may remove this. Part of the excess may well be due to the motion of our own galaxy in space. The Nebula of Andromeda is approaching us."[38] He had no inkling that the great confirmation of the theory of island universes was to come from the very same nebula, although not through the apparently "favorable" direction of its motion. Psychologically, it was equally remarkable that the ominous specter of recessional velocities could recede with such ease from Curtis' mental horizon, when, four pages later, he wrote: "On this theory [of island universes] it is unnecessary to attempt to coordinate the tremendous space velocities of the spirals with those of the average star."[39]

Curtis also felt that by adopting the island-universe theory of spirals, one could avoid "the almost insuperable difficulties involved in an attempt to fit the spirals in any coherent scheme of stellar evolution, either as a point of origin, or as an evolutionary product."[40] In 1920, the question of the evolution of galaxies was qualitatively less explored than that of the stars. In regard to the latter the Hertzsprung-Russel diagram was already available, though nothing was yet known of nuclear transformations in stars. Curtis could, therefore, be forgiven on this score, although his abandonment of a serious problem was another matter. His failure and that of Shapley to mention the General Theory of Relativity in connection with the recessional velocity of most spirals arose in part from the cleavage between observational and theoretical cosmology. The other factor was, as will shortly become clear, that the very few who then understood and discussed Einstein's cosmological equations preferred to argue away the physical reality of the spirals' recessional velocity. Thus, the question of their distance was discussed in isolation from other crucially important points. No trace of them could be found in Curtis' summary of his own position: "The spirals are a class apart, and not intra-galactic objects. As island universes, of the same order of size as our galaxy, they are distant from us 500,000 to 10,000,000, or more, light-years."[41]

By assigning such distances to spirals, Curtis came far closer to the truth than Shapley did. Especially prophetic was the attention Curtis paid to the Andromeda nebula. By his calculations, the Andromeda nebula, to which he assigned a distance of 500,000 light-years, was 17,000 light-years in diameter, rather a good estimate in proportion to its actual distance of 2,000,000 light-years and diameter of about 90,000 light-years. With regard to their respective estimates of the size of the Milky Way, the encounter between Shapley and Curtis was a draw, at least in retrospect. By setting the dimensions of the galaxy at a diameter of 30,000 light-years and a thickness of 5,000 light-years, Curtis fell short of the truth by a factor of about three, and Shapley overshot the target by the same factor. For the moment, however, Shapley seemed victorious. He was one of the rising stars of twentieth-century astronomy, and this effectively determined the direction in which the psychological pendulum of scientific consensus was to swing. Curtis never received adequate credit for his concluding statement, whose truth was fully vindicated by astronomical research within a few years: "The spirals as external galaxies indicate to us a greater universe into which we may penetrate to distances of ten million to a hundred million light-years."[42] Compared to this, Shapley's vision of a larger galaxy was distinctly puny.

On this topic, Shapley and Curtis represented positions based on the latest evidence. Their awareness of its importance was not shared by everyone, not even by such a notable figure of early twentieth-century astronomy as J. H. Jeans. In his Halley lecture of 1922, Jeans confidently estimated the distance of the Andromeda nebula at 5,000 light-years and spoke of spirals as possessing masses enormously larger than our sun.[43] The comparison clearly implied their small size as contrasted with our galaxy. Jeans apparently felt no need to revise his estimate of 1917 when he judged that same nebula to be fifteen times smaller than our galaxy.[44] His references to spirals as "island universes" had, in consequence, a distinctly ambivalent tone.[45] Two years later, a similar hesitation could be detected in the pages of J. Plassmann's monograph on the Milky Way. Until then a firm proponent of the extragalactic status of many nebulae, Plassmann now felt that one must assign unimaginably great brightness to novae in spirals, if these were to be considered really "external."[46]

The Shapley-Curtis debate had, of course, no meaning for the double-tier theory of the Milky Way proposed as an embellishment of Hörbiger's notorious glacial theory of the universe, a major scientific obscurantism of the early twentieth century. The whiteness of the Milky Way was seized upon by an advocate of that theory as evidence of glacial vapor somewhere beyond the orbit of Neptune. This vapor, together with stars in the plane of the Milky Way, was supposed to cause its appearance.[47] Obscurantism on one particular topic always seems to reappear in much the same ironical cloak.

The touch of irony was not completely absent from the most respectable sector of research on the Milky Way and other galaxies. The technical procedure which proved Curtis right proved to be the same observation of Cepheid variables which had enabled Shapley, a few years earlier, to enlarge enormously the accepted estimates of the distances of globular clusters in the Milky Way. The results of applying the same method to the spirals were equally startling, and began to unfold with dramatic speed and directness. During the fall of 1923, an extensive search was undertaken at Mount Wilson, through its newly erected 100-inch telescope, for novae in the Andromeda nebula. The first plates showed two novae and one faint, eighteenth-magnitude object which was classified as a nova. However, comparisons with subsequent plates revealed that the latter was a Cepheid variable with a period of about a month. This put its absolute magnitude at –4, or about 7000 times the brightness of the sun. Its true distance and that of the Andromeda nebula must, therefore, be of the order of about a million light-years, and far outside the Milky Way, even if it were as large as Shapley claimed.[48]

By the end of 1924, Edwin P. Hubble had a list of twelve Cepheids in the Andromeda nebula (M31) and twenty-two in M33. In the late summer of that year, astronomers traveling in and out of Mount Wilson began to hint at some extraordinary development in progress. Thus, Henry Norris Russell of Princeton received word of the new Cepheids when, in October, 1924, he happened to be visiting at Yale where Jeans, then a Research Associate at Mount Wilson, also made a visit. Russell immediately grasped the full import of the news. He also grew impatient with the official silence from Mount Wilson. On December 12, he decided to take up the matter with Hubble in a letter, in which he mentioned his meeting with Jeans and asked Hubble when he would be ready to announce the discovery: "It is a beautiful piece of work," wrote Russell, "and you deserve all the credit that it will bring you, which will undoubtedly be great. When are you going to announce the thing in detail? I hope you are sending it to the Washington meeting, both because we all want to know all about it, and because you ought, incidentally, to bag that $1000 prize."[49] The letter closed with Russell's anticipation of "some grand discussions" at the meeting.

Russell's words could not have been better timed. On the first day of 1925, at the annual meeting of the Astronomical Society in Washington, it fell to Russell to read Hubble's paper on "Cepheids in Spiral Nebulae." Those present, Curtis and Shapley among them, needed no special warning to realize the importance of the moment. The air was full of the tension of the final act of a drama. As Russell read, it became evident that the debate on island universes had come to an end and that a new era in cosmology had begun. Yet, the actual excitement failed to be reflected in the printed record. In the detailed account of the meeting, brief reference was made to the fact that Hubble's paper was "to share in the joint award of the thousand-dollar prize given for an outstanding paper at the Washington meeting."[50] The contents of Hubble's paper received only six matter-of-fact lines: "Dr. Hubble, working with the 100-inch Mount Wilson reflector, had succeeded in resolving portions of two of the spiral nebulae, those of Andromeda and Triangulum, into separate stars, and from a study of the period-luminosity curves of the Cepheid variables in the nebulae had derived distances approaching one million light years for each, thus bringing confirmation to the so-called island universe theory."[51]

The published form of Hubble's paper, which appeared as a three-page abstract in the March issue of *Popular Astronomy,*[52] might have seemed an effort on its author's part to conceal the tremendous cosmological implications of the fact that the two nebulae must be placed at a distance of 285,000 parsecs or 827,000 light-years. Hubble himself gave Russell a good glimpse of his own thinking on the entire matter in a letter of February 19, 1925. He singled out "the flat contradiction to van Maanen's rotations" as the reason of his "reluctance in hurrying to press." Upon examining van Maanen's data, Hubble felt that there was a "magnitude error" in their original evaluation, and he was anxious to show the evidence to Russell during the latter's forthcoming visit to Mount Wilson. The letter made it very clear that he was most elated not by the grand vistas of cosmology, but by the new possibilities for examining questions of detail in the spirals: "Meanwhile a mass of undigested data is accumulating from the observations—star counts and color plate for M33, novae in M31 (6 on one plate), variable in other spirals, evidence of resolution in irregular non-galactic nebulae, etc. *The really big advance,* as I see it, is the possibilities of applying the usual methods of stellar investigations to the spirals."[53] (Italics added.)

Hubble himself made no reference to cosmology or island universes in his detailed accounts of the spirals NGC6822 and M33, published in 1925 and 1926, respectively.[54] Few discoverers, if any, tried more carefully to avoid the most fundamental aspect of their findings than did Hubble. It became his historic privilege to write the concluding page in the long history of man's groping after the truth about the Milky Way. Hubble's page had all the grandeur as well as the baffling myopia of the major steps in that groping. It matched the disinterest with which Galileo and his century left unexplored the sighting of innumerable stars in the Milky Way. It matched the failures of a Newton and a Halley to judge the Milky Way worthy of any sustained reflection. It matched the lack of any serious reaction to the publication of the works of Wright, Kant, and Lambert. No major discovery entered scientific consciousness more surreptitiously than did the decipherment of the true structure of the Milky Way. Herschel became the toast and admiration of the world because of his great reflectors; for his research on the Milky Way he was not given prompt recognition. Herschel united a great observer with a bold theorizer, who was not ashamed to declare that the data of observations were intended for speculations. Of the two errors of speculating too little, or too much, he preferred to be guilty of the latter. Diametrically opposed to this attitude was Hubble, who nevertheless provided the crucial discovery which made Herschel's great dream a reality: the first reliable glimpse into the galactic construction of the heavens.

There were no major demurrers as Hubble disclosed his findings—but proofs were one thing, assent another. The latter could be given wholeheartedly or with a slight air of reserve. Reserve, in turn, could find convenient support in the lessons of history, which always suggest caution. In his *Studies of Anagalactic Nebulae,*[55] Knut Lundmark once more provided good synthesis of the latest research as he did in his lengthy paper which summarized, in 1920,[56] the various efforts made during the previous twenty years to fathom the contours of the Milky Way, and the distances and distribution of globular clusters. In 1926, however, Lundmark prefaced the long list of technical details with a long historical introduction.[57] Although he began with Swedenborg, Lundmark made clear his unfamiliarity with Swedenborg's actual statements on the Milky Way. He was no more fortunate in what he reported of Thomas Wright and Kant. As for Herschel and the nineteenth century, Lundmark fared better. His survey of the recent history of studies on the Milky Way was brought to a conclusion with a paragraph in which he advanced the possibility that the theory of island universes might again fall into disrepute: "During the time passed since William Herschel's days the pendulum of astronomical opinion has swung many times as to the conception of the cosmogonical role of the nebulae. On many occasions the evidences have been rather strong in favor of one or the other theory. For the present it seems rather plausible that the island-universe theory has obtained its final confirmation.—Or, will the pendulum swing again?"[58] Turning to history at this juncture was not without its drawback, for the annals of cosmology could also offer powerful reminders of irreversible changes and advances. After the beacon of Andromeda was once seen, there was no way of extinguishing it. The universe could not shrink again into a one-galaxy, one-island system, surrounded by the "unobservable infinite" outside.

A subtly modified form of the one-island theory of the universe was carefully nurtured by Shapley for some years. In his classic monograph on star clusters, published in 1930, he claimed that researches on extragalactic nebulae during the previous six years "seem to leave no doubt that our galactic system is extremely large compared with typical external systems."[59] These external systems were the spirals of which some rare "gigantic" ones such as the Andromeda nebula had, according to Shapley, a diameter of, at most, one-fifth that of the Milky Way. Such an estimate of the size of Andromeda was as baffling in 1930 as Shapley's other contention that only future research would tell whether the Milky Way was itself a spiral or an irregular nebula like the Magellanic Clouds. Clearly, Shapley preferred to keep the Milky Way and the spirals as different as possible in every respect. Even the steadily increasing number of observed spirals and the likely prospect of finding countless others proved to him the special status of the Milky Way in the universe: "Among the hundreds of thousands, possibly millions, of discoverable external systems or island universes in the oceans of space, our own system tends to be continental in dimensions."[60] Discovery of novae in some spirals had overthrown, according to him, only "the arguments against the Kant-Herschel theory of a plurality of universes," but not the picture of a universe consisting of countless islands and of one big continent. It was this picture of the universe which, he now claimed, represented his position in 1918: "The lack of comparability between galactic system and spiral nebula appears now more certain than before; ours is a Continent Universe if the average spirals are considered Island Universes."[61] Such was a strange reading of the record in which he had played a most active part, but it matched the silent treatment which he gave to Curtis.[62] Shapley also failed to perceive that his "one-continent, many-islands" theory of the universe did an injustice to the cosmology of the General Theory of Relativity.[63] The principle of homogeneity was a foundation of that theory, and in turn demanded that most galaxies be essentially alike.

The General Theory of Relativity was the theoretical one of the two factors that forbade the pendulum to swing back to the one-island theory of the universe. The other factor was observational: the red-shift in the spectra of spirals. Lundmark could have correctly interpreted this observational factor in terms of the theoretical factor when, in 1924, almost simultaneous with Hubble's dramatic disclosure, he considered a possible correlation between the apparent size of spirals and the magnitude of the red-shift in their spectra.[64] He perceived in this merely an indication of the extragalactic status of spirals. The man who had the creative insight to enunciate the concept of a universe of receding nebulae in the four-dimensional space-time manifold of General Relativity was Abbé Georges Lemaître, a Belgian astrophysicist and a former student of Eddington. But once more, the puzzling story of the Milky Way remained consistent. Lemaître's historic paper[65] appeared in the pages of a periodical of limited circulation, and for almost three years the world of astronomy failed to take note of it.

Like most discoverers in science, Lemaître had precursors. The most distinguished of these was Willem de Sitter of the University of Leiden. In three papers published in 1916–1917,[66] he submitted a solution to Einstein's cosmological equations, presenting in a wholly new light the cosmological status of spirals. Computing the radius of the four-dimensional universe by star density,[67] de Sitter obtained a value of 9×10^{11} astronomical units or about 1.44 billion light-years, and 7×10^{19} sun masses for the total mass. Since current estimates put the total number of stars within the Milky Way at about 100 million, the result implied that the total mass of the universe was sufficient for almost a thousand billion (10^{12}) galaxies similar in size and mass to the Milky Way. De Sitter in fact noted that at least some of the spirals were very probably "galactic systems comparable with our own in size,"[68] and that "in the part of space which immediately surrounds our galactic systems, there are many similar systems whose mutual distances are large compared with their dimensions."[69] Confining his horizon to the immediate neighborhood of our galaxy was a position derived in part from his realization of "the uncertainty (which is considerable) of the hypotheses and of the numerical data."[70] Clearly, observational evidence did not yet encourage ideas of a universe of billions of galaxies, in spite of the fact that spirals had already been observed in very large numbers. Practically no observational evidence was available to substantiate another challenging suggestion of de Sitter concerning the change in the frequency of light as it traveled across the four-dimensional space-time manifold. This was the famous de Sitter effect, which he, however, did not ascribe to the possibility that nebulae may have a velocity relative to the Milky Way.[71] According to his equations, the frequency change could be either toward the violet or toward the red. It was in the latter instance that de Sitter noted "the lines in the spectra of very distant stars or nebulae must therefore be systematically displaced towards the red, giving rise to a spurious positive radial velocity."[72] The word spurious was well chosen so far as de Sitter's thinking was concerned. He was reluctant, and rightly so, to conclude that the few measurements of red shifts in the spectrum of some nebulae provided an experimental verification of a universal "slowing down of light" in the finite though unbounded universe of galaxies.[73]

A. Friedmann, the famous proponent of an oscillating solution of Einstein's world model, had far less right than de Sitter to be considered a forerunner of Lemaître. Nebulae were not mentioned in Friedmann's paper,[74] in which he gave a detailed account of the period of oscillation (expansion-contraction), but omitted the questions of its speed at a given moment and of the principal material carriers of the oscillation. A year later, in 1923, Eddington moved nearer to the issue when he noted in his *Mathematical Theory of Relativity* that "a number of particles initially at rest will tend to scatter."[75] Eddington's book even contained the latest list of "radial velocities" prepared expressly for him by Slipher, the leading authority on the subject.[76] In the list of forty-one spirals, all but five appeared to be receding. However, Eddington's derivation of a "universal recession" had some loopholes, and these were quickly brought to attention.

Eddington himself was the first to acknowledge in 1930, when he "discovered" Lemaître's paper, that it gave "a remarkably complete solution of the various questions connected with the Einstein and de Sitter cosmogonies."[77] From Lemaître's formulas it was "at once apparent," to use Eddington's words, "that the Einstein world is unstable."[78] This was a most important point which, as Eddington noted, was not until that time appreciated in cosmological discussions. In 1930, Eddington had reason to register that astronomers were currently much interested in cosmologies based on the General Theory of Relativity. The interest was suddenly aroused, not by Lemaître's paper of 1927, but by an observational exploit which he predicted, in essence, two years ahead of time. "The receding velocities of extragalactic nebulae, are a cosmical effect of the expansion of the universe,"[79] wrote Lemaître, and in Section 4 of his paper he derived the numerical value of the Doppler effect in question from the variation of the radius of the universe, which could contain only a finite number of nebulae. For this feat, strangely enough, he did not receive credit from Eddington, although the latter pointed out with generous frankness that his own "original hope of contributing some definitely new result has been forestalled by Lemaître's brilliant solution."[80]

The observational exploit in question was a feat of Hubble, already famous for his evidence of the extragalactic status of the Andromeda nebula. Since then, he had ascertained the distances of twenty-four extragalactic nebulae. Their recessional velocities appeared unmistakably as functions of their distances: "The results establish a roughly linear relation between velocities and distances among nebulae for which velocities have been previously published, and the relation appears to dominate the distribution of velocities."[81] Hubble, nonetheless saw the "outstanding feature" of his results in "the possibility that the velocity-distance relation may represent the de Sitter effect and hence that numerical data may be introduced into discussions of the general curvature of space."[82] Hubble cannot, of course, be faulted for characterizing, in 1929, as a "de Sitter effect" the recessional spectrum of the extragalactic nebulae he studied. Lemaître's paper had yet to come to the attention of Eddington himself. A year earlier, Eddington had described, in his famous Gifford lectures, the red-shift of the nebulae as the "slowing down" of the vibrations of light as it travelled across the de Sitter space.[83] Hubble also had in mind a static universe of galaxies and not an expanding one. He did not even mention the distance-velocity relationship in his paper, "The Exploration of Space," written in 1929 to inform the general public of his work on the nebulae.[84] He presented his findings as the vindication of island universes, and of the finite though unbounded universe postulated by the General Theory of Relativity.[85] Hubble had no thought of the de Sitter effect when he hinted that further confirmation of the distance-velocity law might "lead to a solution having many times the weight."[86]

Rarely in the history of science has the real significance of a great discovery been left in such vagueness. Avoiding the speculative perspectives of the astronomer's findings was not a momentary attitude for Hubble.[87] In his memorable lecture series on the realm of nebulae, given at Yale in 1935, Hubble even tried to create the impression that speculation on island universes had largely been done by others than astronomers: "Astronomers themselves," he noted somewhat petulantly, "took little part in the discussions: they studied the nebulae."[88] But then what were Shapley, Curtis, Newcomb, Struve, and the two Herschels? The opening concept of Hubble's Halley lecture had, however, a theoretical perception of great significance: "The nebulae are great beacons, scattered through the depths of space . . . Observations give not the slightest hint of a super-system of nebulae. Hence, for purposes of speculation, we may invoke the principle of the Uniformity of Nature, and suppose that any other equal portion of the universe, chosen at random, will exhibit the same general characteristics. As a working hypothesis, serviceable until it leads to contradictions, we may venture the assumption that the realm of the nebulae *is* the universe—that the Observable Region is a fair sample, and that the nature of the universe may be inferred from the observed characteristics of the sample."[89] In doing the sampling he was without a peer.

Hubble's studies of the nebulae culminated in his classification of their shapes. His lectures at Yale were largely devoted to the details of that classification, which consisted of a sequence that began with spherical nebulae developing into increasingly eccentric ellipsoids and branching into normal and barred spirals[90] (Illustration XXb). Hubble, it should be noted, carefully avoided seeing an evolutionary process in the sequence, although it was almost immediately regarded as such by others. This time his distrust of theorizing stood him in good stead, at least in part; subsequent research left the sequence intact but inverted the order. When, in 1961, eight years after Hubble's death, A. Sandage published *The Hubble Atlas of Galaxies,*[91] progress of studies in stellar populations clearly suggested that galaxies start as spirals and assume an elliptical form only after their spiral arms have been dissipated. Or, in Sandage's words: "The continuity of sequence is still present, only the direction of travel is reversed."[92] Still, one must recognize an extraordinary achievement in the survival of Hubble's classification despite later research with the 200-inch telescope at Mount Palomar and through radio astronomy. It truly bears, as was noted by W. W. Morgan, the mark of the "genius of its formulator."[93]

The most provocative aspect of the latest speculations on galaxies does not relate to research on their individual structure and evolution. J. S. Plaskett incorrectly diagnosed the situation when he claimed, in his Halley lecture of 1935, that topics of such general application as the expansion of a universe of galaxies were exceeded in importance "by investigations into the nature of the stellar system in which we are situated."[94] Questions about the structure and dynamics of our galaxy had, of course, their share of excitement, especially since 1927, when Jan H. Oort of Leiden Observatory provided decisive data in support of the rotation of the Milky Way. As a result, such highly interesting details as the revolution of our solar system around the galactic center once in every 200,000 years immediately became evident. Knowledge of the specifics of rotational movement in turn permitted a close analysis of the stability of our stellar system. The results were quickly brought into an impressive synthesis by Eddington in his Halley lecture of 1930.[95] The Milky Way appeared to him as a rotating Catherine wheel, precarious and highly unstable. The chief lesson to be learned was, he said, "to wake us from our dream of leisured evolution through billions [10^{12}] of years," for the Milky Way could not be credited "with so much age and endurance."[96] He conjured up a much shorter future of a few thousand million [10^{9}] years in which "our skies will have lost one of their chief telescopic glories," the Milky Way.[97] He did his literary best "to emphasize our sense of the transitoriness of things." The date was May 30, 1930, and he had just read Lemaître's historic paper. Its words could not have reached a more receptive mind than that of Eddington: "the other galaxies," he remarked, by way of conclusion, "are rushing away at high speed as though our poor system were the plague-spot of the universe."[98]

Clearly, beneath questions about the structure, rotation, and stellar population of the Milky Way and other galaxies, lay the specter of the singularity, both *in space* and *in time,* of a universe of galaxies. Singularity or finiteness in space was a consequence of the finiteness of matter in the universe. Mental capitulation to that singularity was relatively quick, though painful, for the idea of an infinite, homogeneous universe had been idolized, in one way or another, for the previous three centuries. Impelled by spurious piety and poor theology, Newton had described infinite space as the sensorium of God. Less than a century later, the idea of an infinite universe was seized upon as the ultimate order of existence; lack of respect for scientific laws typically accompanied such efforts. Thus, in our day, Bertrand Russell, who was always illuminating, though not convincing, chose to emphasize the possibility of revising the basic laws of science rather than accepting the stringent reasons which allow but a finite mass for the universe.[99]

Singularity in time, or an irreversible, finite time-span for the universe, was, to some, an even more ominous notion. More than pure science almost immediately raised objections to the idea of an expanding universe of galaxies. The Symposium on "The Evolution of the Universe,"[100] held in 1931 at the meeting of the British Association for the Advancement of Science in London, provided especially significant details. Jeans strongly deplored the idea of an expanding universe as one in which the whole life, that is, "productive" life, of the universe was reduced "to a matter of hundreds of thousands of millions of years at most," bringing an "almost complete chaos into the already chaotic problem of stellar evolution."[101] Beyond that "productive" life-span of the universe, Jeans saw a universe becoming infinite in size as time went on. Unbelievably, he also submitted that, after a sufficiently long period, "the different galaxies or star-systems will be scattering away from one another *with speeds greater than that of light,* so that radiation will be unable to bridge the gap between them."[102] (Italics added). Jeans offered no scientific data as he aired his misgivings about the expanding universe of galaxies: "it will have become impossible to see any one galaxy from any other, even by light [which] left it at the very beginning of time; only the mathematician will be able to deduce the existence of the other galaxies in recondite ways—and probably no one will believe him. Then at least these will have justified Herschel's name for them—'island universes'."[103]

In another address to the meeting, Jeans again attacked the expansion of galaxies as a theory which reduces the universe to an "ephemeral concern."[104] He did not seem, however, to express the views of many when he claimed: "The concept of an expanding universe may prove after all to be a false scent, and the truth may lie in some other direction."[105] In support of this, he could offer only two considerations. One was the thousands of billions of years which, in his view, were necessary for the evolution of stars. Jeans' claim certainly clashed with the few billion years of the estimated past of the universe, and also with subsequent findings about the evolution of stars. His other consideration was less specific and, therefore, more difficult to discredit. On the basis that relativity had imposed drastic reformulations of our concepts of space and time, he held out a prospect for similar changes in our understanding of matter and energy.[106] Since the date was almost thirty years after Einstein formulated the total conversion of matter into energy, Jeans must have had something else in mind if his suggestion was to be at all meaningful.

Jeans did not live to see a dubious form of that "something else" spelled out in 1948 by H. Bondi and T. Gold, the original proponents of the steady-state theory.[107] The continuous "creation of hydrogen atoms out of nothing"[108] and, of course, without a Creator, seemed to them the only way to eliminate the great cosmic singularity in space and time: the progressive, permanent separation of island universes from one another. In place of a world of nebulae proceeding as a whole through an irreversible evolution, they proposed a universe finite in space but remaining, as a whole, forever the same. It must, however, be noted that Bondi and Gold paid insufficient attention to the infinite amount of matter that should have accumulated, had the continual "creation" of hydrogen atoms been going on throughout an infinite past. They seemed merely to have in mind their "finite" universe when they noted that "the process of getting rid of both matter and radiation from any fixed volume is by pushing both across the surface bounding this volume; and both are replenished from within."[109] Actually, the only process that could have saved the steady-state universe from the gravitational and optical paradoxes of an infinite and eternal universe was to assume the spontaneous and continual *annihilation* of matter at the identical rate that it was "created." Pushing matter and radiation into a cosmic sink beyond the confines of a relativistic universe was indeed an expedient with no scientific merit. It amounted to re-introducing into cosmology the long discredited dichotomy which cuts the universe in two parts: one, finite and observable, another, infinite and forever unobservable. It is a cut which severs cosmological reasoning at its most vital juncture by the imposition of the debilitating precept that the "infinite" part of the universe can have no observable influence whatever on its puny, observable section.

At any rate, with the alleged "creation" of matter steadily proceeding, the universe could always appear the same with no major singularity breaking its uniformity. The condensation of freshly "created" hydrogen atoms into nebulae was always in progress where the recessional motion of galaxies had emptied sufficient volume for the process. The "creation" was supposed to proceed at a rate of one atom per liter of space in every 5×10^{11} years.[110] Though such a rate appears vanishingly small, the total amount of matter thus "created" at every second is enormous even when considering only that "smallish" portion of the universe which is circumscribed by the farthest visible galaxies receding from us at one-half the speed of light. Strictly speaking, the amount or rate of the "creation" should make no difference so far as its basic logical merit is concerned. Creation of one atom out of nothing and without a Creator is as much sleight of hand as that of a whole galaxy. If this remark be dismissed as metaphysics, one must still confront the metaphysics in Bondi's insistence that the continuous creation of matter is a more perfect scientific principle than the hallowed tenet of the conservation of matter to which classical or modern physics found no exception:

> When observations indicated that matter was at least very nearly conserved it seemed simplest (and therefore most scientific) to assume that the conservation was absolute. But when a wider field is surveyed then it is seen that this apparently simple assumption leads to the great complications discussed in connexion with the formulation of the perfect cosmological principle. The principle resulting in greatest overall simplicity is then seen to be not the principle of conservation of matter but the perfect cosmological principle with its consequence of continual creation. From this point of view continual creation is the simplest and hence the most scientific extrapolation from the observations.[111]

The "perfect cosmological principle," or the contention that, apart from local irregularities, the universe presents the same aspect from any place and at any time, was already the guiding star of Aristotle. It led him to sweepingly bold, deceivingly logical, but wholly erroneous cosmological systematization. As a result, Aristotle had no choice but to mishandle such a vast singularity of the universe as the Milky Way. In the process, he was forced time and again to pay lip service to the role of observation when discoursing about nature. Bondi, too, found it scientific to dismiss the need for experimental verification of the alleged rate of "creation." "It is clear," he wrote in 1952, "that is is utterly impossible to observe directly such a rate of creation. There is therefore no contradiction whatever with the observations, an extreme extrapolation from which forms the principle of conservation of matter."[112] But such an extremist, if not wholly willful, position, could not remain acceptable in an age of atomic energy and space travel. The mythical rate of "creation" had at long last to be submitted to direct test by monitoring devices carried through outer space by artificial satellites. No trace of the alleged "creation" was found.

In a sense this was anticlimatic. The original and principal confirmations sought by the proponents of the steady-state theory rested within the realm of galaxies. Had the theory been correct, the relative frequency of young and old galaxies should have been the same at both small and great distances. Again, the theory demanded a special relationship of the number of galaxies to distance. The partial response obtained to these questions by radioastronomy repeated an old but most instructive story. As the singularity of the Milky Way was a stumbling block to the Aristotelian universe and to the infinite, homogeneous universe of classical physics, so too the steady-state theory was affected by the great singularities implied in an expanding universe of a finite number of galaxies. The parallel provided by scientific history does not end here. There should be considerable food for thought in the failure of the champions of the Aristotelian universe to read that most magnificent handwriting in the sky, the Milky Way. The myopia and neglect of the generations of Galileo and Newton with respect to the Milky Way also offer much material to students of the psychological labyrinths in scientific thinking. Still to be unfolded in detail is the story of the success of the steady-state theory to parade, for the last two decades, as one of the major "rival theories in cosmology."[113]

Part of the answer to this problem can be easily given. The

steady-state theory of the universe seemed to satisfy an age-old craving in man to find a self-explanatory, if not *a priori*, solution to the riddle of the cosmos. Time and again, scientists have come to believe that they had a definitive synthesis within their grasp. The history of the search for the fundmental units of matter and for the true picture of the realm of stars provides ample illustrations. As for the realm of galaxies, the research of the last decade was ushered in by the hopeful statement: "The solution of the cosmological problem is an achievement reserved to great telescopes. As telescopes and techniques improve, astronomers eventually reach a critical barrier of ignorance. In due course this barrier falls. The breach, when open, permits all to follow."[114]

The new techniques of celestial observation certainly broke down many barriers of ignorance. In general, the advances of the last decade uphold an old lesson—the crucial role played by studies of the Milky Way (and of other galaxies) in the quest for an understanding of the universe. From earliest times, sustained attention to the Milky Way could have kept men of science from falling victims to the Aristotelian belief in a perfect, spherical heaven. The belief in the ultimate truth of a homogeneous, infinite universe of stars might also have been less blinding had serious attention been given to the Milky Way. In historical retrospect, it is all too clear that the Milky Way should have been the road toward a more truthful picture of the cosmos. The historical record also shows that, if this road eluded men of science, it was in a sense their own fault. Nevertheless, even today, two centuries after the true picture of the Milky Way has emerged, and almost half a century after its true position in the universe was revealed by the beacon from Andromeda, the study of the Milky Way and of other galaxies remains the principal road toward the understanding of the universe as a whole. Estimates of the size of the universe are a function of counting the nebulae, and accounts of its evolution are intimately connected with the knowledge gained about their evolution. In this respect, the achievements of the last two decades are undoubtedly impressive. Galaxies are now being studied with the most advanced technical and conceptual tools of nuclear and plasma physics. Study of the evolutionary process, which determines the shape and physical properties of a galaxy at a given time, is assuming a precision which the study of the properties of ordinary matter did not possess a few decades ago.

Still, it must be recognized that the actual turn of events cast doubt on some sanguine expectations. The cosmological balance sheet of the last ten years shows decidedly greater increase in the column of new great riddles than in the column of major solutions. The latter may at first loom large, but new riddles soon make them appear less final and fundamental. Astrophysics lacks an answer to the newly found phenomenon of quasars as completely as it still stands baffled before the more than fifty-year-old problem of gravitational collapse, the mysterious fate of stars of exceptionally large mass. Again, there is not a clue to the properties of matter in that superdense state of contraction through which it must pass if the realm of nebulae indeed forms an oscillating universe. Its advocates are obviously uneasy about the great cosmic singularity which, to all evidence, characterizes the universe of nebulae not only in space but in time too. Students of that universe should remain mindful of the lesson that marked the entire history of scientific study of the local nebula, the Milky Way: only a readiness to admit its singularity could at last unveil the true image of what has been a most elusive road for science.

References

[1] *Stellar Movements and the Structure of the Universe* (London: Macmillan, 1914), p. 240.

[2] "Über die räumliche Verteilung der Veränderlichen vom δ Cephei-Typus," *Astronomische Nachrichten* 196 (1913): cols. 201–08.

[3] *Ibid.*, col. 208.

[4] "New Stars in Spiral Nebulae," *Publications of the Astronomical Society of the Pacific* 29 (1917): 180–82; for quote, see p. 182.

[5] *The Hubble Atlas of Galaxies* (Washington: Carnegie Institution of Washington, 1961), p. 3.

[6] "Novae in Spiral Nebulae and the Island Universe Theory," *Publications of the Astronomical Society of the Pacific* 29 (1917): 206–07.

[7] *Ibid.*, p. 206.

[8] "Note on the Magnitudes of Novae in Spiral Nebulae," *Publications of the Astronomical Society of the Pacific* 29 (1917): 213–17.

[9] *Ibid.*, p. 216.

[10] "Globular Clusters and the Structure of the Galactic System," *Publications of the Astronomical Society of the Pacific* 30 (1918): 42–54. Here Shapley put the distance of the most remote cluster at 200,000 light-years. He revised this figure upward shortly after.

[11] *Ibid.*, p. 50.

[12] *Ibid.*

[13] *Ibid.*

[14] *Ibid.*, p. 53.

[15] *Ibid.*

[16] *Ibid.*

[17] "Modern Theories of the Spiral Nebulae," *Journal of the Washington Academy of Sciences* 9 (1919): 217–27.

[18] *Ibid.*, p. 217.

[19] *Ibid.*, p. 219.

[20] *Proceedings of the American Philosophical Society* 57 (1918): 513–20.

[21] *A Study of Absorption Effects in the Spiral Nebulae* (Berkeley: University of California Press, 1918).

[22] "Modern Theories of the Spiral Nebulae," p. 225.

[23] "On the Dark Markings of the Sky with a Catalogue of 182 such Objects," *Astrophysical Journal* 49 (1919): 1–23.

[24] "Modern Theories of the Spiral Nebulae," p. 221.

[25] "On the Existence of External Galaxies," *Publications of the Astronomical Society of the Pacific* 31 (1919): 261–68; for quote, see p. 261.

[26] *Ibid.*, p. 268.

[27] "Modern Theories of the Spiral Nebulae," p. 227.

[28] "On the Existence of External Galaxies," p. 268.

[29] The papers of Shapley, "Evolution of the Idea of Galactic Size," and of Curtis, "Dimensions and Structure of the Galaxy," were published in the *Bulletin of the National Research Council* 2 (1921): 171–93 and 194–217, respectively.

[30] *Ibid.*, p. 191.

[31] *Ibid.*, p. 192.

[32] *Ibid.*

[33] *Ibid.*, p. 193.

[34] *Ibid.*

[35] *Ibid.*, p. 198.

[36] *Ibid.*, p. 214.

[37] *Ibid.*

[38] *Ibid.*, p. 213.

[39] *Ibid.*, p. 217.

[40] *Ibid.*, p. 216.

[41] *Ibid.*, p. 198.

[42] *Ibid.*, p. 217. Curtis' position soon received some confirmation from the Swedish astronomer, Knut Lundmark, who, in 1920, estimated the distance of the Andromeda nebula at 650,000 light-years in his impressive study, "The Relations of the Globular Clusters and Spiral Nebulae to the Stellar System: An Attempt to Estimate their Parallaxes," *Kungl. Svenska Vetenskapsakademiens Handlingar,* Vol. 60, No. 8 (p. 63). But Lundmark also agreed with Shapley that "the wonderful system of spirals . . . obviously seems to be connected with the globular clusters and with the Milky Way" (p. 63), and he strongly doubted that the Milky Way was itself a spiral (p. 62).

[43] *The Nebular Hypothesis and Modern Cosmogony* (Oxford: Clarendon Press, 1923), pp. 15–16.

[44] *Problems of Cosmogony and Stellar Dynamics* [The Adams-Prize-winning Essay for 1917] (Cambridge: University Press, 1919), p. 223.

[45] *Ibid.*, pp. 203, 218 and 220.

[46] *Die Milchstrasse* (Hamburg: Henri Grand, 1924), pp. 75–76.

[47] See H. Fischer, *Rätsel der Tiefe* (Leipzig: Vogtländer, 1923), p. 25.

[48] See Hubble's report on variable stars in NGC6822 and M31 in *Carnegie Institution of Washington: Annual Reports of the Director of the Mount Wilson Observatory. 1922–23 and 1923–24,* pp. 81–114.

[49] The quotation is from the carbon copy of the letter in the Russell papers deposited in the Firestone Library of Princeton University. The same collection also contains the copy of the letter written on October 23, 1924, by Jeans to Hubble, to which Jeans attached a six-page outline of two proofs in support of Hubble's estimate of the distance of Andromeda.

[50] "Thirty-Third Meeting of the American Astronomical Association," *Popular Astronomy* 33 (1925): 159.

[51] *Ibid.*

[52] *Ibid.*, pp. 252–55. Neither the Henry E. Huntington Library in San Marino, Cal., where Hubble's papers are deposited, nor the Goodsell Observatory of Carleton College, Northfield, Minn., which houses the records of the American Astronomical Association, possesses a copy of the full text of Hubble's paper as it was read at the meeting.

[53] This letter is also included in the Russell papers and details of it are printed here by permission of Princeton University.

[54] "N.G.C.6822, A Remote Stellar System," *Contributions from the Mount Wilson Observatory,* No. 304 (1925), and "A Spiral Nebula as a Stellar System: Messier 33," *Contribution* No. 310 (1926).

[55] In *Meddelande fran Astronomiska Observatorium Upsala* (Bulletin of the Astronomical Observatory at Upsala), No. 30 (1926).

[56] See note 42, above.

[57] *Ibid.*, pp. 7–19.

[58] *Ibid.*, p. 19.

[59] *Star Clusters* (New York: McGraw-Hill Book Company, 1930), p. 179.

[60] *Ibid.*, p. 179.

[61] *Ibid.*, p. 195.

[62] The two references to Curtis (p. 151 and p. 166) were perfunctory.

[63] Shapley admitted that study of our galaxy would permit estimating "the radius of the space-time world, the total mass of the universe, the comparability of galaxies" (p. 171).

[64] "Motions and Distances of Spiral Nebulae," *Monthly Notices of the Royal Astronomical Society* 85 (1925): 865–94; see especially p. 867. Lundmark took his cue from Carl Wirtz of Kiel Observatory, who had suggested a year earlier a relationship between the red-shift of the spectra of spirals and the logarithms of their apparent diameters, in his paper, "De Sitter's Kosmologie und die Radialbewegungen der Spiralnebel," *Astronomische Nachrichten* 222 (1924): cols. 23–24. But these and other studies at that time of the cosmological significance of the red-shift represented at best tentative probings in the direction of the idea of an expanding universe.

[65] "Un univers homogène de masse constante et de rayon croissant, rendant compte de la vitesse radiale des nébuleuses extra-galactiques," *Annales de la Société Scientifique de Bruxelles* 47A (1927): 49–59. It is practically known and quoted only in its English translation, "A Homogeneous Universe of Constant Mass and Increasing Radius Accounting for the Radial Velocity of Extra-galactic Nebulae," *Monthly Notices of the Royal Astronomical So-*

ciety 91 (1931): 483–90. It is nowhere been noted that three long notes of the original were not incorporated in the translation. The translation and its publication were done at the behest of Eddington.

[66] "On Einstein's Theory of Gravitation, and its Astronomical Consequences," *Monthly Notices of the Royal Astronomical Society* 76 (1915–16): 699–728; 77 (1916–17): 155–84; 78 (1917–18): 3–28.

[67] *Ibid.*, 78, p. 24.

[68] *Ibid.*

[69] *Ibid.*, 78, p. 25.

[70] *Ibid.*

[71] *Ibid.*, 77, pp. 176–77.

[72] *Ibid.*, 78, p. 24.

[73] *Ibid.*, 78, pp. 27–28.

[74] "Über die Krümmung des Raumes," *Zeitschrift für Physik* 10 (1922): 377–86. Curiously enough, Friedman made the claim two years later, ("Über die Möglichkeit einer Welt mit konstanter negativer Krümmung des Raumes," *Zeitschrift für Physik* 21 [1924]: 326–33), that Einstein's equations did not necessarily impose a finite universe, although he should have known that Einstein's General Theory of Relativity was in part motivated by the recognition of the gravitational paradox of the infinite Newtonian universe of homogeneously distributed matter.

[75] *The Mathematical Theory of Relativity* (Cambridge: Cambridge University Press, 1923), p. 161.

[76] *Ibid.*, p. 162.

[77] "On the Instability of Einstein's Spherical World," *Monthly Notices of the Royal Astronomical Society* 90 (1929–30): 668.

[78] *Ibid.*

[79] See the English translation (note 65, above), p. 489.

[80] "On the Instability of Einstein's Spherical World," p. 668.

[81] "A Relation between Distance and Radial Velocity among Extra-galactic Nebulae," *Proceedings of the National Academy of Sciences* (Washington) 15 (1929): 173.

[82] *Ibid.*

[83] *The Nature of the Physical World* (Cambridge: Cambridge University Press, 1928), pp. 166–67.

[84] *Harper's Monthly Magazine* 158 (1929): 732–38.

[85] *Ibid.*, pp. 732 and 735.

[86] "A Relation between Distance . . .," p. 173.

[87] The exceptions were rare. In 1926 Hubble estimated from his list of nebulae the effective radius of Einstein's finite universe as 3 x 10^{10} parsecs or 600 times the range of the 100-inch telescope at Mount Wilson. See his "Non-galactic Nebulae," *Publications of the Astronomical Society of the Pacific* 38 (1926): 258–60.

[88] *The Realm of the Nebulae* (New Haven, Conn.: Yale University Press, 1936), p. 25.

[89] *Red-Shifts in the Spectra of Nebulae* (Oxford: Clarendon Press, 1934), pp. 3–4. But Hubble held out the possibility that the recession of galaxies was not real (p. 17)!

[90] *The Realm of the Nebulae,* p. 45.

[91] Washington D.C.: Carnegie Institution of Washington, 1961.

[92] *Ibid.*, p. 6.

[93] "Some Characteristics of Galaxies," (The Fourteenth Henry Norris Russell Lecture of the American Astronomical Society, Nantucket, Mass., June 19, 1961) *Astrophysical Journal* 135 (1962): 2.

[94] *The Dimensions and Structure of the Galaxy* (Oxford: Clarendon Press, 1935), p. 3.

[95] *The Rotation of the Galaxy* (Oxford: Clarendon Press, 1930).

[96] *Ibid.*, p. 30.

[97] *Ibid.*

[98] *Ibid.*

[99] *The ABC of Relativity* (New York: Harper & Brothers, 1925), pp. 163–68.

[100] *British Association for the Advancement of Science: Report of the Centenary Meeting, London 1931 Sept. 23–30* (London: Office of the British Association, 1932), pp. 573–610.

[101] *Ibid.*, p. 578.

[102] *Ibid.*, pp. 578–79.

[103] *Ibid.*, p. 579.

[104] "Beyond the Milky Way," *ibid.*, p. 563.

[105] *Ibid.*, p. 564.

[106] *Ibid.*

[107] "The Steady-State Theory of the Expanding Universe," *Monthly Notices of the Royal Astronomical Society* 180 (1948): 252–70.

[108] This clarification of the concept of "creation" in the steady-state theory came four years later in Bondi's *Cosmology* (Cambridge: Cambridge University Press, 1952), p. 144.

[109] "The Steady-State Theory . . .," p. 258.

[110] *Ibid.*, p. 256.

[111] *Cosmology*, p. 144.

[112] *Ibid.*, pp. 143–44.

[113] See the "Third Programme of the British Broadcasting Corporation in 1959," published under the title, *Rival Theories of Cosmology* (London: Oxford University Press, 1960).

[114] A. Sandage, in his Preface to *The Hubble Atlas of Galaxies*, p. 6.

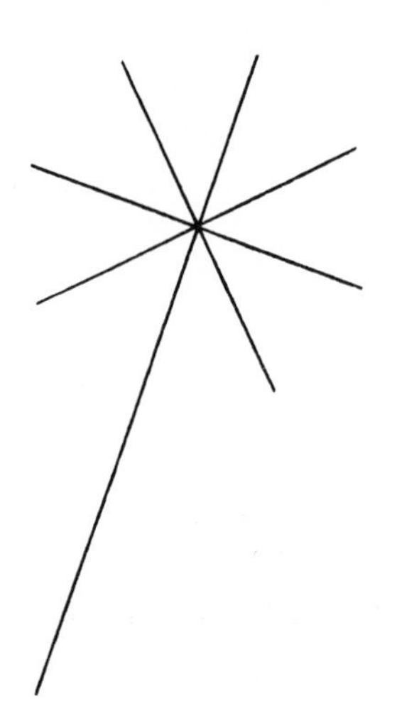

Illustrations

ILLUSTRATION Ia

The theory of the Milky Way attributed by Aristotle to Anaxagoras and Democritus. From Francisco Mateo Fernandez-Bejarano, *Super quatuor libros Meteororum Aristotelis philosophorum principis quaestiones* (Lyons: sumptibus Petri Prost, 1643), p. 72; slightly redrawn. For further details, see pp. 125 and 2.

ILLUSTRATION Ib

The refutation of the theory of the Milky Way attributed by Aristotle to Anaxagoras and Democritus. From Jonas Hertzberger, *Theorian Viae Lacteae* . . . (Leipzig: typis Colerianis, 1663), p. 7; slightly redrawn. For further details, see pp. 131 and 3.

ILLUSTRATION II

Aristotle's theory of the Milky Way. For further details, see p. 5.

ILLUSTRATION IIIa

The parallax problem of a comet (and of the Milky Way) if located below the moon's orbit. From Joannes Dullaert's commentary on the *Meteorologica* of Aristotle, *Habes humanissime lector* . . . (Paris: a Thoma Rees, 1512), f. 31v; slightly redrawn. For further details, see pp. 52 and 24.

ILLUSTRATION IIIb

The parallax problem of the Milky Way if located below the moon's orbit. From John Wilkins, *The Discovery of a World in the Moone* (London: printed by E. Griffin, 1638), p. 174; slightly redrawn. For further details, see pp. 128 and 24.

ILLUSTRATION IVa

The "nebulous star," Praesepe, between Aselli, in the breast of Cancer. Center part of Plate XXV in Johann Bayr, *Uranometria* . . . (A[ugsburg]: M[angus], 1603). Courtesy of the Library of Leiden Observatory. For further details, see pp. 90 and 104.

ILLUSTRATION IVb

The resolution of Praesepe by Galileo into "more than forty starlets." From his *Sidereus nuncius* . . . (Venice: apud Thomam Baglionum, 1610), f. [16c v]. Courtesy of the Rosenwald Collection of the Library of the Institute for Advanced Study, Princeton, N.J. For further details, see pp. 104, 90, and 123.

ILLUSTRATION V

Gassendi's ring theory of the Milky Way and its consequence for the Milky Way's visual appearance. For further details see, p. 122.

ILLUSTRATION VI

Title page of Ludolph Georg Lünde's dissertation on the Milky Way. Courtesy of the Library of the British Museum. For further details see p. 130.

ILLUSTRATION VII

Swedenborg's bar-magnet theory of the Milky Way from his *Principia rerum naturalium* (Dresden and Leipzig: sumptibus Friderici Hekelii, 1734), slightly redrawn. For further details, see pp. 170 and 180.

ILLUSTRATION VIII

Thomas Wright's illustration of the confinement of the stars of the Milky Way between two parallel planes. Plate XXI [XXIII] in his *An Original Theory or New Hypothesis of the Universe* (London: H. Chapele, 1750; facsimile reprint edition edited by M. A. Hoskin, London, Macdonald & Company, 1971). For further details, see pp. 187-88.

ILLUSTRATION IX

The orbit of the sun and of the earth around the center of the Milky Way according to Wright. *Ibid.*, Plate XXI. For further details, see pp. 188-89.

ILLUSTRATION X

The motion of the sun and of the earth around the center of the Milky Way according to Wright. *Ibid.*, Plate XXII. For further details, see pp. 188-89.

ILLUSTRATION XI

Wright's spherical shell model of the Milky Way. *Ibid.*, Plate XXVI. For further details, see p. 189.

ILLUSTRATION XII

Cross sections of the same. *Ibid.*, Plate XXVII. For further details, see p. 189.

ILLUSTRATION XIII

Wright's ring model of the Milky Way. *Ibid.*, Plate XXVIII. For further details, see pp. 190-91.

ILLUSTRATION XIV

Cross sections of the same. *Ibid.*, Plate XXIX. For further details, see pp. 190-91.

ILLUSTRATION XV

The system of Milky Ways according to Wright. *Ibid.,* Plate XXXI. For further details, see pp. 191-92.

ILLUSTRATION XVI

The divine omnipresence in each Milky Way according to Wright. *Ibid.,* Plate XXXII. For further details, see pp. 191-92.

ILLUSTRATION XVII

Herschel's model of the Milky Way. Tab. XVIII, Fig 16, in *Philosophical Transactions of the Royal Society of London,* vol. LXXIV. For the Year 1784. Part I. For further details, see pp. 223-24.

ILLUSTRATION XVIII

Herschel's outline of the boundaries of the Milky Way. Tab. VIII, Fig. 4, in *Philosophical Transactions of the Royal Society of London,* vol. LXXV. For the Year 1785. Part I. For further details, see pp. 228-29.

ILLUSTRATION XIX

The one-island theory of the universe from Simon Newcomb, *Popular Astronomy* (4th rev. ed.; New York: Harper & Brothers, 1882), p. 493. For further details, see pp. 274-75.

ILLUSTRATION XXa

Cross section of the "Stellar System" according to Arthur Stanley Eddington from his *Stellar Movements and the Structure of the Universe* (London: Macmillan and Co., 1914), p. 31. Reproduced with the publisher's permission. For further details, see p. 282.

ILLUSTRATION XXb

The evolutionary sequence of nebulae according to Edwin Hubble from his *The Realm of the Nebulae* (New Haven, Conn.: Yale University Press, 1936), p. 45. Reproduced with the publisher's permission. For further details, see p. 308.

ILLUSTRATION Ia

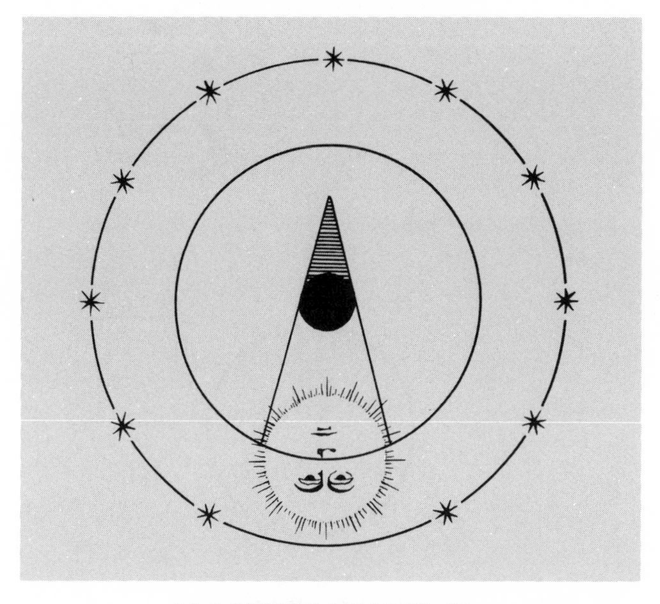

ILLUSTRATION Ib

ILLUSTRATION II

ILLUSTRATION IIIa

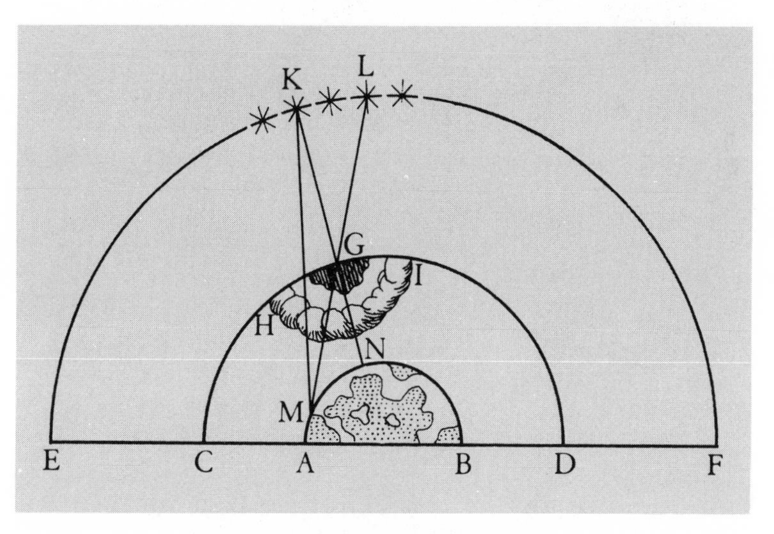

ILLUSTRATION IIIb

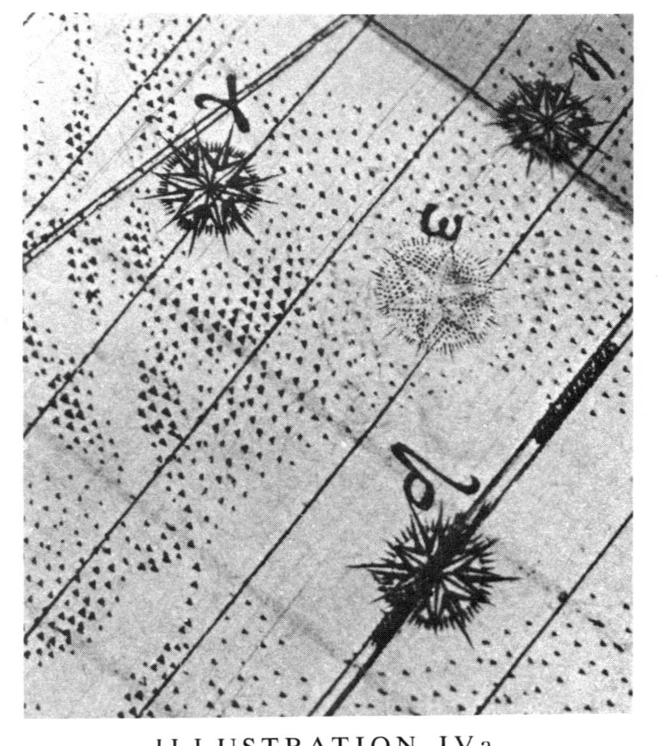

ILLUSTRATION IVa

ILLUSTRATION IVb

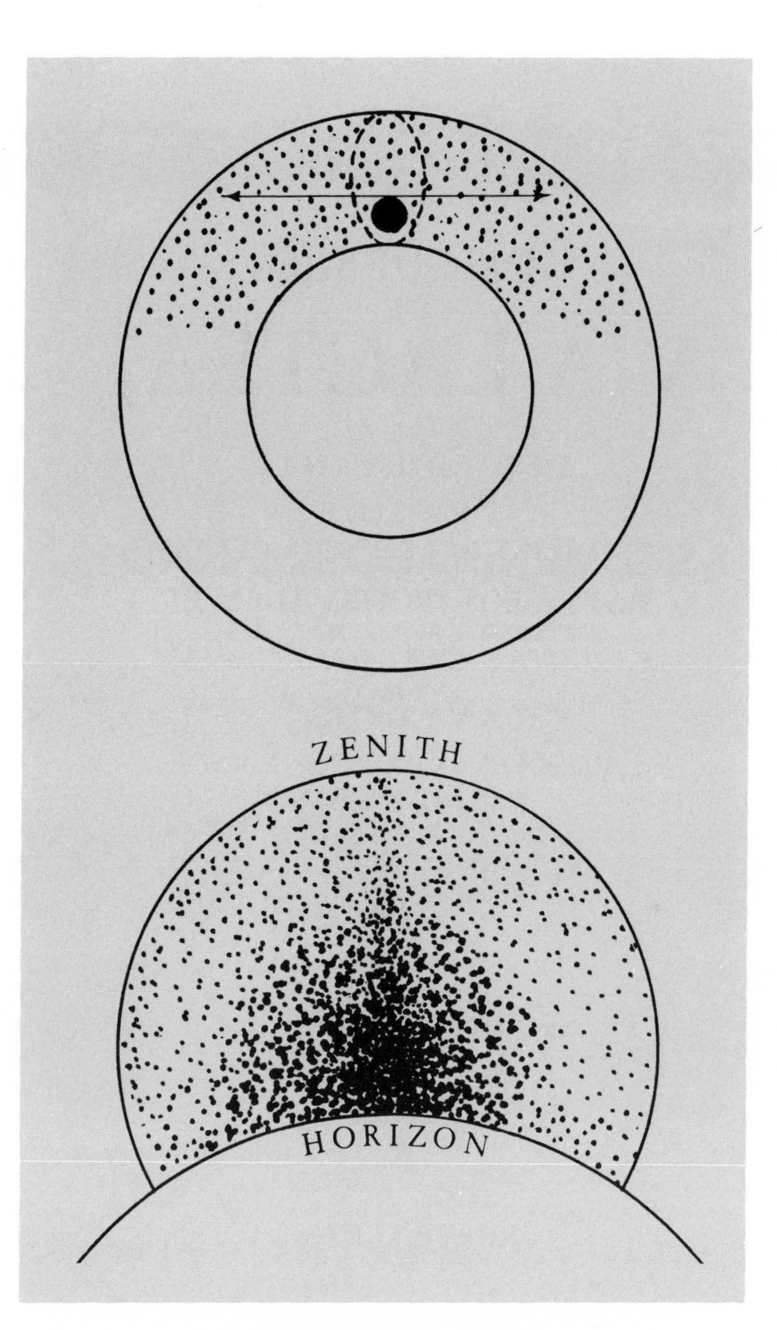

ILLUSTRATION V

DISCURSUS

DE

VIA LACTEA,

QUEM

DEO ADJUVANTE

PRÆSIDE

SIGISMUNDO HOSEMANNO

MATHESEOS PROFESS. PUBL. ET
ORDINARIO IN ILLUSTRI JULIA

EXAMINI COMMILITONUM
SISTIT

AUTHOR

LUDOLPHUS GEORGIUS LUNDE
HANNOVERANUS.

HELMSTADII,
TYPIS JACOBI MULLERI,
ANNO M DC LXV.

ILLUSTRATION VI

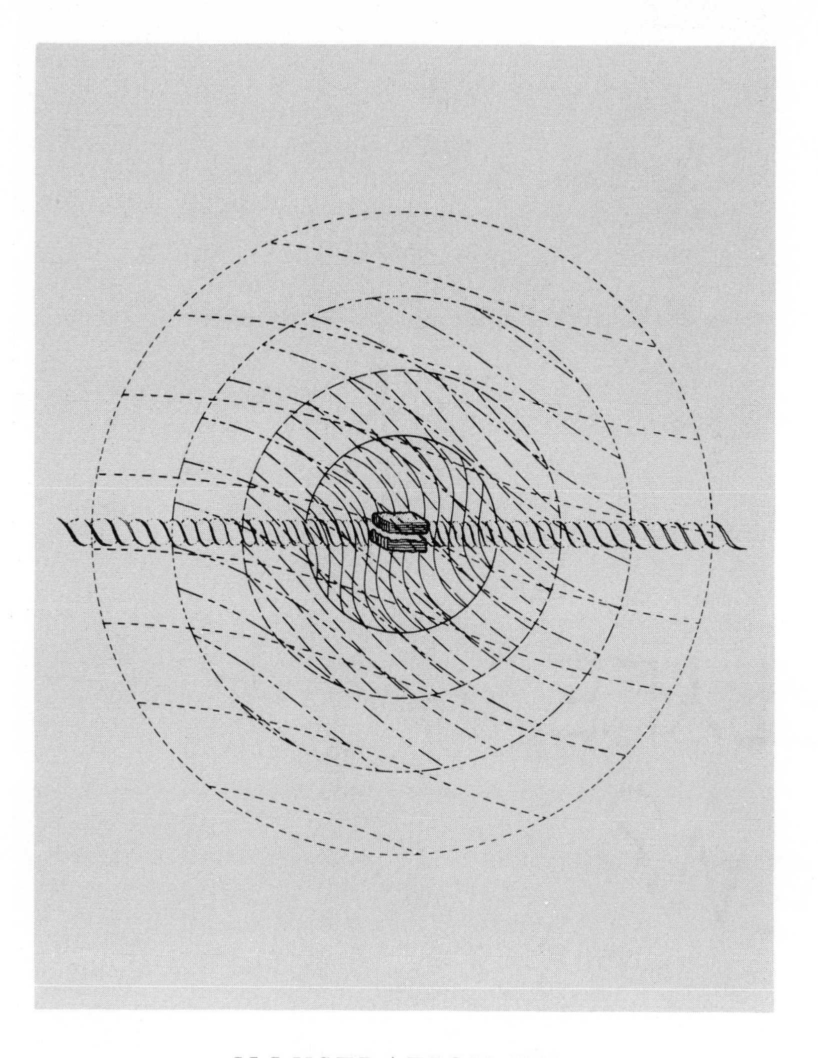

ILLUSTRATION VII

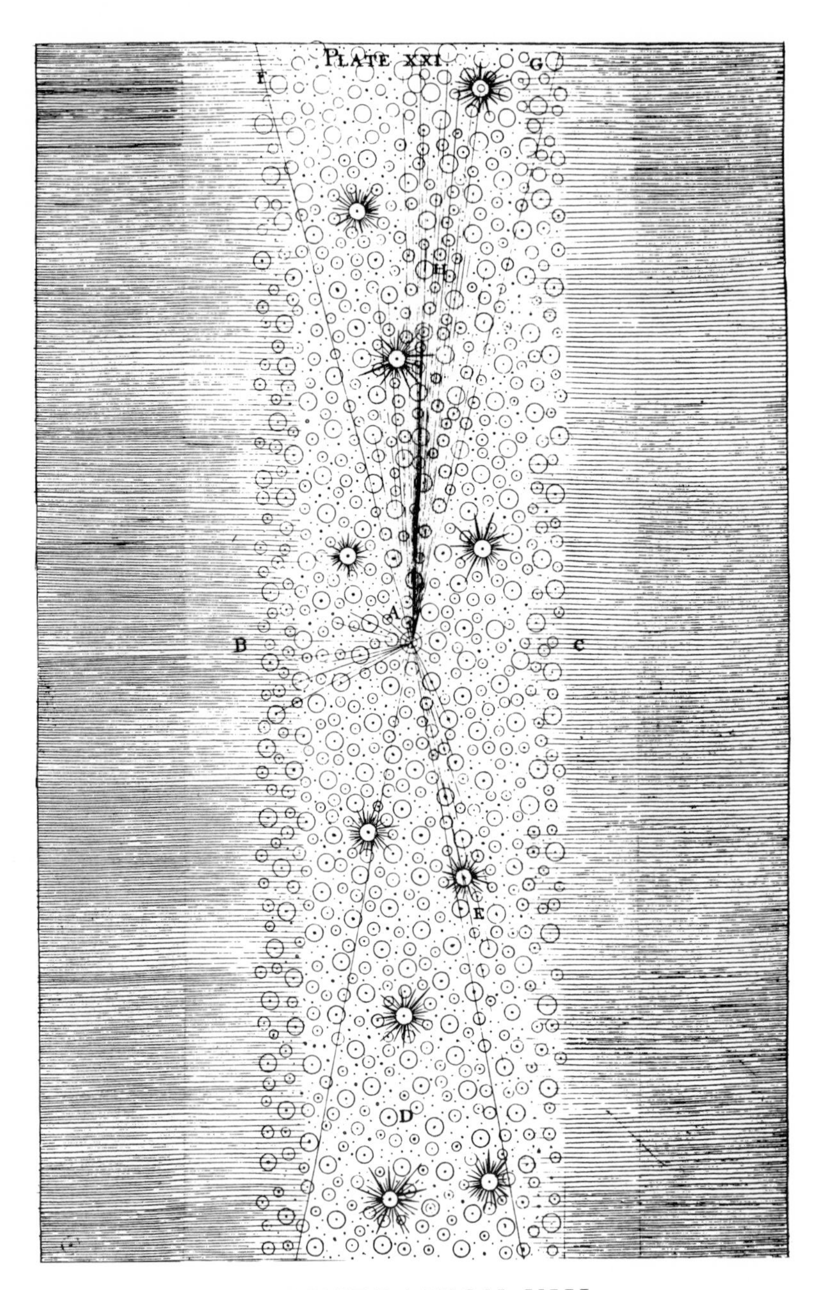

ILLUSTRATION VIII

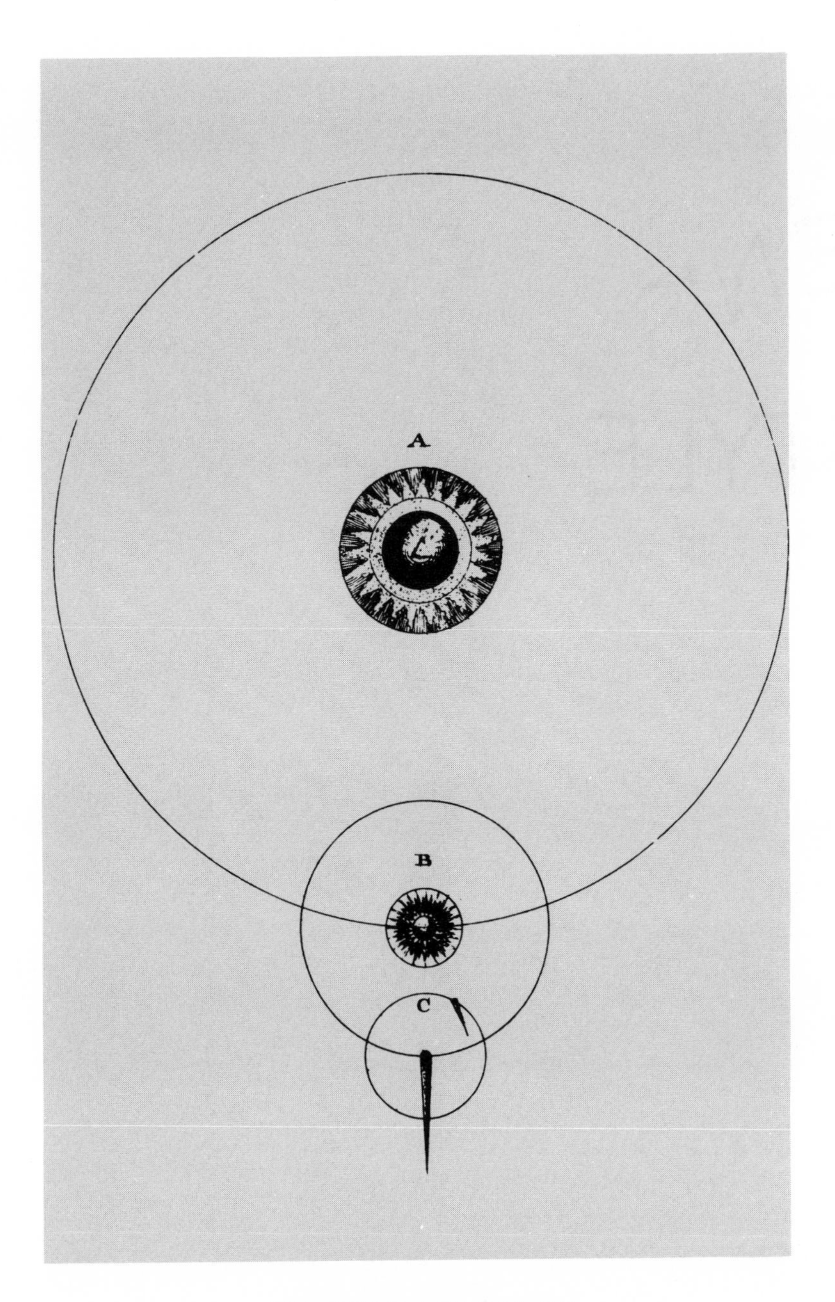

ILLUSTRATION IX

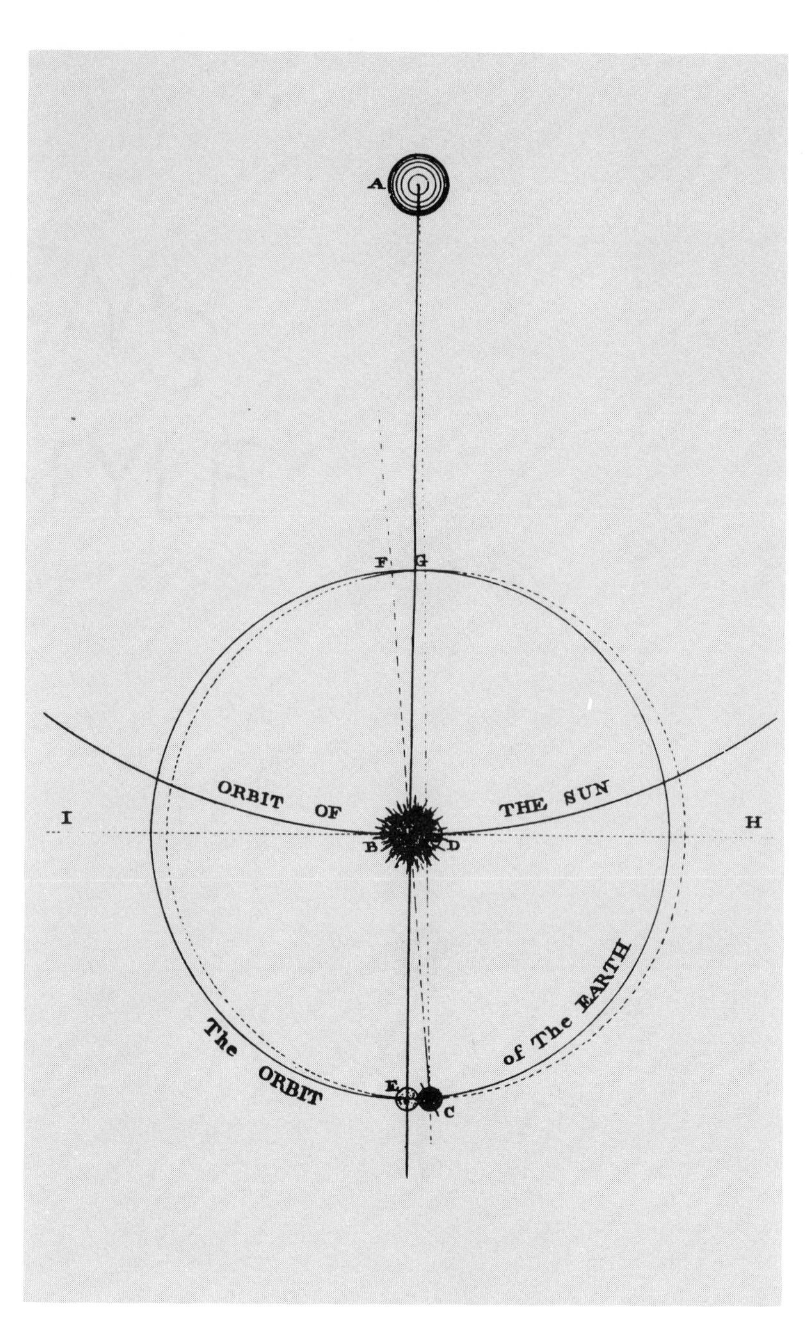

ILLUSTRATION X

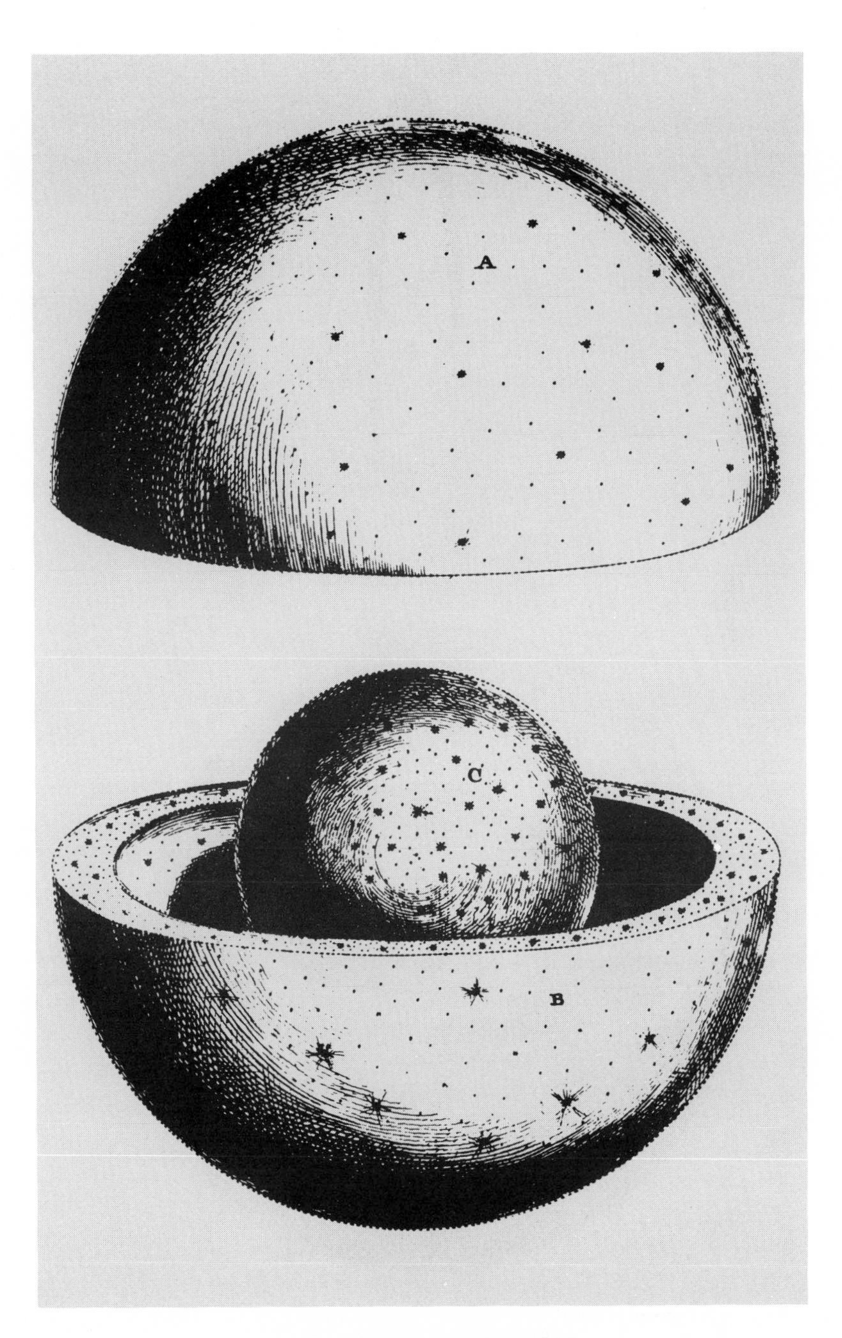

ILLUSTRATION XI

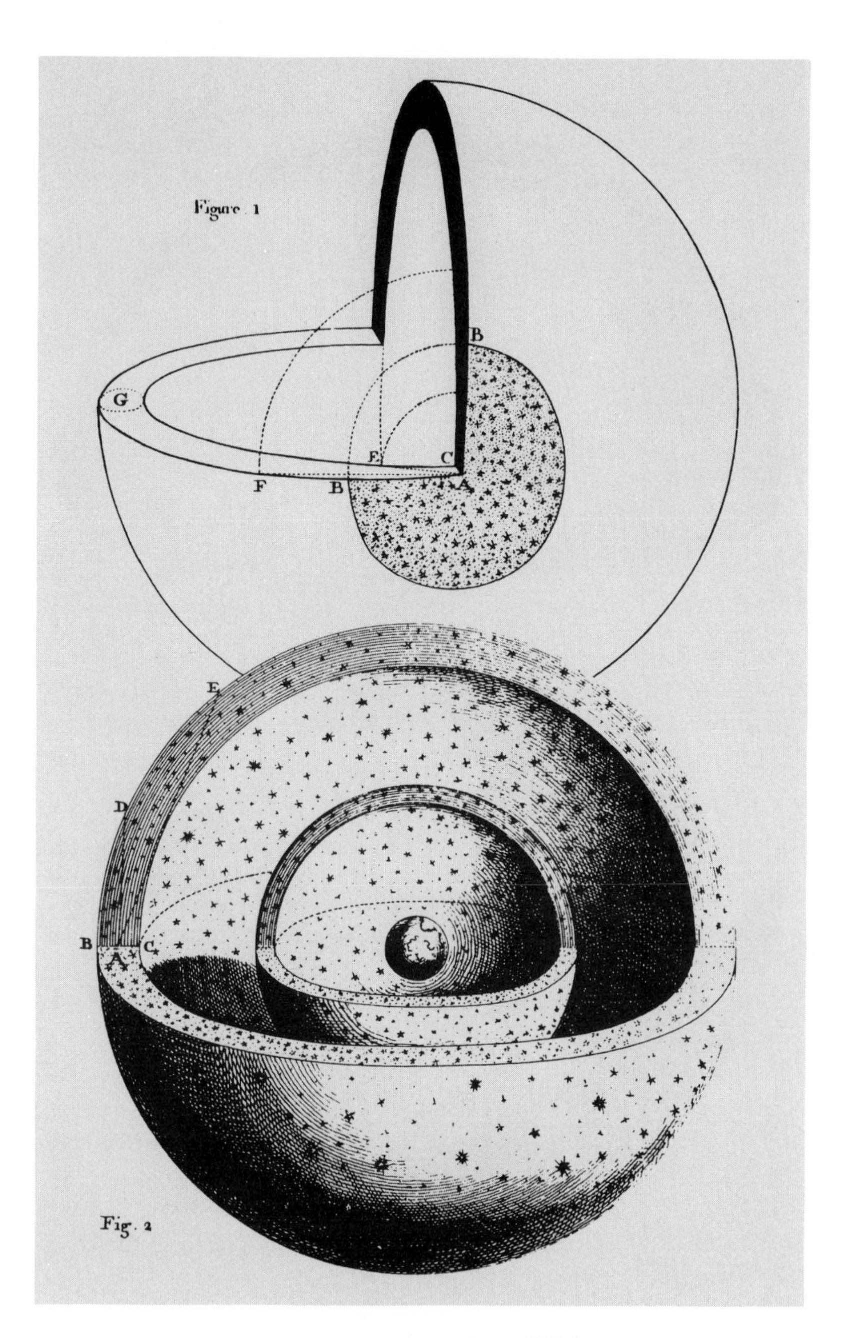

ILLUSTRATION XII

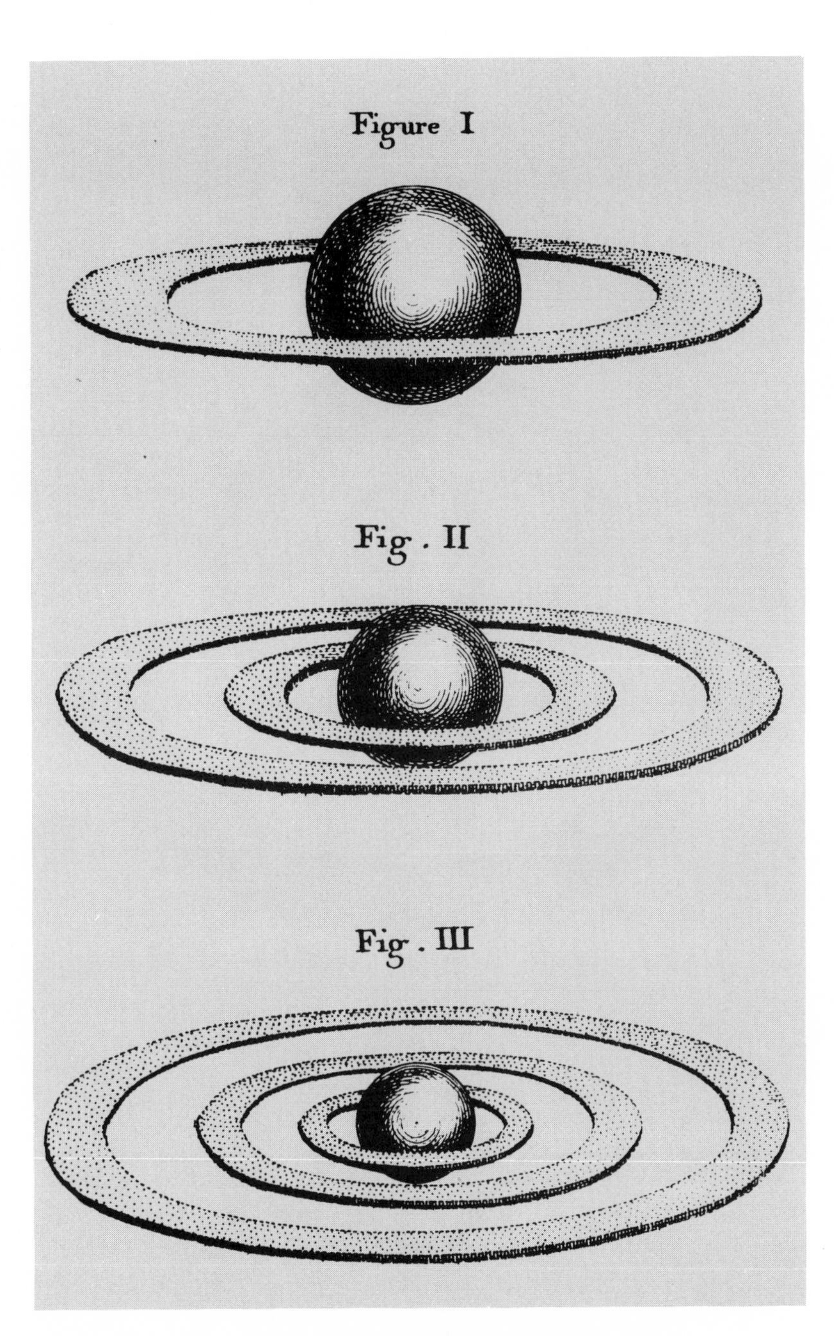

ILLUSTRATION XIII

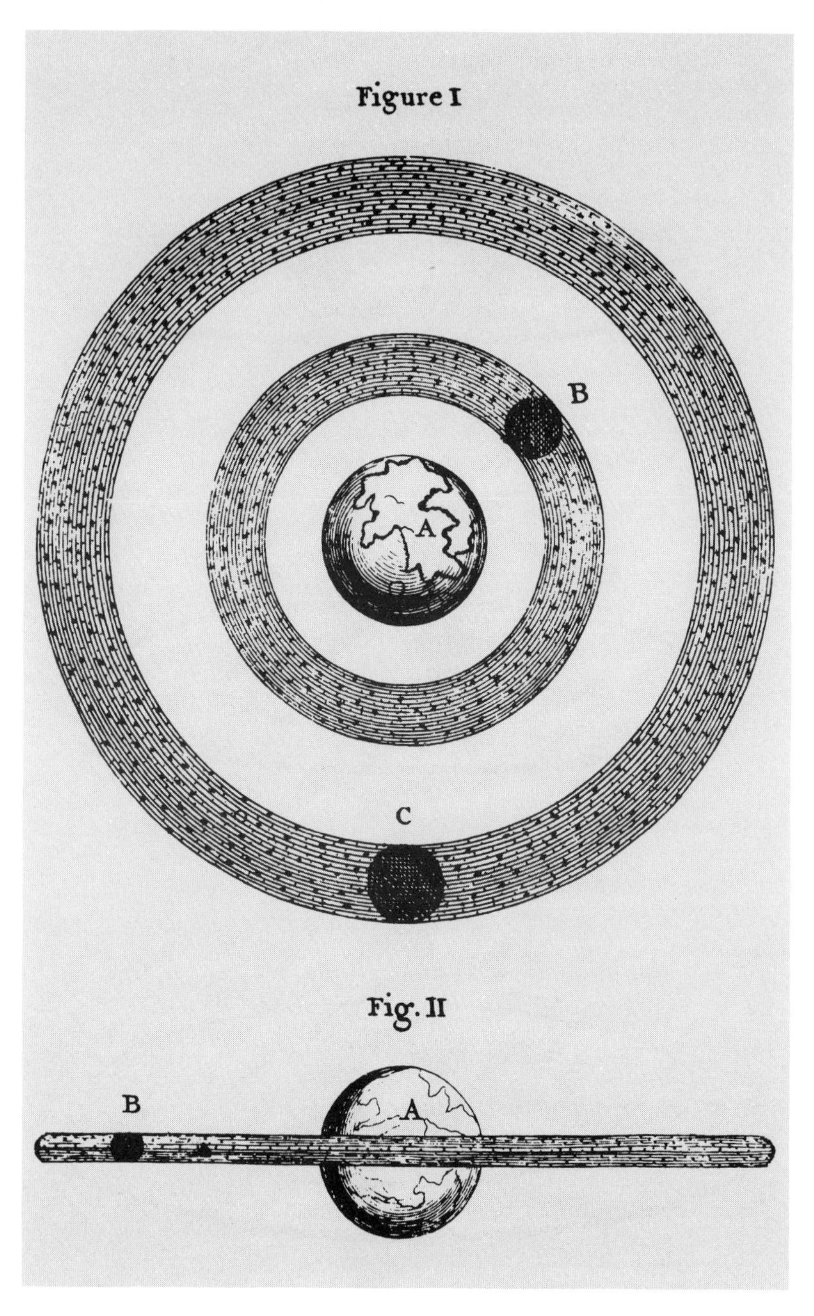

ILLUSTRATION XIV

ILLUSTRATION XV

ILLUSTRATION XVI

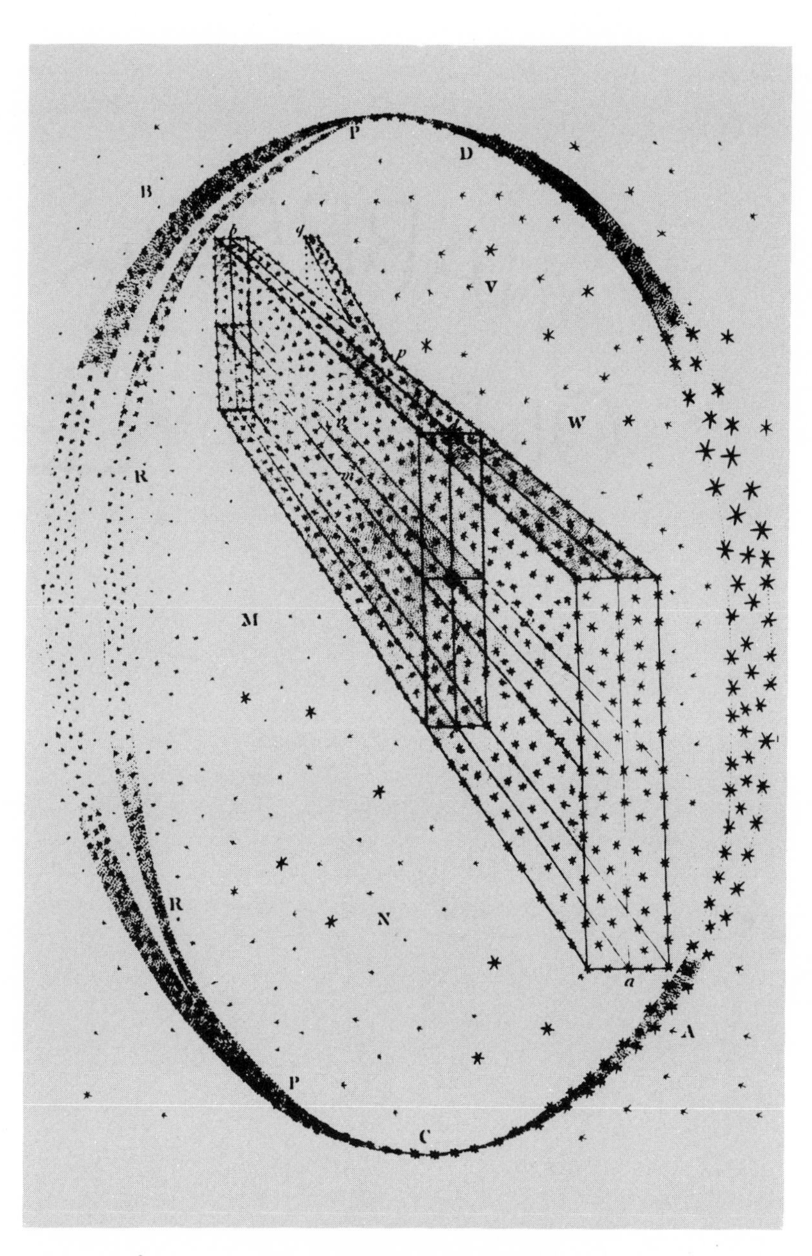

ILLUSTRATION XVII

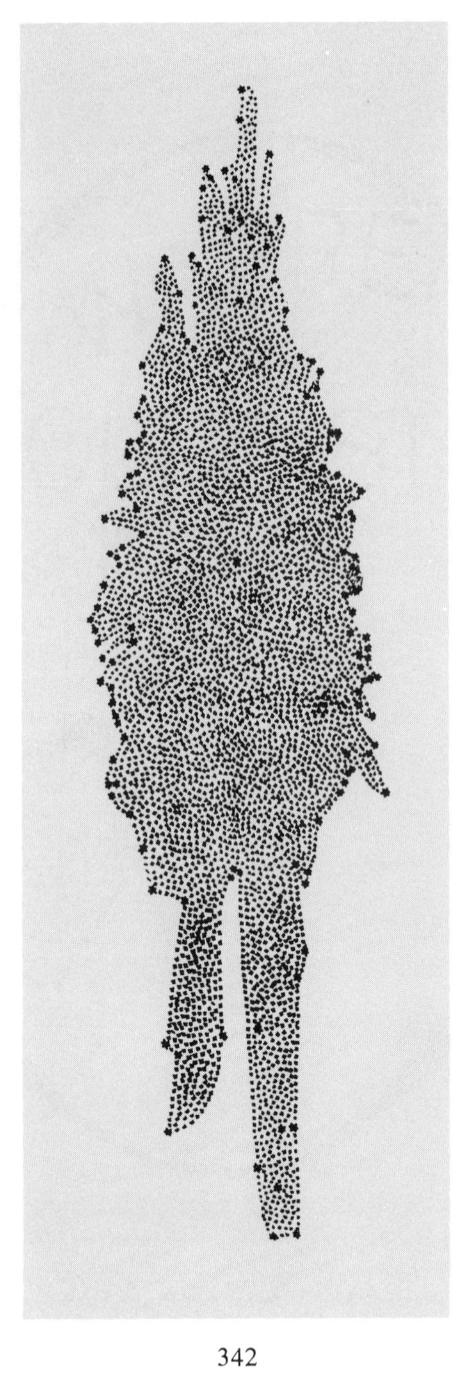

ILLUSTRATION XVIII

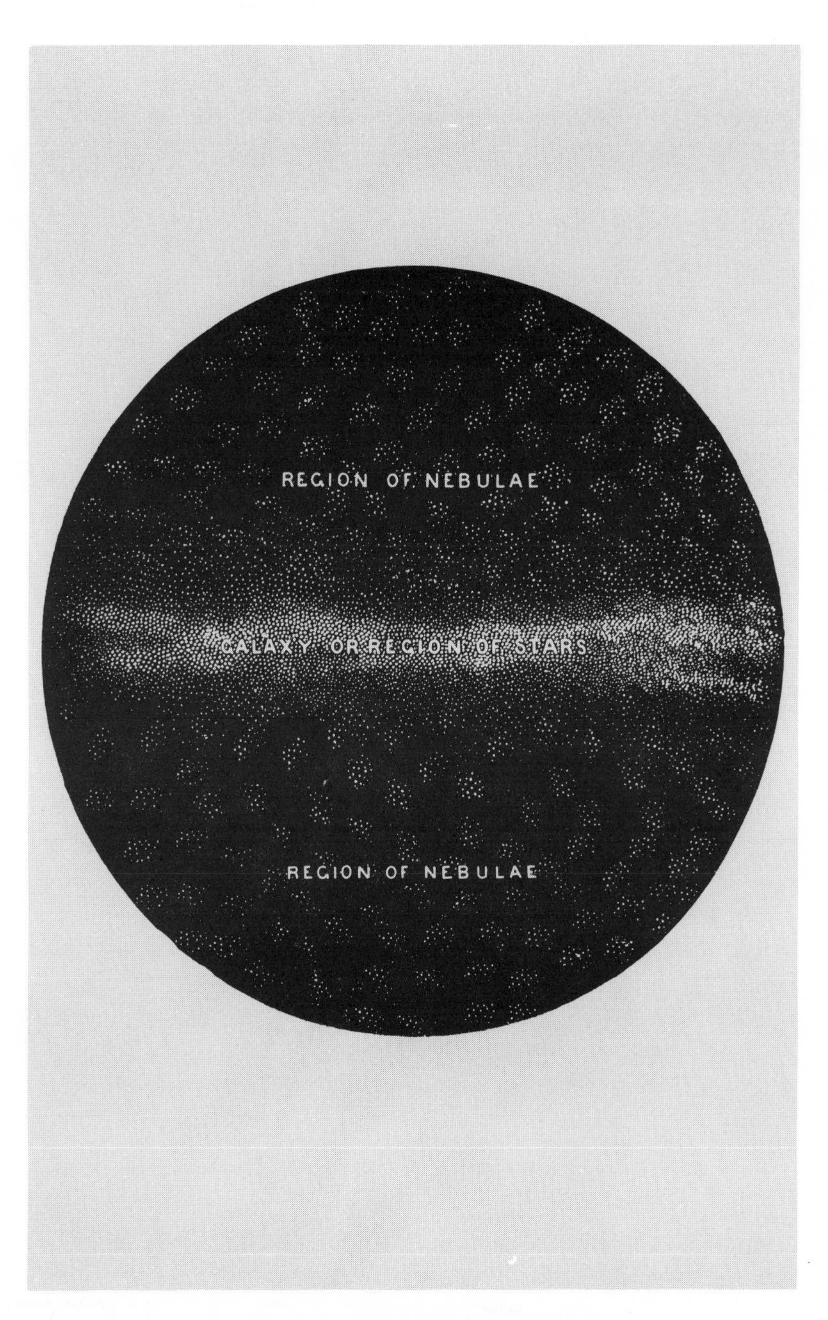

ILLUSTRATION XIX

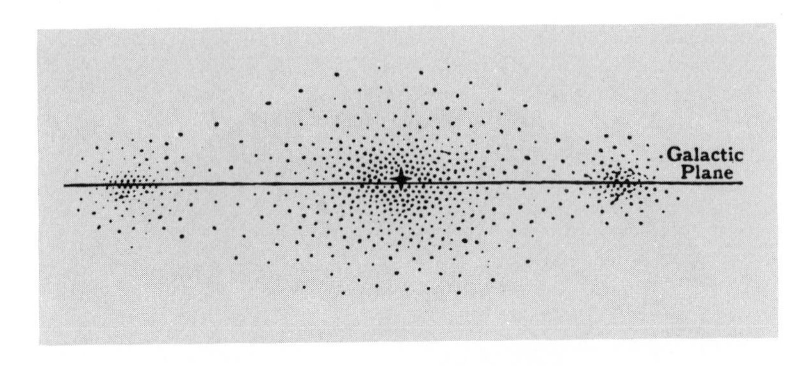

ILLUSTRATION XXa

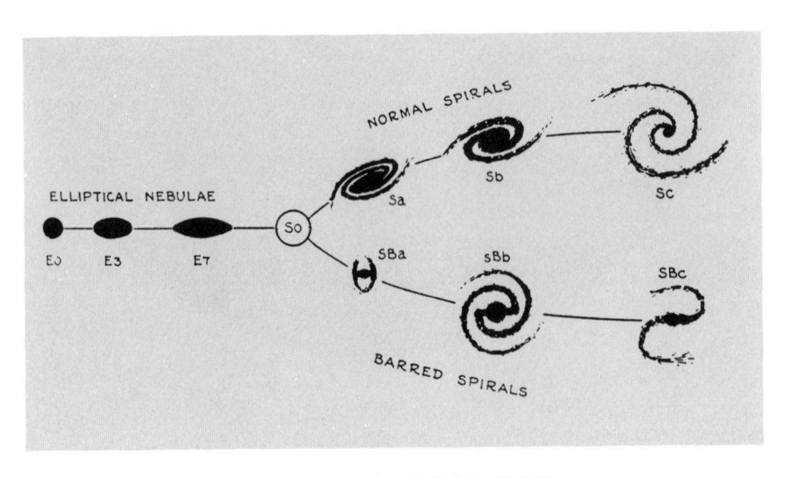

ILLUSTRATION XXb

Index of Names